# CONTENTS

# PATENTS FOR INVENTIONS

*This is a volume in the Arno Press Collection*

# THE SOURCES OF MODERN PHOTOGRAPHY

*Advisory Editors:*
Peter C. Bunnell
Robert A. Sobieszek

*See last pages of this volume for a complete list of titles.*

# PATENTS FOR INVENTIONS

## Abridgments of Specifications
## Class 98, Photography
## Periods 1839 through 1900

Great Britain
Patent Office

Volume I

ARNO PRESS
A New York Times Company
New York • 1979

**Publisher's Note: This book has been reproduced from the best available copy.**

Editorial Supervision: Erin Foley

---

Reprint Edition 1979 by Arno Press Inc.

Reprinted from copies in the University of Illinois Library

THE SOURCES OF MODERN PHOTOGRAPHY
ISBN for complete set: 0-405-09597-X
See last pages of this volume for titles.

Manufactured in the United States of America

---

**Library of Congress Cataloging in Publication Data**

Great Britain. Patent Office.
Patents for inventions.

(The Sources of modern photography)
Reprint of 8 v. published at the Patent Office, London, 1861-1903.
1. Photography--Patents. 2. Patents--Great Britain. I. Title. II. Title: Abridgments of specifications, class 98, photography, periods 1839 through 1900. III. Series.
TR147.G73 1979 770'.2'72 76-23063
ISBN 0-405-09626-7

PATENTS FOR INVENTIONS.

# ABRIDGMENTS

OF

# Specifications

RELATING TO

# PHOTOGRAPHY.

PRINTED BY ORDER OF THE COMMISSIONERS OF PATENTS.

LONDON:
PRINTED BY GEORGE E. EYRE AND WILLIAM SPOTTISWOODE,
PRINTERS TO THE QUEEN'S MOST EXCELLENT MAJESTY.
PUBLISHED AT THE GREAT SEAL PATENT OFFICE,
25, SOUTHAMPTON BUILDINGS, HOLBORN.

1861.

# PREFACE.

The Indexes to Patents are now so numerous and costly, as to render their purchase inconvenient to a large number of inventors and others, to whom they have become indispensable.

To obviate this difficulty, short abstracts or abridgments of the Specifications of Patents under each head of Invention have been prepared for publication separately, and so arranged as to form at once a Chronological, Subject-matter, Reference, and Alphabetical Index to the class to which they relate. As these publications do not supersede the necessity for consulting the Specifications, the prices at which the latter are sold have been added.

The Specifications included in this series of inventions relate:—

1st. To the processes of Photography.

2nd. To the chemical, optical, and mechanical apparatus used in the practice of the art, or its applications.

3rd. To the various applications of Photography.

4th. To the camera obscura, so far as its use for photographic purposes is concerned.

5th. To the stereoscope; this instrument depends upon the results of photography for its efficient action; it gives to photographs the nearest possible approach to reality; therefore, all Specifications relating to stereoscopes are included in this series.

All the quotations from the printed Specifications (included between quotation commas throughout the work) are given in the exact punctuation and orthography therein used; however, to draw attention to any passage more immediately connected with this series of abridgments, portions are sometimes italicised that appear in Roman type in the original.

Berzelius' ammonium theory has been adhered to throughout the text of the work, and the best recognised names for chemical and metallic bodies such as ammoni*um*, platin*um*, have been adopted.

When two words are used as one adjective to qualify a noun, they are connected by a hyphen, thus:—" camel-hair " pencil," " eye-piece frame," &c.

The definition of " Photography " for the purpose of these abridgments is, " the art of copying designs or " images, however they may be produced, by the chemical " action of light upon surfaces prepared to receive that " action."

It is hoped that the publication of these abridgments will prevent the disappointment consequent on repatenting an old invention, and, by setting forth what has been already done in this department of applied knowledge, enable inventors to exert their talents only upon discoveries and applications that are new.

B. Woodcroft.

July 1861.

# INDEX OF NAMES.

[The names of persons by whom inventions have been communicated from abroad, when known, are printed in *Italics*. The Roman numerals refer to the Introduction.]

# INTRODUCTION.

In order to render the Abridgments of Specifications in which Photography is referred to as complete as possible, the following brief summary of the progress of knowledge in reference to this subject is prefixed. By the chronological arrangement of the matter, the treating of each discovery or invention in a separate paragraph, and the citation of the works consulted, it has been sought to make the heads of the subject easy of reference, and to show readily where more detailed information can be found.

Although the action of light on coloured bodies must have been of every-day occurrence, the philosophers of antiquity did not record the fact, and the chemical action of light upon matter would appear to have entirely escaped their observation. The first definite knowledge of this action appears to have been some observations of the alchemists, that "horn silver" (chloride of silver) was blackened by exposure to light. Although the middle ages thus furnished the discovery of one of the facts upon which the scientific principles of photography depend; the others—viz., the proper application of designs or images to sensitive surfaces and the permanent fixing of the pictures so obtained—are belonging to the present era.

B.C.

300. Euclid, about 300 B.C., in his "Treatise on Optics" (26th, 27th, and 28th theorems) proves that he knew that the pictures of bodies seen by both eyes are formed by the union of two dissimilar pictures formed by each eye. (See Brewster on the *Stereoscope*, p. 6; also Euclid's *Optics*, Edit. of Pena, pp. 17, 18, Paris, 1577; or *Opera*, by Gregory, pp. 619, 620, Oxon. 1703.)

A.D.

170. Galen, about A.D. 170, in the twelfth chapter of the tenth book of his work, "De Usu Partium Corporis Humani," described the phenomena attendant upon looking at

A.D.

bodies with both eyes, and alternately with the right and left eye. (See Brewster on the *Stereoscope*, pp. 6, 7; also *De Usu Partium Corporis Humani*, edit. Lugduni, 1550, p. 593.)

1646. LICETAS and KIRCHER, in 1646, noticed the phosphorescent influences of solar rays. (See Hunt's *Manual of Photography*, p. 335.)

1722. PETIT, in 1722, noticed that solutions of nitrate of potash and muriate of ammonia crystallized more readily in the light than they did in darkness. (See Hunt's *Manual of Photography*, p. 4.)

1777. SCHEELE, in 1777, observed the dissimilar powers of the rays of the spectrum in darkening nitrate of silver; he found the violet ray to have the most blackening effect. (See Hunt's *Manual of Photography*, p. 4.)

1786. SCHEELE, in 1786, observed that nitric acid was decomposed by the chemical action of light. (See Hunt's *Manual of Photography*, p. 335.)

1791. SENEBIER, in 1791, discovered that yellow wax was bleached by the action of light. (See Hunt's *Manual of Photography*, p. 335.)

1795. FISCHER, in 1795, observed the action of light upon the ferrocyanates of iron. (See Hunt's *Manual of Photography*, p. 334.)

1801. J. W. RITTER, in 1801, pointed out "the separate existence of chemical rays in the spectrum which extend "beyond the most refrangible or violet rays." (See *Encyclopædia Britannica*, 8th edit., Dissertation on Optics, pp. 921, 922.)

Mr. THOMAS WEDGWOOD and Sir HUMPHREY DAVY, in
1802. June 1802, published their method of copying paintings, &c., by means of the action of light upon nitrate of silver. Paintings on glass were copied by placing white paper or white leather, sensitized with this salt, behind the painting, and exposing the arrangement to solar light. This method was also employed to delineate profiles or shadows of figures, the woody fibres of leaves, the wings of insects and the images produced by the solar

A.D.

microscope. Muriate of silver was found to be more sensitive than the nitrate, and both were more readily acted upon when moist than when dry. These pictures could not be fixed. (See Nicholson's *Journal of Natural Philosophy*, &c., 8vo. series, vol. III., p. 167; also *Encyclopædia Britannica*, 8th edit., art. Photography, p. 544.)

1802. Sir HUMPHREY DAVY, in 1802, noticed the effect of light upon the puce-coloured oxide of lead. (See Hunt's *Manual of Photography*, p. 334.)

1802. Mr. ROBERT HARUP, in 1802, showed that several salts of mercury were reduced by light. (See Hunt's *Manual of Photography*, p. 5; also Nicholson's *Journal* for 1802.)

1803. WOLLASTON, in 1803, discovered the action of light upon gum guaiacum. (See Hunt's *Manual of Photography*, p. 335.)

1803. Dr. YOUNG, about 1803, proved the interference of the obscure chemical rays. (See *Encyclopædia Britannica*, 8th edit., Dissertation on Optics, p. 922.)

1810. SEEBECK, in 1810, observed the production of colour on chloride of silver by the various rays of the spectrum; the violet rays rendered it brown, the blue produced a shade of blue, the yellow preserved it white, and the red gave the salt a red colour. (See Hunt's *Manual of Photography*, p. 8.)

1812. BÉRARD, in 1812, showed that the chemical rays are polarized by reflection. (See *Encyclopædia Britannica*, 8th edit., Dissertation on Optics, p. 922.)

1814. M. NICEPHORUS NIÉPCE, in 1814, "attempted to fix the "pictures produced in the camera obscura, and to copy "engravings by means of light transmitted through "them upon substances made sensible to its action;" he used a tablet of copper coated with silver, and called his process "*Heliography*." (See *Encyclopædia Britannica*, 8th edit., art. Photography, p. 545.)

1824. M. DAGUERRE, in 1824, began experiments with the view of fixing the pictures in the camera. (See *Encyclopædia Britannica*, 8th edit., art. Photography, p. 545.)

A.D.

1827. M. Nicephorus Niépce, in December 1827, submitted some pictures taken upon silvered copper plates smeared with the bitumen of Judæa to the Royal Society. The resin is rendered insoluble in certain essential oils by the action of light, on subsequent treatment with the oleaginous solvent, the shadows dissolve away, and the lights are represented by the unaltered resin remaining on the plate. In order to produce a better effect, M. Niépce darkened the silver surface with a film of iodine. (See *Encyclopædia Britannica*, 8th edit., art. Photography, p. 545; also Hardwich's *Manual of Photographic Chemistry*, pp. 7, 8.)

Mr. William Henry Fox Talbot's researches in pho-
1833. tography date from October 1833, and were suggested by an unsatisfactory attempt to use the camera lucida for the purpose of sketching. (See Supplement to the *Penny Cyclopædia*, art. Photography.)

1834. Mr. William Henry Fox Talbot, in 1834, took pictures in the camera by the action of light upon paper washed with nitrate of silver, and succeeded in fixing them; he called this art "*Calotype.*" (See *Encyclopædia Britannica*, 8th edit., art. Photography, p. 545.)

1835. Mrs. Somerville, in 1835, made some experiments on the permeability of different bodies to the chemical rays. Copper-green glass intercepts, and the emerald transmits them; red glass stops most of them, whilst the garnet transmits them. (See *Encyclopædia Britannica*, 8th edit., Dissertation on Optics, p. 922.)

1838. Mr. Wheatstone, on 21st June 1838, read a paper to the Royal Society, in which he described an instrument for uniting two dissimilar pictures of solid bodies, called a "Stereoscope." This instrument is known by the name of "The Reflecting Stereoscope," as the images are made to coincide by means of mirrors. The observer looks into two plane vertical mirrors (one being opposite to each eye) inclined to each other at an angle of 90°; the drawings, taken according to the laws of perspective, are fixed to two upright planes equally inclined to the mirrors.

A.D.

(See Brewster on the *Stereoscope*, pp. 18 and 58–61; also *Phil. Trans.* 1838, pp. 371–394.)

1839. M. DAGUERRE'S invention (the "Daguerreotype") was communicated to the Academy of Sciences on the 7th January 1839. (See *Penny Cyclopædia*, art. Photogenic Drawings.)

1839. Mr. WILLIAM HENRY FOX TALBOT, on the 30th January 1839, communicated his discovery to the Royal Society. The paper was sensitized by means of nitrate of silver, and the image fixed by common salt. (See *Encyclopædia Britannica,* 8th edit., art. Photography, p. 545.)

1839. The Rev. J. B. READE, in April 1839, delineated, by the agency of light, objects of natural history from their images taken by the solar microscope. Nitrate of silver solution sensitized writing paper which was washed with an infusion of nut-galls just prior to use, and was employed wet. Hyposulphite of soda fixed the picture. (See *Encyclopædia Britannica,* 8th edit., art. Photography, p. 545.)

1839. Mr. MUNGO PONTON, on the 29th May 1839, announced to the Royal Scottish Society of Arts that bichromate of potash might be used in solution to sensitize paper. The dark orange tint of the parts exposed to light remains on the immersion of the picture in water, only the yellow portions being dissolved out. (See *Encyclopædia Britannica,* 8th edit., art. Photography, p, 545.)

1839. M. DAGUERRE in conjunction with M. ISIDORE NIÉPCE, on the 15th June 1839, received a pension from the French Government for their invention of the "Daguerreotype." (See *Encyclopædia Britannica,* 8th edit., art. Photography, p. 545; also Hunt's *Manual of Photography*, p. 21.)

1839. Sir JOHN HERSCHEL, in 1839, used glass plates in conjunction with precipitates of silver to obtain photographs upon. (See Hunt's *Manual of Photography*, pp. 276–279 and 333.)

1839. Mr. ELLIOT, in 1839, constructed an "ocular" stereoscope, consisting of a wooden box, without lenses or mirrors, in which two dissimilar pictures as seen by each eye,

A.D.

were placed. The images were united by the eyes alone. (See Brewster on the *Stereoscope*, pp. 18–22 and 56–58.)

1839. Mr. T. B. Jordan, in 1839, used photography for automatic registration. The barometer, thermometer, and magnetometer were thus made self-registering. Mr. Jordan also devised a "Heliograph" for numerically registering the intensity of an incident beam. (See Hunt's *Manual of Photography*, pp. 285–289.)

1839. Dr. Fyfe, in 1839, used the following sensitizing process:—The paper is soaked in phosphate of soda solution, dried, treated with nitrate of silver solution, and "again put "through the salt, by which any excess of silver is con"verted to phosphate." (See Hunt's *Manual of Photography*, pp. 174 and 332.)

1840. Sir John Herschel, in 1840, made his important researches on the chemical properties of the solar spectrum. He operated upon sensitized paper; the solutions used were, nitrate of silver, nitrate of silver mixed with chloride of sodium, and nitrate of silver mixed with hydrobromate of potash. In the first instance the chemical spectrum was 1·57 times the length of the luminous spectrum, in the second instance 1·81 times, and in the third nstance 2·16 times. (See Sir David Brewster's *Optics* in Lardner's Cabinet Cyclopædia, pp. 104, 105; also *Phil. Trans.* for 1840, p. 26.)

1840. Sir John Herschel, in 1840, procured upon photographic paper a coloured image of the solar spectrum. (See *Encyclopædia Britannica*, 8th edit., art. Photography, p. 552.)

1840. Sir J. F. W. Herschel, on 20th February 1840, read a paper to the Royal Society on an "Actinograph" or self-registering photometer. (See *Phil. Trans.*, vol. 130, 1840.)

1840. Sir John Herschel, about 1840, discovered "that certain "of the persalts of iron when exposed to sunlight in "contact with organic matter, were reduced to the state "of protosalts." (See Hardwich's *Manual of Photographic Chemistry*, p. 182.)

1840. Dr. Draper, in 1840, published his process in England, for taking Daguerreotype portraits. (See Hunt's *Manual of Photography*, pp. 281–284.)

A.D.

1840. Mr. JOHN GODDARD, in 1840, proved that the sensibility of the Daguerreotype plate was greatly promoted by exposing it to the vapours of iodine and bromine in succession. (See Hardwich's *Manual of Photographic Chemistry*, p. 182.)

1840. M. E. BECQUEREL, in 1840, used a combination of bichromate of potash and iodide of starch to sensitize paper with. After exposure to light, the dry photograph is steeped in an alcoholic solution of iodine, then washed and dried. (See Hunt's *Manual of Photography*, pp. 142, 143, and 335.)

1840. Mr. HUNT, in 1840, introduced the use of protosulphate of iron as a photographic agent for developing the image. (See Hunt's *Manual of Photography*, p. 109.)

1840. BAYARD, in 1840, discovered the use of bromide of silver in photographic operations. (See Hunt's *Manual of Photography*, p. 332.)

1841. M. CLAUDET, in 1841, employed chloride of iodine as a means of accelerating the action of sensitive surfaces. M. Claudet also invented a "photographometer," a "dynactonometer," and a "focimeter." (See Hunt's *Manual of Photography*, pp. 92, 288, and 291.)

1842. Messrs. HUNT and TOWSON, in 1842, made a sensitizing solution containing fulminate of silver; this agent was remarkable for its high degree of sensibility to light. (See Hunt's *Manual of Photography*, pp. 174, 175, and 332.)

1842. Sir J. F. W. HERSCHEL, on the 16th June 1842, read a paper to the Royal Society on the action of the rays of the solar spectrum on vegetable colours and on some new photographic processes; among the processes described were the "Chrysotype" and the "Cyanotype." (See *Phil. Trans.* vol. 132, 1842.)

1843. M. DAGUERRE, in 1843, made certain improvements in polishing and preparing Daguerreotype plates. The chief feature was heating pure water over the plate. (See Hunt's *Manual of Photography*, pp. 177–179; also *Comptes Rendus* of 13th March 1843.)

A.D.

1843. Mr. HUNT, in 1843, used a mixture of bichromate of potash and sulphate of copper as a sensitizing solution. The picture is developed by means of nitrate of silver, and fixed by washing in pure water. This process is called "Chromatype." (See Hunt's *Manual of Photography*, pp. 143–145, and 334.)

1844. M. DAGUERRE, in 1844, published an extremely complex "instantaneous" Daguerreotype process, in which the sensitizing agent was a mixed iodide of gold and platinum. (See Hunt's *Manual of Photography*, pp. 181–184; also *Comptes Rendus* for April 1844.)

1844. Mr. CUNDELL, in 1844, invented an important improvement upon Mr. Fox Talbot's "Calotype" process. The chief feature consists in using common salt mixed with the iodide of potassium to "iodize" the paper; the details of the process are also different from Mr. Fox Talbot's method. A *negative* picture is produced. (See Hunt's *Manual of Photography*, pp. 56–62; also the *Philosophical Magazine* for May 1844.)

1844. Mr. HUNT, in 1844, published his process called "Energia-"type," or "Ferrotype." The paper is sensitized by means of succinic acid, common salt, gum arabic, and nitrate of silver. The picture is developed by a mixture of protosulphate of iron and gum arabic, and fixed by hyposulphite of soda. (See Hunt's *Manual of Photography*, pp. 145–147.)

1844. Mr. HUNT, in 1844, published a process which he called the "Fluorotype." The sensitizing solution contains bromide of potassium and fluoride of sodium, a solution of nitrate of silver is then applied. In the developing process protosulphate of iron and weak muriatic acid are used; the image is fixed by means of hyposulphite of soda. (See Hunt's *Researches on Light;* also Hunt's *Manual of Photography*, pp. 83 and 257.)

1844. Dr WOOD, in 1844, invented a process called "Catalyso-"type," from its dependence on catalysis for its action. The paper is sensitized by means of weak hydrochloric acid, a mixture of syrup of iodide of iron and tincture of iodine, and a solution of nitrate of silver. When left in

A.D.

the dark, after exposure in the camera, a negative picture becomes gradually developed. The picture may then be fixed by washing it with a solution containing bromide and iodide of potassium. (See Hunt's *Manual of Photography*, pp. 147–150, and 334.)

1845. M. FIZEAU's method of etching photographic impressions by means of an acid menstruum was presented to the British Association in 1845. (See Supplement to the *Penny Cyclopædia*, art. Photography.)

1846. Mr. R. J. BINGHAM, in 1846, published his improvements on the Daguerreotype process. These consisted in the substitution of certain compounds of bromine, chlorine, and iodine with lime for the bromine or other accelerating solution. A dry accelerating mixture unaffected by heat is thus secured. (See *Philosophical Magazine* for October 1846; also Hunt's *Manual of Photography*, pp. 92–94.)

1848. M. NIÉPCE DE ST. VICTOR, in 1848, substituted for the paper used by Mr. Fox Talbot, a film of albumen spread upon glass. (See *Encyclopædia Britannica*, 8th edit., art. Photography, p. 546.)

1849. Sir DAVID BREWSTER, in 1849, exhibited his lenticular stereoscope to the British Association. (See Brewster on the *Stereoscope*, p. 29.)

1851. Mr. ARCHER, in the autumn of 1851, published his collodion process, he at the same time proposed the substitution of pyro-gallic acid for the gallic acid previously employed in developing the image. M. LE GREY originally suggested the use of collodion for photographic purposes. (See Hardwich's *Manual of Photographic Chemistry*, p. 10.)

1851. Messrs LANGENHEIM, in 1851, introduced "Hyalotypes" into England. These were positive pictures, copied on glass from negatives; they were adapted to the slides of magic lanterns. (See Hunt's *Manual of Photography*, p. 84.)

1852. M. ADOLPHE MARTIN, in 1852, used a collodion process in which the sensitizing operation is performed by means of nitrate of silver, iodide of ammonium and nitric acid.

A.D.

Protosulphate of iron developes the image, which is changed from negative to positive by a bath of double cyanide of silver and potash. (See Hunt's *Manual of Photography*, pp. 124, 125.)

1852. Mr. STEWART, in 1852, employed a negative paper process (wet or dry) in which the following solutions were used:—Iodide of potassium, aceto-nitrate of silver, gallic acid (for developing), and hyposulphite of soda. To iodize the paper the air-pump was used. (See Hunt's *Manual of Photography*, pp. 68–71; also the *Athenæum* for December 1852.)

Mr. HARTNUP, Dr. EDWARDS, and Mr. J. A. FORREST, in
1854. 1854, attempted to obtain photographs of the moon. (See *Liverpool Photographic Journal* for 1854, p. 14.)

1854. Mr. SHADBOLT, in March 1854, executed and exhibited microscopic photographs from the 20th to the 40th of an inch diameter. These pictures are made with a structureless collodion. (See *Encyclopædia Britannica*, 8th edit., art. Photography, p. 552.)

1856. M. M. MELINE and MONTREUIL, in 1856, published a dry collodion process, the principal feature of which was the washing the free nitrate of silver away from the sensitized collodion before exciting the plate, and allowing the plate to dry. (See *Liverpool and Manchester Photographic Journal* for 1857, pp. 7, 8.)

1857. M. TAUPENOT, about 1857, used albumen as a varnish to preserve sensitive collodion plates. (See Hunt's *Manual of Photography*, pp. 119 and 336.)

1857. M. JULIEN BLOT, in 1857, improved upon M. Taupenot's process, and prevented the blistering of the albuminized collodion film by a process in which dextrine was used. (See Hunt's *Manual of Photography*, p. 336; also *Cosmos* of 17th April 1857.)

1857. Messrs. SPILLER and CROOKES, in 1857, used nitrate of magnesia and nitrate of zinc to preserve sensitive collodion plates by their deliquescent properties; the plates thus being kept slightly moist. (See *Phil. Mag.*, May 1857; *Photographic Journal*, vol. I., p. 223, and vol. II., p. 6;

A.D.

also *Encyclopædia Britannica,* 8th edit., art. Photography, p. 548.)

1858. Mr. Llewellyn, in August 1858, published his oxymel process, in which bromide of potassium, alcohol, oxymel, citric acid, and nitrate of silver are used. (See *Encyclopædia Britannica,* 8th edit., art. Photography, p. 549.)

1858. Mr. R. M. Grier, in 1858, published the following process:—" A plate coated with asphalt varnish, known in " Prussia by the name of *eisenlack,* diluted with rectified " benzole or benzine (first product obtained by distillation of coal tar at a low heat), and when barely dry, and " still slightly sticky, was placed in contact in the pressure frame with a negative on albumen, and exposed to " the direct rays of the sun for half an hour. On removal from the pressure-frame, the plate was breathed " on over its whole surface, until the image became distinctly visible, the parts changed by the action of light " absorbing moisture, and those covered by the blacks of " the negative remaining unchanged, and repelling it. " In this state, the image being distinctly visible, it was " quickly covered by the bronze powder known as *aurum* " *musivum,* which at once changed the almost invisible " image to a direct positive on black and gold, the gold " adhering to the parts that had been protected from the " light, and not adhering to those where the actinic rays " had effected such a change in the molecular structure " of the film as rendered them capable of absorbing " moisture, thus producing a complete picture detailed in " all its parts." (See *Journal of the Photographic Society,* 22nd Nov. 1858, p. 87; also *Encyclopædia Britannica,* 8th edit., art. Photography, pp. 546, 547.)

1859. M. Van Monkhoven, in 1859, presented a memoir to the Académie des Sciences, on a new method of photography by means of the solvents of cellulose; he employed the ammonio-cupric solution of Dr. Schweitzer. (See the *Photographic News,* edited by W. Crookes, for April 15th 1859, pp. 61, 62.)

1859. Messrs. Bunsen and Roscoe, on the 26th May 1859, read a paper before the Royal Society, respecting their "photo-chemical researches." The standard flame for their purposes was that of carbonic oxide, and by its means they established a unit of chemical light. A thin film of cloud was found to increase the chemical effect, and thick clouds to extinguish the chemical light of diffuse daylight. In the higher latitudes it was found that the chemical action of diffuse daylight greatly exceeded that of direct sunlight. The chemical brightness of the sun, in the latitude of Heidelberg, was ascertained to be 36·6 times as great as that of burning magnesium wire when the sun's zenith distance is 67° 22. By means of a quartz prism and other appliances the chemical action of the various parts of the solar spectrum was measured. (See *Phil. Trans.* for 1859, pp. 879—926.)

---

# PHOTOGRAPHY.

A.D. 1839, August 14.—N° 8194.

BERRY, MILES (*a communication. Invented by Messrs. Louis Jacques Maude Daguerre and Joseph Isidore Niepce, junior*).—This invention is commonly known under the name of "Daguerreo-"type."

"The reproduction of the images received at the focus of the "camera obscura, is effected on plates or surfaces of silver, which "may be plated on copper." The silver surface of the plate is first polished and cleaned; secondly, a coating sensitive to light is applied to the silver surface; thirdly, the prepared surface is submitted to the action of light in the camera obscura, "so that "it may receive the images;" fourthly, the photographic picture, which is invisible when the plate is taken from the camera, is brought out; fifthly and lastly, the sensitive layer is removed from the plate, thus fixing the picture.

1st process.—Polishing the plate is effected by means of pounce pumice powder, or calcined Venetian tripoli, which is rubbed on the plate by means of finely-carded cotton dipped in olive oil. The plate is then cleaned by dusting the powder over the surface and rubbing it with dry cotton. It is further prepared by rubbing with cotton moistened with dilute nitric acid, cleansed again by means of powder and cotton, and submitted to considerable heat until "the surface of the silver has obtained a whitish tint or "coating." The plate is next cooled rapidly by placing it in contact with a cold substance, cleaned as before, and treated with dilute nitric acid as before, three separate times, the last time being just before use; it is finally entirely cleaned from pounce dust by cotton.

2nd process.—The layer sensitive to light is formed upon the plate, after it has passed through the first or cleansing process, by coating it with iodine by spontaneous evaporation in the dark.

The plate is first fixed upon a thin board by means of suitable metallic bands and nails; it is then exposed to the vapour of iodine (being for that purpose enclosed in a suitable box, in the bottom of which is a dish containing iodine,) until it has attained a golden yellow tint. The board and plate must then be introduced into a frame fitting the camera obscura, and having shutters or doors to exclude light until the plate is exposed in the camera.

3rd process.—The prepared surface should be submitted to the action of light in the camera immediately after being sensitized. This is done by adjusting "the focus of the camera obscura so "that the objects be represented perfectly clear and distinct;" a frame containing ground glass is moved forward or backward for this purpose. The glass is then removed, the frame containing the board and plate is substituted for it, and the shutters opened, so that the plate can receive the impression of the image of the objects chosen; the plate remains in this position until the requisite number of minutes have elapsed. The doors are then closed, the frame is removed from the camera, and the plate immediately subjected to the next operation.

4th process.—To bring out the picture the plate is removed from the frame and exposed in the dark to the vapour of mercury. A box is provided with a cup at the bottom, containing mercury the board and plate are introduced into the upper part of the box, and shut up. The mercury is then heated, and the plate left in until, by inspection through a glass by means of a candle, the development of the picture is found to be complete. The plate may now be deposited in a suitable dark box, thus being excluded from the light until it is convenient to fix the picture.

5th and last process.—To remove from the plate the coating of iodine, and thus to fix the picture, a solution of "sea salt" may be used, but a weak solution of hyposulphite of soda is preferred. The plate is first dipped into distilled water, then moved about in the saline solution until the yellow colour of the iodine is entirely removed, again plunged into water, and finally subjected to the action of a continuous stream of hot water falling on an inclined plane carrying the plate, thus cleansing it perfectly; it is then ready for mounting by being placed in a pasteboard case, and covered with glass, thus preserving the silver surface from being touched and from tarnishing.

The Specification describes these processes at great length, and Drawings of the apparatus are given in detail.

[Printed, 11*d*. See Repertory of Arts, vol. 13 (*new series*), p. 176; vol. 6, (*enlarged series*), p. 256; vol. 8. (*enlarged series*), p. 47; and vol. 14, (*enlarged series*), p. 293. London Journal (*Newton's*), vol. 16, (*conjoined series*), p. 1; vol. 21, (*conjoined series*), p. 57; vol. 28 (*conjoined series*), p. 368; vol. 31 (*conjoined series*), p. 64; and vol. 34 (*conjoined series*), p. 438. Mechanics' Magazine, vol. 31, p. 464; vol. 32, p. 77; and vol. 47, p. 45. Patent Journal, vol. 3, p. 157; also vol. 7, p. 125. Inventors' Advocate, vol. 2, p. 163. Common Bench Reports, vol. 3, p. 97; also vol. 8, p. 165. Carrington and Kirwan's Reports, vol. 2, p. 667.]

## A.D. 1840, June 13.—Nº 8546.

BEARD, RICHARD (*a communication*). — This invention consists of:—

1st. "A mode of taking likenesses and representations of "nature, and of drawings by reflecting images on to suitably "prepared surfaces," by means of concave reflectors. Instead of the camera obscura a "rectangular" box is used, inside of which and at the end thereof is fixed a concave reflector; the prepared surface is placed opposite the reflector, and is capable of being moved towards or away from it for the adjustment of the focus. The box has its end opposite to the mirror open; this end is placed opposite to the object to be photographed, so that a clear image is reflected on to the prepared surface.

The following points are mentioned as auxiliaries to the use of this apparatus in the case of taking portraits, &c.:—A head-rest attached to the chair of the person sitting; the glazing of the inclined glass roof of the room with blue glass; the use of a large concave reflector, to reflect the light of the sun on to the person sitting, the rectangular box above mentioned being then placed behind the large reflector, which has a hole through it towards the outer edge, also "a surface of tissue paper varnished with "boiled oil" stretched across its front; a screen of ground glass placed behind the person sitting; a frame containing a white surface placed behind the screen, so as to receive the shadow of the person. By the combination of the screen and frame, varying effects of light and shade are produced in the picture. In the case of plaster busts or other white objects, it is sometimes advantageous to use a brown, blue, or black back surface.

2nd. "A mode of preparing silver surfaces by pressing them "face to face between hardened rollers, when they are to receive "images." Two plates of copper coated with silver are cleaned with cotton and dilute sulphuric acid; their silver surfaces are then placed in contact, and they are passed between a pair of smooth hardened rollers; they are then annealed by heat, permitted to cool, and the whole process repeated until the silver surfaces are highly polished and equal in appearance all over. Tripoli and dilute nitric acid are then lightly rubbed over the silver surface with cotton, and then rubbed off with dry cotton. Lastly, the silver surface is rubbed over with velvet impregnated with charcoal, and is then ready to undergo the iodine process.

3rd. A mode of treating silver surfaces used to receive images, by submitting them to the action of iodine and bromine or bromic acid combined, and, by preference, either nitric acid or sulphuric acid and water is also combined with the iodine and bromine. The silver plate is placed on a glass slide which works in a box, at the bottom of which is a vessel containing the mixture of iodine and acid, or iodine, bromine, and acid, or iodine and bromine; the silver surface is placed downwards, and in a few seconds is ready to receive the image, or to be put into a dark case ready for use.

[Printed, 7*d*. See Repertory of Arts, vol. 15 (*new series*), p. 137, and vol. 1 (*enlarged series*), p. 185; London Journal (*Newton's*), vol. 18 (*conjoined series*), p. 112; and Inventors' Advocate, vol. 4, p. 34.]

### A.D. 1841, February 8.—Nº 8842.

TALBOT, William Henry Fox.—The first part of this invention is generally known as "the calotype process."

The first part of the invention is arranged under the following heads:—

*The preparation of "iodized paper."*—The best writing paper is washed successively with solutions of nitrate of silver and of iodide of potassium, being dried after each washing; it is kept in a dark place until wanted for use, when the next process is performed.

*The preparation of "calotype paper."*—A solution of the "gallo-"nitrate of silver" is formed, by first adding strong acetic acid to a solution of nitrate of silver, and to this compound solution adding a saturated solution of gallic acid or the tincture of galls diluted with water. The "iodized paper" is washed over with the "gallo-nitrate of silver" thus prepared, dipped in water and dried; it is then ready for use in the camera obscura.

*To entirely bring out the invisible or faint image.*—When the

the light thus received. The reflected rays pass through an adjustable tube containing two sets of lenses on to the surface on which the copy is to be made; by adjusting the relative distances of the tube and the reflecting and receiving surfaces, various sizes of copies may be produced. By inclining the reflecting and receiving surfaces to each other, distorted pictures may be copied in true perspective.

3rd. Apparatus for exhibiting photographic pictures on a screen in a dark room. A lime light illuminates a photographic picture made on the surface of a concave metallic mirror; the light from all parts of this picture passes through "a set of achro-" matic glasses," and the image is received on to a suitably placed screen. To enable the dark parts of the picture on the mirror to "become the least polished and the lightest the most so, with the " intermediate gradations," some sulphuric acid is placed in the mirror and heated until the desired change takes place; the mirror is then washed, cleaned, and dried.

4th. Copying photographic pictures on metallic plates by means of the light reflected from their polished surfaces. The light is converged, by means of a lens or lenses, on to the surface of the plate (previously prepared by means of sulphuric acid as in the 3rd improvement), passed through a tube containing a double set of lenses and received on the sensitive surface, which is placed parallel to the original surface in order that it may receive a true image; a small likeness may thus be copied on to paper so as to be the size of life. The apparatus described in the 3rd improvement may be used by placing the prepared surface to receive the copy in the place of the screen.

[Printed, 8*d*. See Repertory of Arts, vol. 2, (*enlarged series*), p. 275; and London Journal (*Newton's*), vol. 25 (*conjoined series*), p. 330.]

## A.D. 1843, June 1.—N° 9753.

TALBOT, William Henry Fox.—This invention relates to the following improvements :—

1st. "To give increased whiteness to calotype and other pho-" tographic pictures, and at the same time make them more " permanent," they are plunged into a hot solution of "hyposul-" phite of soda (or any other soluble hyposulphite)," then removed, washed, and dried. After this process the picture may be waxed "by causing melted wax to penetrate into the "pores of the paper, the object of which is to give increased "transparency."

2nd. To make calotype or other photographic paper more sensitive, it is exposed to the action of a warm plate of iron during the formation of the picture.

3rd. The preparation and use of "io-gallic paper." "Iodized "paper" is washed with a saturated aqueous solution of gallic acid. When wanted for use in the camera, it is washed with a solution of nitrate of silver, which renders it sensitive to light.

4th. The preparation of dry "calotype paper." "Iodized "paper" is washed with the "gallo-nitrate of silver," containing only a small proportion of the nitrate instead of equal parts as used in making moist "calotype paper." This paper is still sensitive to light and can be used in a copying frame, but it "can "easily be dried at a gentle fire without being spoiled thereby."

5th. A photographic copy is taken in the usual way, except that the light is allowed to act upon it twice the usual time. It comes out too dark, is washed, and then plunged into an iodide of potassium solution, which brightens the picture and causes the lights to assume a pale yellow tint; by exposing the picture to the light this effect may be increased. "The picture is then "washed, and then plunged into hot hyposulphite of soda" (as in the 1st improvement) until the lights are of a white colour.

6th. To obtain pleasing effects, a paper photograph is waxed, which makes it more transparent; a sheet of white or coloured paper is then fastened behind it.

7th. Obtaining enlarged photographs. An enlarged negative calotype picture is obtained by throwing a magnified image of a small photograph upon "calotype paper," by means of lenses; positive copies are obtained in the usual manner. Large portraits can thus be taken without the sitter being obliged to sit an inconveniently long time.

8th. "The application of photography to printing by arranging "moveable letters or figures so as to form pages, and making "photographic images thereof." Letters (cut out from some pages of letter-press) are composed and fixed in proper order upon a sheet of white paper; a photographic negative is taken from this, and the requisite number of positive copies taken from it. Another method consists in using a camera obscura to throw upon sensitive paper the reduced image of a tablet, on which large letters have been suitably arranged in grooves.

9th. "Photographic publication," or the multiplication of positive photographs by a certain combination of processes. To form the negative copy, writing paper is dipped into salt water, wiped, and dried; it is then washed with a solution made by precipitating nitrate of silver by means of ammonia and re-dissolving the precipitate, and dried; this prepared paper (called "copying paper") is then used in the copying frame. A negative of distant objects may be obtained with a camera upon "calotype paper." The negative copy upon "copying paper" is fixed by being washed with warm water, placed in a solution of hyposulphite of soda, and all removed that is soluble in water by plunging it into two or three baths of warm water consecutively; it is then dried, and, if necessary, waxed. The "calotype" negative may be fixed as set forth in the 1st improvement. From either of these negatives, the requisite number of positives is obtained upon copying paper by means of the copying frame, and fixed in the same way as the negative copy was.

The terms "calotype paper" and "iodized paper" are used in the same sense as that employed in N° 8842. "A general re-"ference is had to what has been explained and made known" in N° 8842.

[Printed, 4*d*. See Repertory of Arts, vol. 3 (*enlarged series*), p. 47; London Journal (*Newton's*), vol. 23 (*conjoined series*), p. 430; and Artizan, vol. 2, p. 23.]

## A.D. 1843, November 21.—N° 9957.

CLAUDET, Antoine François Jean (*a communication*).—This invention "consists in rendering the daguerreotype picture "susceptible of producing by printing a great number of proofs "or copies, thereby transforming it into a complete engraved "plate."

The plate is first delicately engraved, the "biting in" is then completed, and, lastly, the plate is protected "from the effects of "wear produced by the operation of printing."

*The preparatory engraving.*—The plate is first cleaned by immersion in a weak solution of nitric acid, washed, immersed in "alcoholized potash," then in the biting acid (called "normal "acid"), washed, and then treated with "ammoniacal chloride of

" silver," again washed, heated in caustic potash and left to cool. By repeating these operations a second biting is produced, and by again repeating them (stopping, however, before the ammoniacal chloride of silver is used), a third biting is produced. The plate is then dried, polished with pumice stone, and the chloride of silver filling the black parts removed by means of the normal acid and the ammoniacal chloride of silver solution. The "normal " acid" is composed of nitric acid, "nitrate" [nitrite?] "of " potassa," and common salt, the whole much diluted with water; other solutions used intermediately to the principal ones above mentioned are, a weak solution of caustic potash, a weak solution of ammonia, and distilled water.

*The completion of the " biting in."*—The plate is inked as copper-plate printers do, but with a siccative ink, dried, and polished on the white parts; it is then electro-gilt, cleansed by means of boiling caustic potash, and the lamp black (if any) removed by rubbing the plate with a crumb of bread. The plate is then bitten in by aquafortis (which does not attack the face covering of gold) "according to the ordinary process used by engravers."

*The protection of the plate.*—The surface of the plate is electro-coated with a very thin pellicle of copper; it may then be used to print from. When this thin coating begins to show signs of wear it must be removed and a fresh electro-coat applied in its place. To remove the pellicle, a dilute solution of nitric and nitrous acid, or a caustic solution of ammonia, may be used.

[Printed, 4*d.* See London Journal (*Newton's*), vol. 25 (*conjoined series*), p. 111.]

## A.D. 1849, April 17.—N° 12,575.

ALLIOTT, Alexander.—This invention relates to improvements in various kinds of registering apparatus.

The Specification and Drawings describe and show "apparatus " for ascertaining and for marking or registering the force or " pressure of wind, of water, and of steam, the weight of goods " or substances, and the velocity of carriages; also an apparatus " for ascertaining, under certain circumstances, the length of " time elapsed after carriages have passed any given place, and " for enabling the place or direction of floating bodies to be " ascertained."

The "improvements in apparatus for enabling the place or " direction of floating bodies to be ascertained, consist in the " employment for that purpose of the mariner's compass, *in " combination with chemically prepared paper*, moved by suitable " clockwork." The following is an apparatus by which the direction in which a vessel is sailing may at every moment be registered :—The registering paper is " prepared by some suitable " photographic process," and ruled with five lines by which eight divisions of the compass are shown ; it is caused to travel uniformly in a line parallel with the sides of the vessel, immediately below the compass card, by means of rollers actuated by clockwork. The light passing through two small holes in the compass card marks the sensitive paper according to the course of the vessel. " A separate scale divided into parts of equal length, correspond- " ing to the rate of motion given to the paper, must be employed " to measure the length of time the vessel sails in any particular " course." " During the night some artificial light must be " used."

[Printed, 2*s.* 6*d.* See Mechanics' Magazine, vol. 51, p. 381; and vol. 52 p. 361; also Patent Journal, vol. 8, p. 55.]

### A.D. 1849, December 19.—N° 12,906.

TALBOT, William Henry Fox, and MALONE, Thomas Augustine.—1st. "The use of plates of unglazed porcelain to re- " ceive the photographic images." The porcelain plate is prepared for use by giving it a coating of white of egg, and gently drying it at a fire ; the plate being thus prepared, the remainder of the process and manipulation does not differ from that ordinarily employed in making photographic pictures upon paper. The process described is nearly the same as that set forth in N° 8842.

2nd. " A method of converting or changing negative photo- " graphic images into positive ones." To produce the negative picture, a clean plate of glass is covered with a film of albumen by pouring a solution of albumen over the glass, and drying it ; the film is then iodized by exposure to the vapour of iodine, dipped into nitrate of silver solution, placed in the camera, and treated with an aqueous solution of gallic acid, which developes the latent picture. To convert this picture into a positive picture, a solution of nitrate of silver is allowed to stand some time on the plate ; the resulting picture is still negative if viewed by transmitted light, but, if viewed by reflected light, is positive. To view the positive image well, the glass should be placed upon a dark surface. The positive picture thus obtained must be fixed in the usual way. Other surfaces are stated as being advantageous for this process, such as talc, varnished or oiled paper, dark coloured porcelain, and other dark surfaces. Instead of albumen, gelatine or other substances of animal or vegetable origin may be used.

3rd. The use of varnished or other transparent paper "to "support a film of albumen for photographic purposes." The varnished paper is coated with albumen on one side, and rendered sensitive to light by exposure to the vapour of iodine, as in the 2nd improvement. This paper is useful for "photo-"graphic pictures giving panoramic views of scenery, which are "produced upon a curved surface by a movement of the object "glass of the camera."

4th. "A method of obtaining more complete fixation of pho-"tographic pictures upon paper." In addition to the usual fixing process, the picture is dipped into a boiling solution of caustic potash. To give an agreeable tint to the picture it may then be exposed to the vapour of sulphuretted hydrogen.

5th. Forming photographic pictures upon steel plates. A mixture of solution of iodide of potassium and albumen is uniformly spread over the plate, and dried. The plate is then washed over with an alcoholic solution of "gallo-nitrate of silver," and is ready to receive the photographic image, which is to be fixed by hyposulphite of soda.

A Disclaimer was enrolled May 1, 1855, by William Henry Fox Talbot, in which the 1st, 3rd, 4th, and 5th portions of the invention are disclaimed; also that part of the 2nd improvement which refers to other surfaces besides glass, is disclaimed.

[Printed, 5*d*. See Repertory of Arts, vol. 16 (*enlarged series*), p. 97; Mechanics' Magazine, vol. 52, p. 518; and Patent Journal, vol. 9, p. 143.]

### A.D. 1851, June 12.—N° 13,664.

TALBOT, William Henry Fox.—1st. A method of producing photographic images upon glass plates. Pictures thus obtained are called "amphitypes," in consequence of their appearing either positive or negative, according to the light in which they are held. A clean plate of glass is covered with a film of albumen, which is dried, dipped into a nitrate of silver solution, and dried again, washed, coated a second time with an albuminous film, treated with proto-iodide of iron, containing excess of iodine, acetic acid, and alcohol, and dried. When required for use, the plate is dipped into a solution of nitrate of silver, containing acetic acid; the photographic image is then impressed. To develope the latent image the plate is dipped into a solution of sulphate of iron. To fix the picture, it is dipped into hyposulphite of soda solution, washed, cleaned with cotton and water, and dried. The surface may then be coated with a protective covering of varnish or albumen.

An arrangement for using this invention when the operator is on a journey is fully set forth. A plate prepared for use in the camera is dropped into a glass cell in the camera, which is supplied from the outside with nitrate of silver solution; the image is then taken, and the cell containing the plate is consecutively supplied with the proper quantities of sulphate of iron solution, distilled water, and hyposulphite of soda solution, and is then removed from the camera, having been in the camera (and therefore not exposed to extraneous light) from the time of sensitizing the picture to its fixation. Instead of pouring the contents of the bottles into the glass cell, stop-cocks and supply pipes may be used. The glass cell has an exit pipe.

2nd. Taking "photographic pictures of objects which are in "rapid motion." The glass plate being prepared as set forth in the 1st improvement, and properly arranged in the camera, the moving object is illuminated by the discharge from a large electrical battery. The instantaneous image thus impressed is treated as in the 1st improvement, and the result is a distinct positive image of the moving body upon the glass, "the rapidity of the "motion not affecting the accuracy of the delineation."

A Disclaimer was enrolled March 8, 1854, by William Henry Fox Talbot, in which the arrangement for using the 1st part of this invention when the operator is on a journey is disclaimed; also the whole of the 2nd part of the invention is disclaimed.

[Printed, 5*d*. See Repertory of Arts, vol. 19 (*enlarged series*), p. 41; and Mechanics' Magazine, vol. 55, p. 497.]

A.D. 1852, September 23.—N° 14,302.

TARDIEU, Jacques Léon.—"Certain improvements in the "colouring of photographical images."

The paper photograph, after being varnished, is mounted either on a frame, or on a sheet of glass. Oil colours are preferred to be used; they are laid on at the back of the photograph thus rendered transparent, "the transparency of the paper permitting "the colours to appear in all their brightness on the paper side of "the photographical image."

"The final operation for the preservation of the images" consists in glueing (or otherwise sticking) "on the colour," one or several sheets of paper, tissue, suitable fabric, pasteboard, wood, or other suitable substance of slight thickness. Cloth or canvas, covered with paint in the same manner as canvas for paintings, may be used to glue the photograph on, pressure being used to attach the surfaces, and a proper varnish afterwards applied to the surface of the image.

[Printed, 3½*d*. See Mechanics' Magazine, vol. 58, p. 296.]

---

# PATENT LAW AMENDMENT ACT, 1852.

## 1852.

### A.D. 1852, October 2.—No 179.

NEWTON, Frederic.—"Improvements in the apparatus to be "employed for producing photographic pictures."

This invention consists in "an improved arrangement of apparatus for taking photographic pictures without the use of a dark "room;" the sensitizing, developing, washing, and fixing operations being either performed in the camera itself, or in a box attached to it, without handling the plate or introducing the hands into the camera.

One method consists in introducing the glass plate covered with collodion "into a glass vessel placed in the camera, and supplied "in turn with the proper solutions for rendering the plate, first,

" sensitive, then to develope and finish and fix the picture." This method is set forth only in the Provisional Specification.

A second method consists in suspending the glass or other plate over an aperture at the bottom of the camera, so that the vessels containing the chemical solutions may be successively introduced from below, and the plate thereby submitted consecutively to their action.

A third method consists in making the vessels slide in grooves under the camera, so as to come successively under the plate, which is consecutively dipped down into them, the plate being for that purpose attached to a rod working in a stuffing box at the top of the camera; or the stuffing box may move in grooves, the troughs being fixed.

A fourth method consists in placing the troughs inside the camera, and having a top chamber, into which the plate can be drawn out of the respective solutions by means of a rod and stuffing box.

A fifth method consists in placing the sensitized plate in a dark chamber with a moveable shutter in the camera, impressing the image, and then transferring the dark chamber to the top of a box which contains troughs of the several solutions; these troughs are successively brought under the plate, which is consecutively dipped into them, and is thereby acted upon so as ultimately to produce the finished picture. The cover of this box is so made (with hinges) that, when it is shut, sheets of vulcanized India-rubber come down upon the troughs, and preserve the solutions in them perfectly air-tight.

[Printed, 7½*d.*]

## A.D. 1852, October 29.—N° 565.

TALBOT, William Henry Fox.—This invention refers to engraving pictures, obtained by photographic processes on steel plates; the engraved plates can be used for printing from.

A steel plate (prepared for the use of engravers) is dipped into a solution containing vinegar and sulphuric acid, coated with a mixture of gelatine and bichromate of potash, impressed with the photographic image by means of the copying frame, and washed. The film of gelatine is originally yellow; a yellow image on a brown ground is produced by the action of light, and the washing removes all the bichromate of potash and the greater part of the gelatine from the parts of the plate upon which the sun's rays have not acted; the consequence of this is that the image is whitened. The photographic part of the process is completed by dipping the plate into alcohol, and placing it to dry in a vertical position.

The plate is then etched by means of a solution composed of a certain proportion of "bichloride of platina," containing a little free acid and water. The plate is placed horizontal, and has some of this solution poured upon it, "and quickly diffused and spread "over the whole plate with a camel's hair brush;" the etching is finished, when a very distinct and regular black image of the object is obtained.

"Photographic veils."—To produce the effect of engraved lines or of uniform shading, the image of a piece of folded gauze, or other suitable material, is impressed upon the gelatine prior to the image of the object required being formed.

Plates of zinc or lithographic stones are also readily engraved by this process.

[Printed, 4½*d.*]

## A.D. 1852, October 30.—N° 575.

LUCENAY, Pierre Bernardet de.—"The production of "photographic images by means of artificial light."

To obtain an image of a "stereotype" [stereoscopic or photographic?] plate on unpolished glass. The artificial light is reflected by a mirror, so as to pass through lenses, through the "stereotype" [?] "plate, which is to be copied," then through "an ordinary daguerreotype objective" on to the glass plate where the required representation is produced.

Instead of receiving the image on a sheet of unpolished glass, it may be received on photographic paper, or on a glass plate coated with collodion, and sensitized so as to fix the image thus produced.

"The light is produced either by the galvanic battery" or by "refractory bodies heated by lighted oxigen, or by the combustion "of pyrotechnic compositions, or by any other suitable means." When parallel rays of light are required, a parabolic reflector is used, the light being placed in its focus.

"The light is rendered diffuse for portraits. In this case the " interposition of blue cobalt glasses will be required, and the " reflectors will be a simple screen of a dull ground placed behind " the luminary." This application of the invention is not mentioned in the Complete Specification.

[Printed, 5½*d.*]

A.D. 1852, November 9.—N° 693.

MABLEY, WILLIAM TUDOR.—"Ornamenting glass and other " transparent or partially transparent substances for windows " and for other purposes," by means of photographic designs formed thereon.

A design for a window or lamp may be produced by copying the object required upon glass by the collodion or other process, fixing the image, and applying a transparent varnish to protect it from injury; this, however, will be a negative representation, and not suitable for many designs.

To produce a positive design, a negative photograph (prepared by superposition or by the camera) is used as the original to form a positive picture on the glass by means of the collodion process. One negative will serve to produce a number of positives.

Positives may also be produced on a different scale from their negatives by means of the camera, which also affords the means of combining various negatives into one design. The effect is heightened by placing ground glass or enamelled glass behind the positive picture. The boldness of the picture may be increased by the usual means, and is sometimes essential, in the application of photography to this invention.

To facilitate ornamentation, the sensitive film, with the image on it, may be transferred from one surface to another; parts of surfaces only may be ornamented in this manner; also curved surfaces may be thus made receptive of photographic images. By this means designs may be made up of several photographs, and thus large designs produced that it would not be possible to photograph at one operation.

The designs may be protected from injury by varnish or by a sheet of plain glass.

[Printed, 3½*d.*]

A.D. 1852, December 29.—N° 1196.

POWER, JAMES.—This invention consists of "silvering all "sorts of metals and of glass," "by a solution of silver, with "the aid of electricity, and of covering the said coating of silver "with a coating of copper, in the usual way, and in certain cases" of detaching "the silver and copper plate from the glass, to serve "for daguerreotype, photograph, and other similar purposes."

The solution of silver consists of nitrate of silver, nitrate of ammonium, and "nitrate of silver of alcohol (spirits of wine gun-"powder proof)" [the Provisional Specification says, "nitrate of "alcool (spirits of wine gunpowder proof)"]; to this is added a certain quantity of "resinous spirit," consisting, by preference, of gum galbanum and spirits of wine.

For laying a very thin coating of silver upon glass to give it a conducting surface, the above solution has a few drops of spirits of cloves mixed with it; the glass plate is then transferred to another silvering bath (formed as described above), connected with a galvanic battery, then a coating of copper given by the usual galvanic means.

These silvered plates may be said to be electro-cast upon glass; "chrystal, porcelaine," and other non-metallic substances susceptible of receiving a perfect polish, may be used. The metallic plate, when detached from the glass, is of the most perfect polish, "requiring neither flattening nor polishing, and is most beauti-"fully adapted to the use of daguerreotype and photographic "portraits, views, &c[a]."

[Printed, 4½*d.*]

---

## 1853.

A.D. 1853, January 25.—N° 193.

MAYALL, JOHN EDWIN.—This invention "relates to the pro-"duction of imitation crayon drawings or portraits in or by the "photographic process by the aid of a mechanical contrivance "interposed between the object and the camera."

This contrivance consists of a disc, with a central aperture in it of the form of a large star, which disc is slowly revolved by means of clockwork. "The central or free open portion of this star is "sufficiently large to admit the rays from that part of the object "which is to be shewn in strong light, or as a firm sharp image, "whilst the rays from those exterior parts which are to be "gradually shaded or deepened off to a dark or light background "are partially intercepted by the converging points of the star. "With this apparatus the intensity of the light is gradually "diminished, and the pictures taken in conjunction with it "possess the required softened off crayon effect. The apparatus "is applicable to every kind of camera, and by placing it at "different distances from the camera different portions of the "image may be softened off."

"The pure stellar shape of aperture is not essentially necessary "for the end in view, inasmuch as other shapes would answer for "producing the graduation of the rays."

[Printed, 5½*d.*]

## A.D. 1853, March 23.—N° 711.

CLAUDET, Antoine François Jean.—"Improvements in "stereoscopes." This invention consists:—

1st. In "the producing of pleasing and novel optical illusions "by means of a peculiar construction and arrangement of some of "the parts, which are made moveable so as to impart to the "picture the appearance of moving figures."

To enable the picture to "be seen in a natural position, and not "reversed, as is usually the case," a mirror is placed in the stereoscope at an angle of 45° to the plane of the picture, and the eye pieces are placed at the back of the stereoscope, so that the line of sight is parallel with the plane of the picture.

In another improvement, a slide, adapted to the eye pieces, moves backwards and forwards, and thus opens and closes each eye piece alternately. The pictures are made differently, so that the objects are not similarly placed (but differently viewed) in each, as in the case of ordinary stereoscopic views. One view shows a person with his hat on, for instance; the other the same with his hat off; by the action of the slide it appears that the figure is taking off his hat. "An increased effect and great "variety will be produced by combining a number of pictures "together in the form of a cross, so arranged as to exhibit alter-

" nately under each eye from different positions, so that the " observer may, in succession, see eight different pictures, all in " different attitudes." Two crosses are used; they are mounted on horizontal axles in the body of the stereoscope, and are rotated by a coiled spring, free to act by the movement of the slide.

2nd. "In so constructing and combining the several parts of " which the stereoscope is composed that the instrument may be " collapsed and brought into a small compass, so as to be ren- " dered more conveniently portable than ordinary rigid stereo- " scopes."

The mirror stereoscope mentioned in the first part of the invention is rendered portable by making the bottom of the stereoscope correspond to the bottom of the case formed by the instrument when folded; the ends of the instrument form the lids, the mirror with its "lids" fold up against the back of the instrument, which back is hinged to the bottom, and the eye pieces are made collapsible by means of helical springs.

In another construction, " immediately the fastening which " holds the instrument in a closed state is undone, all the parts " will be at once made to assume their proper position for viewing " a picture." The bottom of the closed instrument corresponds to the bottom of the instrument when in use, and the lid to the back; the said back supports the eye-piece flap, as well as another flap which forms the partially open front of the stereoscope. The eye pieces are as in the folding mirror stereoscope. The instrument is ready for use when the cover is opened, the flaps being made to open by an elastic strap. A modification of this construction consists in supporting the eye-piece flap by a central partition, which springs into its proper position by an elastic spring. Another modification consists in altering the external form of the instrument so that the eye pieces need not be collapsible.

" Another form of making the instrument consists in attaching " the eye pieces to two flaps, which have holes to look through " made therein, and are hinged to a central vertical partition, " which is made moveable in a vertical guide, so that it may be " moved up or down for the purpose of altering the focus when " required." This instrument will open without assistance from the observer. The top and bottom (together) of the case form the bottom of the instrument when in use. In a modification of this form the vertical guide forms the back of the instrument and also the top of the closed box; the eye-piece flap is in one piece, and has an overhanging flap for a shade.

[Printed, 6½d.]

## A.D. 1853, May 21.—N° 1260.

SCOUTETTEN, HENRI JOSEPH.—This invention is entitled " An improved plastic compound applicable to various ornamental " and useful purposes," and (amongst other things) it relates to rendering paper impermeable, which impermeable paper "is " suitable for photography." This application of the invention is not mentioned in the Provisional Specification.

" When it is desired to render paper or stuffs impermeable, the " caoutchouc and the gutta percha must be separately dissolved " in sulphuret of carbon, in the proportion of eight of gutta " percha for one hundred of sulphuret of carbon, well purified; " the solution is left to rest during eight days, and the white of " eggs is added to it; when the impure matters are deposited it " is poured forth to obtain an almost colorless liquid. Paper or " stuff may be then steeped in this liquid, and drawn from it by " passing them between cleansing rollers, which equalize the layer " of the matter."

[Printed, 2½d.]

## A.D. 1853, July 8.—N° 1629.

BRETT, JACOB (*a communication*).—(*Provisional Protection only.*) "This invention consists in combining together into one " apparatus two photographic cameras in such a manner that two " stereoscopic pictures may be taken at the same time and on the " same plane. The instrument is arranged in such manner that " the angle at which the two pictures are taken may be varied at " pleasure, but at the same time retaining the distance between " the pictures on the plane always constant."

" The invention also consists in obtaining from a negative pho- " tographic picture a positive picture of either a larger or smaller " size than the original negative, by so arranging the apparatus " that the light which passes through the negative picture (which " is preferred to be on glass) is caused to pass through a series of " lenses which bring the light to a focus at the other end of the "apparatus, where a prepared surface is placed to receive the "positive image, the size of which will vary according to the "distance at which the negative picture is placed from the "lenses."

[Printed, 2½*d.*]

A.D. 1853, October 31.—Nº 2510.

GOËTHEL, Christian, and ZIMMERMAN, Charles Moritz.—(*Provisional Protection only.*) "Improvements in stereo-"scopes."

"The improved portable stereoscope" "consists principally of "three parts hinged together so as to allow of the whole being "folded up."

The Drawing shows an instrument with a moulded base, which constitutes the "centre piece" or back of the stereoscope when in use. Two vertical planes are hinged to this base; one carries the lenses, the other the picture to be viewed. These planes may be held in a vertical position by ties, and are also capable of being folded down (one over the other) on to the base.

"The distance of the verticals is regulated according to the "focal distance of the lenzes. This arrangement allows of a more "minute examination of the details of the object, and the whole "when folded up occupies no more room than a small sketching "block."

[Printed, 4½*d.*]

A.D. 1853, November 3.—Nº 2552.

DUPPA, Bryan Edward.—"Improvements in coloring pho-"tographic pictures."

These improvements "consist of rendering the front surfaces "transparent with suitable varnish, and applying colors to the "back surfaces, by which means the laying on of the colors "will not require that artistic skill heretofore necessary, and "results are obtained which cannot be obtained by the coloring "on the front surface."

A coating of wax, mastic, or other varnish is given to the front surface of a paper photograph; this process renders the paper transparent. Oil colours are then suitably applied to the back of the picture by means of a brush; the colour being applied uniformly (in most cases), the lights and shadows are entirely produced by those of the photographic picture itself. The back of the picture is then cemented on to a surface of white lead spread over cardboard, and, finally, the front of the picture is again varnished.

The image shown by the photographic picture may be reversed by colouring the "face" or front of the photograph, the paper being rendered transparent as explained above; "the photo-"graph image will then be seen through the paper, and the "colours through both."

[Printed, 2½*d.*]

A.D. 1853, December 8.—N° 2856.

LAVERDET, MARCEL GUSTAVE.—(*Provisional Protection only.*)
"This invention relates to a mode of coloring photographic "pictures, so as to give them a life-like or natural appearance, "without impairing the exactitude of the photographic image, as "is the case with the ordinary method of coloring, and without "requiring the manipulation of a skilful artist."

The photograph is rendered transparent "by coating the back "with varnish, and causing it to penetrate the pores of the fabric "containing the picture." When the varnish is dry, oil colours are laid on the back of the picture, "so as to imitate as closely as "possible the natural tint of the object represented.

"These colors may be diluted with rectified essence of tur-"pentine so as to obtain the requisite tints, care being taken to "saturate the paper, so that the color may penetrate through to "the other side.

"The picture is then allowed to dry, and is afterwards varnished "on the front side. If the tones are not correct they may be "removed by means of essence of turpentine, and others laid on "without injuring either the picture or the paper

"It is only now necessary to place the photographic picture "upon a white ground and the operation is complete."

[Printed, 2½*d.*]

## 1854.

### A.D. 1854, January 18.—N° 125.

BOURQUIN, Jean Pierre. — (*Provisional Protection only.*)

This invention relates :—

1st. To a method of making troughs "for holding the various "chemical solutions required in photographic processes." The troughs are made "of a square, rectangular, or other conveniently "shaped framework of wood, in which a glass bottom is cemented "in any suitable manner by means of a cement which is not "liable to be acted upon by the chemical solutions to be em- "ployed."

2nd. "To a method of levelling the troughs or vessels, so that "the same depth of liquid may be maintained in every part of "the trough." Three or more levelling screws are adapted to the framework of the trough; these screws "pass through screwed "shoulders or brackets, fixed on the tray, trough, or vessel, and "form the feet whereon the vessel rests."

[Printed, 3*d.*]

### A.D. 1854, February 20.—N° 407.

URIE, John.—"Improvements in photographic pictures."

To give photographs on glass or other transparent plates, "the "effect of standing out in high relief from the apparent image "surface," "the back of the glass or the surface opposite to that "on which the image is taken" is "coated with a black varnish "or other pigment; care being taken to confine this reverse "coating to the exact area opposite to or covered by the picture, "or that portion of the picture to which the appearance of relief "is to be given." Paper or other dark material may be used to give the effect of relief, "and two or more kinds of reverse "coatings may be used." In some cases the background may be coloured and shaded, "or the part uncovered by the actual image "may be treated with fluoric acid." Different degrees of relief may be given by using two or more thicknesses of glass or other transparent material, "and by putting the dark reverse coating

" for one part of the picture at a greater or less distance from the " picture surface than that for another part."

By another process, plaster of Paris or other plastic substance is poured or deposited upon the surface of a positive image on glass, and, when solidified, the substance is detached from the glass, " taking with it the positive image film."

[Printed, 4*d.*]

## A.D. 1854, April 7.—N° 808.

FONTAINEMOREAU, Peter Armand le Comte de (*a communication*).—(*Provisional Protection only.*) " Certain improve- " ments in the preparation of photographic paper."

" The invention consists in the construction of an apparatus " by which photographic paper is instantaneously obtained, and " which shews visibly the progress in coloration of the proofs, " producing at the same time the degree of heat favorable to the " operation.

" It is composed of pans with edges perfectly flush, resting on " a plate of glass, to allow the liquid chemical solution spread on " the surface of the paper to combine with it, by absence of any " intervening atmosphere, light, dust, &c.

" Secondly, of a frame with sliding door, which perfectly en- " closes the moistened paper, so as to preserve it from the effects " of light, though the operation takes place in open day."

[Printed, 3*d.*]

## A.D. 1854, April 7.—N° 809.

SAUGRIN, Louis François.—(*Provisional Protection only.*) " Improvements in the construction of stereoscopes."

This invention " consists in constructing stereoscopes with " moveable lenses and collapsable springs, and rendering them " more portable than heretofore."

This instrument is made so as to be entirely compressible into the stereoscope case; for that purpose the piece carrying the lenses is connected to the bottom of the case by means of a " rectangu- " lar spiral spring." The bottom of the case forms the bottom of the instrument, and has an aperture " to show the object by trans- " parency." The lenses are " set in a sliding piece, to facilitate the " change of glasses." " Black stuff is stretched " around the "nose piece," "so as to prevent the reflection of the face on the "object."

A spring "which can be entirely compressed," and applicable to a stereoscope, is shown in the Drawing, and consists of two flat semicircular springs in the case, and one abutting on them to support the eye piece. Another form of spiral spring is also shown.

[Printed, 5*d.*]

## A.D. 1854, April 7.—N° 822.

NEWTON, William Edward (*a communication*). — "Im-"provements in producing stereoscopic pictures, and in the appa-"ratus for exhibiting such or similar pictures."

1st. The method of producing stereoscopic pictures "consists in "taking the two pictures from the extreme ends of the hypothe-"nuse of a right angle triangle, one of the sides of which is hori-"zontal and the other vertical, each of the sides being two and a "half inches, or thereabouts, in length." One method of practically accomplishing this object, is to take the first picture with the camera risen up two and a half inches from its table, by means of studs or pins, and to take the second picture with the camera placed flat on its table, the pins passing through apertures distant horizontally from its former position two and a half inches. Other methods can be used "which shall allow the camera to be put into the two positions required."

2nd. The method of exhibiting stereoscopic pictures consists in giving them "a panoramic motion into and out of the field of "vision by means of machinery." In a reflecting stereoscope the pictures are brought in succession into the field of view by means of screw and screw-wheel machinery. The pictures are arranged in sets, one set on each side of the instrument; each picture has a rack attached to its under side, by means of which spur wheels on the screw-wheel shafts give the desired "panoramic" motion; by means of pins on vertical shafts (worked by the screw-wheel shafts) the pictures have also a lateral motion, to enable them to place themselves in the case in proper order. The machinery is worked by a crank handle in front of the instrument.

[Printed, 1*s.*]

A.D. 1854, April 13.—N° 868.

DEVINCENZI, Guiseppe.—" A method or methods of produc-" ing engraved, figured, and typographic surfaces for printing " and embossing from, and for ornaments, also certain machinery " employed therein."

The following processes treated of in this Specification have relation to photography :—

Impressions are produced on metallic surfaces " with some " greasy matters," by means of a photographic process, either direct " or on a lithographic stone, to transfer from afterwards an " impression of any bituminous or resinous substance." Other means of producing these " impressions " are set forth at length.

To engrave these " impressions " they are varnished on the surface by a process similar to that of inking, and engraved by chemical or electro-chemical action ; or the " impressions " may be simply warmed and dusted over with bituminous or vitreous matter, which is melted, and figured surfaces are thus obtained.

Another process consists in producing the " impressions " upon an electro-deposited layer, and subjecting the resulting surface to the solvent action of the electric current in a suitable liquid. Another method is to electro-coat all such portions of the " im-" pressions " " as have no impressions with a layer of a different " metal," then to " remove the material in which the impressions " are taken," and either to use the resulting surface as a deep engraving by electro-etching it in a liquid which does not act upon the layer, or to use it as an ornamental surface.

Engravings on non-metallic bodies may be produced by means of the above-named " impressions " in combination with the chemical action of acids.

The " impressions " obtained as above may also be used " to " produce electro-plate reproductions or stereotype plates."

Besides the above-mentioned processes to which the method of producing " impressions " by means of photography are applicable, other modes of producing ornamental surfaces are set forth at great length in the Specification; also many applications of this invention are explicitly stated. An electro-magnetic " printing " machine," to work with a limited number of types is described in the Specification and shown in the Drawings.

[Printed, 1s. 6d.]

A.D. 1854, April 27.—N° 956.

JOHNSON, JOHN HENRY (*a communication from Henry Beaud*).—" Improvements in polishing and flattening metal plates," more especially Daguerreotype plates.

The plates are first polished on a flat table, which is provided with a ledge and moveable side. The plate is retained firmly on the table by being nipped between the ledge and the moveable side, the moveable side being provided with guide plates and a screw for that purpose. In this position the plate is well polished by the aid of polishing pads and a suitable mixture of impalpable powder and liquid. A pair of the plates are then laid back to back, and a block or " squeezing plate " is placed against the polished surface of each plate, the plates being thus placed between two blocks. The blocks and plates are then fastened together, so as to form one solid piece, by means of a hinged frame carrying a clamp screw, and are subjected to the pressing action of a pair of rollers, " through which the plates are passed two or three times." " When perfectly flattened, the finishing polish is effected by rub-" bing with the hand, or by suitable polishing mechanism."

It is important that the pressing surfaces of the blocks should be as well and as carefully polished as the surfaces of the plates, before being used.

[Printed, 6*d*.]

A.D. 1854, May 15.—N° 1080.

SAUGRIN, LOUIS FRANÇOIS.—" Improvements in apparatus " for the production of stereoscopic and photographic pictures."

The Specification and Drawings describe and show a camera having three object lenses, and which is capable of taking three stereoscopic views at one set of operations, when used in connection with a second camera, in which the arrangement of lenses and plates is reversed. The three lenses are placed over one another in a vertical line, and have hinged fronts, so that they may have such an inclination as to enable a true picture to be taken by each. The lenses are not placed in the centre line of the camera, but are nearer to one side than the other.

In taking a picture, " the sides of the apparatus, where the " lenses are nearest to the edge, are placed towards each other." To obtain " the angle of two objects upon the same plate," the operator shuts "the right slide of one of the apparatus and the "left slide of the other apparatus;" then "the frame containing "the plate of the left apparatus" is withdrawn, and placed "in "that of the right one, and vice versa." Three pictures, vertically over one another, are first made in one camera, and their stereoscopic copies are made in the other camera. "To obtain "the six proofs double by one operation," three additional lenses are added to each camera.

A similar camera to that described above, but with two lenses instead of three, is described and shown; also another camera, with two lenses placed side by side.

[Printed, 10*d.*]

## A.D. 1854, May 15.—Nº 1086.

EAST, FREDERICK.—"Taking photographic views and portraits "in the open air, by means of vertible machinery attached to "a cubical box, by which the changes are seen, and the light and "the time of exposure regulated."

A camera is described in the Specification, and shown in the Drawings, called "the camara chora," and consisting of the following parts:—A removable chest, which fits into the camera; this contains two water baths, a rack, a bottle case, and a drawer. The camera stand, which consists of a number of thin pieces of wood, jointed at the centres and ends; these "form three legs "when drawn out, and three thin squares when closed up;" they fold under each other "into a frame fixed round the bottom of the "camera, and are thus out of sight." A "diverting bath" to excite the plate, at one end of which is a cylinder, from which the exciting solution runs into the bath, when the bath is turned on the axis of the cylinder downwards; this and the following apparatus are interior fittings of the camera. The developing apparatus, consisting of a flexible case, from which the developing fluid is forced (by the compression of a screw) into a "meter;" when its knob is turned the fluid runs over the plate, and developes the picture. The fixing apparatus is on the other side of the camera to the developing, and consists of a similar arrangement; a thin plate of metal ("the developing tray") in connection with a wire is used to spread the fluid over the plate, in the developing and fixing processes. The focussing apparatus, consisting of a rack and pinion arrangement attached to a plate holder; a socket containing a half zone, in which the lens is so mounted that it can be turned upon its axis; a set of "light "regulators," admitting different colours and qualities of light; a set of diaphragms; and a "paraphon" or shutter; the light regulators, diaphragms, and shutter, are placed within the mounting of the lens, and turned by keys fixed round the head of the mounting; the focussing is observed with a tube at the back of the camera, provided with a "valve" or shutter. A "monitor," (or time-beater), consisting of wheel-work, actuated by a main spring, and regulated by a fly; a wheel "with elastic points" strikes the time, when required, upon "vibrating wires of different "tones;" a dial outside the camera marks the time for exciting the plate; and the exact time of exposure is ascertained by means of a connection from the fly to "the paraphon valve," in the mounting of the lens. Certain outer fittings consist of a "lunette," or "peep-hole," to look into the camera, having a shutter; a "diaphane" or window (with a shutter) to admit such rays of light as will not injure the solution; a handle to carry the instrument, a tap, a waste bag, and "locks and keys "with various other fastenings."

To work the instrument.—The camera is prepared by fixing the legs, taking out the chest, putting the waste bag on the tap and opening the tap. The prepared plate is then placed in the lip of the plate holder, the focus taken by the above described focussing apparatus, and the plate excited by immersion in the "diverting" bath. The picture is then taken, using such diaphragms and "light regulators" as may be necessary, and closing the "paraphon." In developing the picture, light of the requisite colour is admitted through the "diaphane," the plate dropped into the developing tray, and the developing fluid allowed to run over the plate. To fix the picture, the fixing fluid is allowed to run over the plate in a similar way to the process for developing; the operator, during developing and fixing, watches the process through the "lunette." The picture is then taken out of the camera, and washed, rinsed, and dried in the chest. Finally, the waste bag is unhooked, the tap closed, the chest put into the camera, and the legs folded under the camera, preparatory to leaving the field.

[Printed, 1*s.* 8*d.*]

A.D. 1854, May 22.—N° 1139.

SPENCER, JOSEPH BLAKEY, and MELHUISH, ARTHUR JAMES.—(*Provisional Protection only.*) " Improvements in pho-" tographic apparatus."

By means of this invention a series of photographic pictures may be obtained, in succession, upon a long sheet of sensitive paper, " the parts of the paper or sensitive surface not for the " time being in use being rolled up within the frame of the " camera." " The frame in which the prepared surfaces are em-" ployed is fitted up with two rollers;" " in using the apparatus, " after one picture has been taken, that part of the prepared " sensitive surface is wound up on to one of the rollers, and a " fresh quantity of the prepared surface, suitable for receiving " another picture, is unwound off the other roller, and so on, " till all the prepared paper or surface on the roller has been used. " Two rods or tubes of yellow glass are employed to retain the " part of the sensitive surface, which for the time being is brought " into position in the correct plane or position." It is preferred " to focus directly on to the prepared surface when using waxed " paper (instead of focussing on to a plate of ground glass), a plate of yellow glass having previously been placed in front of the lens to prevent the light from injuriously affecting the surface, and also having placed another plate of yellow glass behind the paper with a similar object.

[Printed, 3*d.*]

A.D. 1854, June 23.—N° 1384.

DREYFUS-WERTH, SALOMON, and MEUNIER, PIERRE.—(*Provisional Protection only.*) " A new or improved system of " applying designs to all kinds of fabrics, and of surfaces of " wood, marble, and stone."

This invention " consists in a new or improved system for ap-" plying designs to or ornamenting the surfaces set forth in the " title." For this purpose " the chemical action of light upon " colours" is applied " directly to the surfaces above mentioned. " The surface to be acted upon is first rendered sensitive by " subjecting it to one or more baths of any suitable substance, " such as solutions of chloride of sodium, nitrate of silver, hypo-" sulphite of soda, chromate of potash, and sulphuric acid.

" The fabrics or other surface, after being subjected to one or " more of these baths, is well washed and dried; the designs to " be reproduced on the surface are then laid or put on it in a " suitable manner, and the whole or part of the surface thus " treated exposed to the action of solar light during a few " moments. The surface having thus received the image or " outline of the design, may have any required tint given to it by " the usual processes."

[Printed, 3*d*.]

A.D. 1854, July 26.—N° 1638.

CUTTING, JAMES A.—This invention refers "to the art of " taking pictures photographically on a film of collodion upon " the surface of a sheet of glass;" the process is called "ambro-" type."

The invention is divided into the following parts:—

1st.—Preparing the collodion.—Gun cotton "is plunged into " strong alcohol, which effectually deprives it of the water which it " contains, without exposing it to the atmosphere for the pur-" pose;" this is an essential part of the invention. From the alcohol the gun cotton is taken immediately to a mixture of sulphuric ether and alcohol which dissolves it, thus forming the collodion.

2nd.—Beautifying the picture and sensitizing the plate.—The collodion, made as described above, is allowed to settle, and decanted. Iodide of potassium dissolved in alcohol is then added to the collodion. " After being well shaken" the mixture has gum camphor added to it; this is the second essential feature of the invention, and greatly increases the distinctness and beauty of the picture to be produced. This mixture is poured over the glass so as to form "a colorless transparent film," and, when it has only partially set, it is placed in a solution of nitrate of silver until the ether has escaped from the collodion; the plate is then ready to be placed in the camera.

3rd.—The completion of the process.—After being exposed in the camera the plate has its latent picture developed " by the " application of a solution of protosulphite of iron, acetic acid, " and nitric acid;" it is then washed and its picture fixed by means of hyposulphite of soda solution. The plate is then washed and dried, thus completing the chemical portion of the process. The third important point in this invention, however, consists in applying "a coating of balsam of fir to the surface of the glass " upon which the picture is made, the balsam being confined to " the picture plate by a secondary plate of glass;" for this purpose the balsam is "applied in a line along one edge of the glass," and one edge of the other plate is applied to the edge of the first, the two plates are then pressed gradually together and the superabundant balsam thus removed; this process greatly enhances the beauty and distinctness of the picture, and "secures the whole " from the action of air, moisture, and dust."

[Printed, 4*d*.]

### A.D. 1854, August 1.—N° 1696.

MERRITT, THOMAS EDWARD.—"Improvements in apparatus " for taking photographic pictures in the open air."

1st. A dark chamber or box "capable of containing any re- " quired number of prepared papers or glasses" is fitted at the hinder part of the camera. "A screw of quick pitch," passing through the back of the box, brings "the prepared papers or " glasses forward in succession into the focus of the camera." A second chamber is fitted beneath the camera and the box, in which is placed "a tray deep enough to hold the prepared papers " or glasses edgewise, and divided into as many compartments " as the box is capable of holding papers or glasses;" or the tray may have ears that slide in grooves at the bottom of the camera, thus dispensing with the second chamber. A slit is formed through the bottom of the dark chamber; this can be closed by means of a slide, and is employed to allow each paper or glass to fall into the tray after having been used. The tray is shifted "as " it is required to receive a paper or glass, so as to bring forward " a fresh compartment for it to fall into." A number of pictures can thus be taken without being exposed to light, and they may remain in the tray "until a convenient time for developing " them."

2nd. A small frame of tin, or other suitable material, to hold collodionized glass by means of a spring or springs; this arrangement may also be employed to hold the papers prepared for use in the camera mentioned in the 1st improvement. The paper is prepared by being placed between two pieces of glass, "which are "held together by a strip of paper pasted at their edges."

3rd. "A close box for containing one or more prepared papers "or glasses for taking photographic pictures in the open air with "a camera of the common description."

Besides the above particulars, the Specification and Drawings describe and show "an improved bath for developing collodion "pictures taken in the camera" described in the 1st improvement. The bath is adapted to the camera by means of ears, and consists of two cells communicating with one another by means of a perforated partition. The solution is poured into one cell and the partition admits it to the photograph in the other. There is a lid to keep light out of the bath and a corked spout for drawing off the used liquids.

[Printed, 1*s.* 4*d.*]

## A.D. 1854, August 11.—N° 1752.

MONSON, Edward.—This invention "consists of a machine or "apparatus" "for manufacturing or getting up, cleaning, polish- "ing and buffing daguerreotype plates."

The plate or plates are fixed upon a horizontal platform by means of a holder. This platform rests upon a second platform, and each platform is capable of a rectilinear sliding motion, "the "two sliding motions being at right angles to one another." By means of a rapid alternating motion given to the upper platform, and a slow direct motion given to the under platform, the whole of the surface of the plates is brought under the action of the tools for hammering, burnishing, or polishing them. The hammer is made to have a suitable vertical and alternating motion by means of a cam arrangement in combination with a helical spring; the cam raising the vertical shaft that carries the hammer, and the spring (in combination with the weight of the hammer and shaft) forcing the hammer on to the plates. The burnisher or buff is pressed on to the surface of the moving plate by means of a weight at the end of a lever, and besides can have a rotary motion on its axis given to it when required. It is preferred to hammer a number of plates first and then to buff them, but the hammering and buffing may be done at the same time. The buff is fixed, so as to prevent its rotation, for finally polishing the plate. "To prevent the plates from being spoiled, by dust being ham-"mered therein," a blowing apparatus is used.

[Printed, 6*d.*]

A.D. 1854, November 9.—N° 2373.

PRETSCH, PAUL.—This invention "consists in adapting the "photographic process to the purpose of obtaining either a raised "or a sunk design on glass or other suitable material or materials, "covered with glutinous substances mixed with photographic "materials, which aforesaid design can then be copied by the "electrotype process, or by other means, for producing plates "suitable for printing purposes, or can be applied for producing "moulds applicable for obtaining plates."

To obtain a raised design, the following process is employed:— A solution containing glue, nitrate of silver, iodide of potassium, and bichromate of potash is poured over the plate on which the design is required. When the coating thus formed is completely dry, the subject to be copied is laid on the prepared surface, and the whole transferred to the photographic copying frame and exposed to light. After a sufficient exposure the plate is removed "and washed either with cold water or a solution of borax or of "carbonate of soda, as may be necessary." The picture thus obtained in relief is treated with spirits of wine, dilute copal varnish, and a weak solution of tannin; when sufficiently raised it is washed and dried. The picture may now be copied by the electrotype, stereotype, or other like process.

To obtain a sunk design:—The previous process is employed, substituting gelatine for glue and drying the plate after washing it with spirits of wine; the picture is then ready to be copied.

Printing ink may be applied to the design developed as above, and thence zincographic or lithographic impressions obtained.

The pictures have flat surfaces when taken from the copying frame. By using more sensitive compounds as the iodide or bromide of ammonium instead of iodide of potassium, designs can be obtained by the camera.

[Printed, 3*d.*]

A.D. 1854, November 18.—N° 2447.

LUFF, HENRY JAMES.—(*Provisional Protection only.*) "Im-"provements in the mode of attacking hostile bodies, also appli-"cable to the obtainment of plans of forts." Photography is mentioned as a means of obtaining the plans of forts.

These improvements "consist in the mode of attacking hostile "forts and towns, or bodies of troops when stationed near to the "sea, by having, in connexion with a vessel at sea or in a harbour, "a balloon having a telegraphic communication with such vessel, "such balloon being freighted with explosive compounds of any "description most suitable with the purpose. The balloon having "ascended from the vessel, and being then in connexion there-"with, can readily direct the movements of such vessel, whilst "such movements by suitable connexions can again operate in "the position of the balloon, so as to bring the same readily into "the most favorable position with regard to the fort, army, or "other object to be acted upon, when the explosive compounds "may be easily let down in the exact position where the greatest "injury can be inflicted. It is also to be observed, that by "similar means, plans of fortifications, &c., may readily be "acquired by the photographic process. It is obvious, that "where inland forts or armies are to be attacked, the balloon "may be in connection with any apparatus upon land, so as to "effect the same results."

[Printed, 3*d.*]

A.D. 1854, December 13.—N° 2620.

FONTAINEMOREAU, PETER ARMAND LE COMTE DE (*a communication*).—This invention "consists in fixing, vitrifying, "and coloring photographic images taken by the collodion pro-"cess upon enamel, metal, stone, porcelain, glass, china, and all "kinds of earthenware."

"Upon a plate of glass is poured a composition consisting of "rectified sulphuric ether, rectified alcohol, gun cotton, and "hydrate of ammonia; after leaving this layer to dry for a few "seconds, the plate is plunged in a solution composed of dis-"tilled water, crystallized nitrate of silver, and rectified alcohol, "with or without a solvent. The plate is then ready to receive "the photographic image in the usual manner; after which it is

" introduced into a bath composed of filtered water, sulphate of " iron, acetic and sulphuric acid. After having washed the proof " several times with fresh water, it is plunged into a bath com- " posed of cyanide of potassium and distilled water, and then " submitted to the action of a solution of bi-chloride of mercury, " hydrochloric acid and distilled water. This operation being " terminated, a very sharp permanent proof is obtained, which " may be separated from the glass upon which it has been made, " by plunging it during a few seconds into water and taking it " off with blotting paper. This image may be applied to a sur- " face of metal or glass, &c., of any form by means of a thin " solution of gum or other adhesive matter." It is then baked in a porcelain furnace. The images may be coloured by applying vitrifiable colours either in a state of powder " or mixed with " spirits of turpentine, lavender, &c.;" these colours are also baked in.

[Printed, 3*d.*]

## A.D. 1854, December 29.—N° 2745.

THOMPSON, FREDERICK, and WAGSTAFF, WILLIAM.—(*Provisional Protection only.*) " Improvements in photography."

" The invention consists in marking on the back of a glass " plate on which a picture or photograph has been taken, or upon " one or more seperate glass, metal or other plates placed behind " the same, the outline or form of the object or objects in the " picture or photograph, with any sort of varnish, cement, or " other material that can mark or define a line, or with paper, " cloth, velvet, silk, gutta percha, wood, or other suitable material, " and filling the space or part thereof within such outline with " one or more coats of the above, or other suitable material, " thereby producing an appearance of relief or rotundity. And " the same effect may be produced by eating away with any acid " or other suitable material the whole or part of the said outline, " or any part of the back of the glass or picture, or upon any " seperate plate of glass, metal, or other suitable material, and " coloring or coating the part affected by the said acid or other " suitable material, or with varnish or other material before " described.

" The invention farther consists in producing a colored back- " ground for photographic pictures, by placing behind the plate

" on which the picture is produced, or upon or behind one or " more seperate plates of glass or other suitable material, tinted " paper, cloth, velvet, silk, gutta percha, wood, stained glass, or " other suitable material, and which background will add to or " assist the appearance of relief or rotundity before described."

[Printed, 3*d.*]

---

## 1855.

### A.D. 1855, February 20.—N° 371.

SCHOTTLANDER, Henry.—The title of this invention is, " Improvements in ornamenting looking glasses."

Various methods of painting, transferring, or printing ornamental designs " upon the back of a looking glass not yet " silvered," and afterwards silvering the same in the ordinary way, are set forth. " Through this process the drawing, design, " &c. appears between the silver and the glass either in black or " colours, as the case may be, and becomes closed in hermetically. " The parts not covered with colour become thereby ordinary " looking glasses, and form the basis or ground of the design, " &c."

The patentee further states :—" I further deposit photographies " direct upon glass, colour the same if required, and silver them " afterwards, thereby transforming them partly into looking " glasses, the photography appearing between the glass and the " silver, in manner as heretofore described, by which process the " same are protected against any effects of the atmospheric air, " and consequently retain their freshness permanently. In order " to detach the outlines of a photography upon glass from the " chemical substances which surround them, I apply the described " process of scraping, provided that the design does not contain " too many minute parts; otherwise I cover the photography " itself with varnish, and wash the surpassing chemical sub- " stances off with water or other dissolving liquids."

The process of " scraping " above alluded to is thus described: —" The drawing or design themselves are formed into looking "glasses in imitation of inlaid silver by covering the whole "surface of a glass on one side with paint, and scraping the "colours off those parts where I desire to have looking glass."

[Printed, 3*d.*]

A.D. 1855, February 27.—N° 426.

BERCHTOLD, Alfred Jean.—(*Provisional Protection only.*) The title of this invention is, "Certain improvements in apply- "ing the photographic engraving on metals or other materials."

The patentee states:—"I take a negative or positive photograph "either on paper, or on collodion, or albumen, or other suitable "material, and I produce a grain upon this photograph by "printing the grain in black or any other colour upon it from a "plate or block or other printing surface, or by perforating or "making strokes, lines, or dots upon it by a roller or other "instrument, worked by the hand or by machinery.

"The design is then transferred by the action of light to a "varnished surface of metal or stone or other suitable material by "the process already known, and the varnish is afterwards treated "by solvents, and the surface etched by acid, as already known. "By this means the strokes, lines, or dots, or grain, are produced "upon the metal, stone, or other suitable material by the action "of the light, and not by the mechanical application of powdered "resin, as heretofore."

"The printing, or marks, or perforations may be made on a "separate sheet of paper or other suitable material, instead of "being made on the photograph itself. The perforations may "pass completely through the paper, or only partially through it."

[Printed, 3*d.*]

A.D. 1855, March 5.—N° 488.

GARNIER, Arsène Louis.—The title of this invention is, "An "improved process for producing photographic pictures, which I "intend to denominate 'Système Garnier de Photochérographie "'coloriée.'"

The invention consists in:—

1st. "Employing a textile or woven fabric instead of paper as "the surface on which the picture is to be produced."

2nd. A certain process "for neutralizing any alkaline salts, and "for removing any impurities or injurious matters contained in "the fabric or paper employed." For this purpose the fabric is immersed consecutively in boiling water, alcohol, and potash.

Having cleansed the fabric according to the 2nd part of the invention, it is cut into pieces of suitable sizes coated with a paste of rice flour, allowed to dry, and immersed in a mixture of virgin wax, Venice turpentine, and linseed oil; the fabric has thus a body and a finely polished surface, upon which a positive photograph may be obtained in the usual manner, and fixed in a bath "of ammoniacal hyposulphites." "To remove any alkaline salts "which may remain after the ordinary washings," the picture is immersed in pure alcohol and hot water. The picture is then ready for the reception of oil colour (which is applied at the back), and can be attached to a suitable flat and even surface and varnished on the face.

The 2nd part of the invention can also be applied to paper photographs after they have been obtained and fixed in the usual manner. The picture is then treated with starch, waxed, coloured on the back, mounted, and varnished.

[Printed, 4*d.*]

## A.D. 1855, March 8.—N° 515.

CLAUDET, ANTOINE FRANÇOIS JEAN.—"Improvements in "stereoscopes."

1st. "Giving to the interior of the casing or box of the stereo-"scope such a form that no part of it can be reflected from the "pictures or the frame upon which they are to the eye of the "observer." "A perpendicular from the tangent at any point "which could otherwise be reflected of the curve of the interior "of the casing must make with the slide such an angle as will "not reflect it to the eye of the observer." The casing of the stereoscope shown in the Drawings has its vertical sections towards the end that receives the pictures composed of curves of contrary flexure.

2nd. Placing within the outside casing two "chambers," each chamber being larger than, but in continuation with, and concentric with each eye tube. These chambers are suitably blackened and have openings corresponding to the form of the picture; they thus "allow the pictures to be seen freely, and wholly hide the "frame."

3rd. An "adaptation of the lenses so as to obviate the necessity "of any adjustment to suit observers whose eyes are at different "distances apart." Each eye piece is made out of a half lens, "the only transparent part of the lenses through which the ob-"server can view the pictures" being a thin central strip in the longitudinal direction of the instrument; "the other portions of "the lenses are blackened and rendered opaque." If the meniscus form of lens be used, "the concave side should be towards the "eye, in order to the prevention so far as possible of reflection "from the lens."

4th. A "mode of adjusting the foci of the lenses to the focal "length usually required by the varieties of long, common, and "short sight." The sliding parts of the eye tubes are connected by a cross piece which carries "a small arm, upon which is "mounted an index." A graduated scale, fixed on the outside of the casing, enables the index to be suitably placed for correct vision.

5th. Modes of "mounting several slides on a revolving frame, "and exhibiting them to view in succession." According to one method a frame revolving on an axis has as many sides to it as it is required to carry slides. To give rotation to the frame a button or "screw" on the axis is turned by hand, and a ratchet wheel and pall prevents the frame revolving while the picture is being viewed. According to another method machinery is used; by pulling out a bar that carries a rack working into a wheel on the frame, a shutter is made to cover the interior of the eye pieces during the motion of the frame by the rack. On the release of the bar, a "spiral" [helical?] spring in combination with a weighted cord, brings it (the bar, but not the frame) back to its original position, thus opening the shutter and exposing a new slide to view. By fixing the slides on an endless band more slides than four can be arranged in the instrument, and when an apparatus is required to show pictures to several persons at once several sets of apparatus may be mounted in a polygonal box.

6th. "The use for each eye piece of the central part of a whole "lens, instead of a half lens, so that the pictures are seen through "the lenses at or very near to their centres." In this case "the "axes of the eyes of the observer must converge in order to pro-

" duce coalescence." To adapt this part of the invention to the 3rd improvement, the excentric parts of the longitudinal strips must be rendered opaque.

[Printed, 1*s.* 5*d.*]

A.D. 1855, March 17.—Nº 598.

PETITJEAN, TONY, and PÊTRE, LOUIS (*a communication*).—(*Provisional Protection only.*) " Certain improvements in the " manufacture of daguerreotype plates and of electro-plated sheets " of metal, part of which improvements may be applied to the " production of polished surfaces on metallic articles."

This invention " consists of the following method for giving a " perfectly smooth and polished surface to the plates or sheets of " electro-plated copper."

A piece of glass is taken, of the size and shape of the plate to be produced, and prepared " to receive the action of the galvanic " battery ; " it is then electro-silvered and electro-coppered. Instead of being electro-coppered, the plate may be electro-coated with iron or other metal or with " as many layers of different " metals as we please. This method of giving several layers of " metal we think especially useful for daguerreotype plates, as such " plates would form a kind of voltaic pile, and render them more " sensible to the chemical action to which they are afterwards ex- " posed. As to the electro-plated sheets of metal used for reflec- " tors and other purposes, we think it sufficient, after giving a coat " of silver as before described, to give a single coat of iron or " other metal, but we will not limit ourselves in that respect.

" By using proper moulds, metallic articles may receive a high " polish by the above-described process."

[Printed, 3*d.*]

A.D. 1855, March 27.—Nº 674.

BOURNE, JOHN COOKE.—This invention relates to portable photographic apparatus, and consists in :—

1st. Constructing photographic apparatus " in such a manner " that the box or case in which the camera is contained may serve " as a foundation on which the camera may be supported when in " use, and when out of use the front part of the camera may be " folded into the back part, and be placed in a compartment within " the box."

2nd. "Constructing the box or case so that when opened it may "form the base for a tent, which is constituted by means of a lath "fixed on a centre or pivot at each angle of the inside of the box, "the several laths being raised and united at the top in pairs, so as "to form, as it were, two gable ends, in order to support a covering "of suitable fabric, within which the operations of preparing "the sensitive surface, developing the photographic picture, &c. "may be performed, the camera remaining suspended from the lid "of the box during such operations. The said fabric or covering "is confined to the box or foundation by means of an elastic cord, "and it has suitable openings in it to admit of the head and hands "of the operator being introduced as required."

3rd. "Arranging and constructing the means of adjusting the "position of the lens of the camera, so as to ensure its movement "and ultimate position in a direction always in that of a curve, "the radius of which is the focal length of the lens, the rays of "light passing through the centre of the lens at right angles to "the surface thereof will fall upon the centre of the picture."

A camera is described and shown, consisting of a "slide case" connected to the lens end of the instrument by means of a dark chamber formed of cloth or other suitable material that will enable the lens end of the instrument to be collapsed against the slide case, so that the camera may be packed in a box. The box contains the camera, a triangular table to which the legs are fixed, and on which the box is fastened by a central bolt and nut, a "triangular developing stand, constructed on the principle of a "book hinge," chemicals, plates, washing trough, also laths and waterproof covering for forming the tent. Besides the above there are the legs of the instrument, which are retained on the pegs of the triangular table by means of elastic bands.

In using the apparatus the triangle and legs are set up, then the box is fixed on to "the table so formed," and the camera is fixed and adjusted to the lid of the box. A "swivel foot" and "hinged "tail board" allow of the fixing, adjustment, and inclination of the slide case as required; the lens end of the instrument is adjusted horizontally along curved slots, and vertically upon a standard that is bolted to the box lid. When the objects to be taken are at unequal distances from the camera, they may all be nearly focussed for vertical lines by means of a slot in the lens standard in combination with an inclination of the slide case upon the vertical axis of the swivel foot; for horizontal lines this adjustment is made by raising or depressing the lens on the standard, and suitably inclining the slide case by means of the "hinged tail-" board." The plate is then prepared under the tent (the camera during this operation being suspended downwards, from the lid of the open box), the tent removed, the box closed, the slide substituted for the focussing glass, the picture taken by exposure of the plate, the tent re-formed, and the picture developed in the usual manner.

[Printed, 1*s.* 8*d.*]

## A.D. 1855, April 7.—N° 770.

ROLLASON, ALEXANDER.—This invention "consists of im-" provements in transferring to paper, linen, cardboard, bone, " ivory, wood, metal, or stone, the film of collodion or albumen " used in collodiotype or albumenized plates." By this invention a photograph may either be removed "from the glass or plate on " which it may have been produced; or, by transferring the plain " film on to certain of the substances above named," a new base or medium on which photographic pictures may be taken may be produced.

After having produced the picture in the ordinary way, it is immersed in dilute nitric acid and allowed to dry, coloured (if requisite), and varnished with "asphaltum or Brunswick black " dissolved in mineral naptha to about the consistency of cream."

The film is then transferred from the glass to the substance destined to receive it, by means of a mucilage composed of gum arabic and honey; during the separation of the film, "water, " spirits of wine, or other limpid fluid" is introduced between the film and the glass.

"The irridescence" that appears after the transfer may be removed by rubbing the film over with cotton wool and oil, thus " leaving a slight stickiness, to which the dry colours known as " 'mansions' and many other dry colours will adhere." In some instances this last operation may be omitted, and the picture completed by means of water, oil, or varnish colours.

In producing the new medium above referred to, the film is transferred "before the picture is taken instead of after."

[Printed, 4*d.*]

A.D. 1855, June 7.—N° 1303.

ORANGE, André.—(*Provisional Protection only.*) "Improvements in obtaining representations for commercial purposes of "articles for sale."

"This invention relates to the obtainment of the representa-"tions of articles intended for sale, such representations being "used in lieu of samples in negotiating with purchasers: such "representations are produced by any of the well-known photo-"graphic processes, either on paper, glass, or other material, "when the articles to be represented are in their finished con-"dition, varnished, polished, burnished, or otherwise treated, so "as to reflect the light. It is preferred to take the representa-"tions of them before the final finishing stage, or when they "present a dull appearance, or such as is suitable for producing "a well-defined outline in the camera; or the finished articles "may be washed over or prepared in some suitable manner, so as "to present the required dull appearance."

[Printed, 3*d.*]

A.D. 1855, July 7.—N° 1523.

GEDGE, John (*a communication from Louis Samson*).—(*Provisional Protection only.*) "Improvements in photographic "glasses."

"The 'panes' or 'plates' to be made are of two kinds, one "vitrified and the other not; the latter are composed of two "transparent or translucid glasses, colored or not. The proof is "obtained on one of these glasses by means of a very delicate "preparation of albumine and collodien, permitting it to retain "great transparence, so as to enable it to support with facility "the subsequent operations. To obtain 'positive' proofs on the "above" the "types" are arranged "in a framing so as to form "a group; this frame is placed behind a second, furnished with "drawings or designs corresponding to the types aforesaid "forming a whole, the parts of which may be varied at pleasure. "The transfer apparatus is furnished with powerful lenzes, and "being placed in position, receives the glass prepared as aforesaid "with albumine and collodien, on which the light will act in a "direct manner, traversing the designs and types aforesaid. The "production is now 'positive;' it may be taken of any size by "merely advancing or withdrawing the apparatus."

"To vitrify these panes, they may be submitted to a high "temperature, so as to volatilize the albumine or collodien; the "silver (reduced) will penetrate into the glass (liquified on its "surface), leaving thereon a scarcely visible design or image of a "dirty yellow color. To turn this into 'bistre,' it is necessary to "combine therewith that which is obtained by the vitrification of "colcothar or oxide of iron, and for this purpose the employment "of protosulphate of iron will be found most serviceable as a "means of developing the image or design, as the oxide of iron "and silver and a small quantity of sulphur of silver, which are "formed during the vitrification, gives to the proof a sufficient "strength of color."

[Printed, 3d.]

## A.D. 1855, July 13.—N° 1581.

GAUDIN, Pierre Ignace Alexis.—(*Provisional Protection only.*) The title of the invention is, "Improvements in baths "used for photographic purposes."

The inventor further states:—"My improvements in baths "used for photographic purposes relates to baths of nitrate of "silver or collodion, into which the paper or glass is immersed "previous to receiving the photographic impression, and consists "in maintaining and the means of maintaining the bath at a "proper temperature for the purpose. To effect this I surround "the bath with a double casing of metal or other suitable "material, into which I place a freezing mixture or other cooling "material, in order to cool and reduce the bath to the proper "temperature when it is naturally too warm for the purpose; "and on the other hand, when too cold by reason of the natural "coldness of the surrounding atmosphere, I place warm water "within the double external casing of the bath, in order to raise "the temperature to the required degree."

[Printed, 3*d*.]

## A.D. 1855, August 11.—N° 1824.

PRETSCH, Paul.—This invention relates to "obtaining cylin- "drical and other engraved and ornamented surfaces" "by

" photographic and other agency," and refers to further improvements in the application of the processes set forth in N° 2373, A.D. 1854, (which see).

Copper or other suitable plates are engraved by the processes set forth in N° 2373 (A.D. 1854), and are employed "for the forma-" tion of cylinders to be employed in calico and similar printing, " embossing, or other purposes; or cylinders may be formed " directly by the electrotype process by means of suitable tubular " or other arrangements" of the engraved plates, "to serve " as moulds, and the cylinders produced therefrom may be " strengthened by the insertion of metal rollers, cast metal, and " similar methods.

"When it is desired to ornament manufactured articles with " engraving," according to this improved process, "the same can " be variously applied to flat, curved, and other surfaces, and " when required the engraved surfaces may be covered with gold, " silver, or other metals or materials, or may be inlaid with metals " or other materials."

[Printed, 3*d.*]

## A.D. 1855, August 14.—N° 1842.

SHEARS, George.—"An improved construction of stereo-" scope."

This invention "has for its object to construct these instruments " in such a manner as will admit of their being folded up and " packed away in a shallow box when not in use." Each side is in two pieces, connected together "by a butt hinge" and "a " vulcanized india-rubber spring," "in such a manner, that when " the stereoscope is opened these springs will retain the parts in " position, but will yield to pressure, and allow the parts to fold " the one over the other when it is required to close the instru-" ment. The cross piece at top which carries the lenses is jointed " by hinges to the folding sides, and these sides are hinged to the " side or bottom of the stereoscope box." All unnecessary light may be shaded from the picture under inspection by connecting the side pieces by cloth or other flexible material. "For the " purpose of exhibiting transparent pictures, the bottom of the " stereoscope box may be formed of glass, or this bottom may be " made to consist merely of a hollow frame and covered with an

" opaque door, which may be removed or folded back when the " light is required to be introduced to the back of the picture. " The lid of the box will assist in forming with the flexible " material before mentioned a dark chamber for the reception of " the transparent picture." The " shallow box " itself forms the bottom of the instrument, and is attached to the lid by a leathern joint or hinge; when not in use, the instrument is compressed within the box, and the lid is closed by hooks and eyes.

[Printed, 7*d.*]

A.D. 1855, August 24.—Nº 1914.

ARCHER, Frederick Scott.—This invention relates to " the " manufacture or production of transparent or semi-transparent " tough and flexible photographic pictures, capable of being " used for the production of other or positive pictures by coating, " or covering, or combining a collodion film with a film of gutta " percha, and removing the combined films from the glass or " other surface which supports the collodion during the operation " of taking the picture."

" The negative picture is produced in the ordinary manner " upon the collodion film on a sheet of glass, and it is fixed and " dried in the ordinary manner." The picture is then treated with a solution of gutta percha in benzole, either by pouring the solution over the plate or dipping it (the plate) into the solution; " after draining off the excess of the solution," the plate " is dried " by a gentle heat, and leaves a nearly transparent film of gutta " percha upon the collodion. If the film is not sufficiently thick, " this operation is repeated one or more times, until a sufficiently " thick film of gutta percha is formed." The whole is then immersed in water, " which causes the collodion to separate from " the glass," and " the film or sheet of gutta percha with the " collodion film firmly adhering or combined with it," is removed.

The glass plate may be first coated with a film of gutta percha, then with a collodion film, and when the picture is produced and fixed, with another film of gutta-percha; on removing these films from the glass plate, a collodion film coated on both sides with gutta percha, is obtained. Instead of using the second film of gutta percha, the picture may be simply varnished with any suitable varnish.

[Printed, 3*d.*]

A.D. 1855, September 19.—N° 2112.

CORNIDES, Louis.—This invention is entitled "Certain im-"provements in obtaining impressions of prints or drawings, and "in transferring printing, and colouring or ornamenting the "same on glass or other surfaces;" one of these improvements relates to an application of photographic processes. The improvements are:—

1st. A method of transferring "lithographic or other impres-"sions" on to a surface of glass, coated with "gelatine or "animal glue." After the impression is transferred, it is "fixed" by means of "a solution of any of the known salts of alumina."

2nd.—"Coloring the gelatined surface of glass with or without "printing or transferring impressions thereon after the same has "been steeped in the before-mentioned fixing bath." "Negative "impressions" are transferred to the coated glass surface, and coloured so as to produce a positive picture by transmitted light.

3rd.—The use of finely pulverized coloured glass applied to the coated surface of glass by means of stencil plates.

4th.—"The use of the gelatined glass surface for the purpose "of making photographic impressions." Positive impressions are obtained from photographic negatives on the gelatine surface, which is sensitized for the purpose. The positive picture thus produced is subjected to the action of the fixing bath mentioned in the 1st improvement; it may also be subjected to the "color-"ing or coating processes mentioned in this Patent."

5th.—Colouring the transferred impression by means of a coat of fine metallic powder, and transferring the same to the gelatined surface.

6th.—Protecting the printed or coloured surface of gelatine by means of a "solution of explosive cotton in pyroxylic spirit, com-"bined with gums or resins."

Nos. 2066 (A.D. 1854), and 745 (A.D. 1855), are referred to in the description of this invention.

[Printed, 4*d.*]

A.D. 1855, September 25.—N° 2139.

CLIVE, Joseph Charles.—(*Provisional Protection only.*) This invention "relates principally to such photographic pictures as "are taken upon glass, and are called collodion positives;" its object is "the production of 'backgrounds' to portraits and " groups, which said 'backgrounds' shall be well defined."

The inventor further states :—" My invention consists in taking " the portrait or group on one side of a sheet of glass, and after " having removed from the glass the whole of the background, I " take upon the other side of the sheet of glass any scene which I " wish to be the background, or I take the said background on a " second sheet of glass, and place the same at the back of the " first. The pictures are thus superposed, the portrait or group " being uppermost, and the picture constituting the background " undermost."

" By the method described, figures or objects taken at different " times and at different places can be brought into one picture."

[Printed, 3*d.*]

## A.D. 1855, October 24.—N° 2381.

MAYALL, John Edwin.—This invention relates to the application and use of artificial ivory, " for receiving the photographic " pictures, as a substitute for the metal or glass plates, or pho- " tographic paper ordinarily employed for that purpose," such material " having all the properties and beautiful finish of ivory."

The methods of making artificial ivory for photographic purposes are as follows :—

1st method.—Tablets or slabs composed of gelatine or glue in its natural state " are immersed in a bath of alumina, which is " held in solution by sulphuric or acetic acid ;" " a complete " combination takes place between the alumina and the gelatine " or glue." The tablets remain in the bath until they are sufficiently thick; they are then removed, allowed to dry or harden, and dressed and polished by any known process. The material thus produced is " known in France as Pinson's artificial ivory."

2nd method.—Alumina may be mixed directly with gelatine or glue, and sheets manufactured therefrom.

3rd method.—Equal portions of bone or ivory dust and albumen or gelatine are worked into a paste, and afterwards rolled out into sheets. The sheets are then allowed to harden, " and " are cut into slabs or tablets of the required size."

4th method.—Finely powdered baryta and albumen are " well

" worked together, and rolled out into slabs." Preference is given to this method.

The slabs produced by either of the above means are " carefully " scraped, to give them a perfectly even surface," washed with alcohol, and prepared in the ordinary way to receive positive pictures. " The pictures having been printed, the entire slab " is immersed in weak nitro-sulphuric or nitro-hydrochloric acid, fixed, and dried under pressure on a marble slab.

[Printed, 3*d.*]

A.D. 1855, November 13.—N° 2555.

MAWSON, John.—(*Provisional Protection only.*) " Improve-" ments in cameras for taking photographic pictures."

The inventor states :—" My improvements in cameras consist " in forming them in such a manner that they are capable of " being closed up so as to occupy less space than ordinary, and " also in so arranging the front of the camera, or bracket carrying " the lens, that a readier method is afforded for focussing or " setting the instrument. I also place at the back of the camera " a shutter or screen, which serves as a permanent focussing cloth, " which, when not in use, folds down and fits into the body of " the camera. I form the body part of my improved camera of " cloth, the ordinary grooves for the reception of the focussing " glass and dark chamber being made in the framework of the " camera. In front of the body of the camera I place a conical " tube of pliant or flexible cloth. This tube is connected to the " back of the bracket carrying the lens by means of an elastic " neck, which fits into a deep groove in the bracket. The bracket " carrying the lens is fixed to a slide, which fits into a groove or " sheath in the framework of the camera, and is pushed out or " drawn in by a regulating screw. The sheath or base of the " camera is fitted with a slotted plate, by means of which the " main frame of the camera may be turned square upon the " base, or otherwise adjusted without detachment therefrom. " Where it is desired to render the camera more portable I dis-" pense with the ordinary tubing or casing of the lens, and fix " the lens permanently in the bracket before mentioned, and " attach in the front thereof a flexible tube, fitted with a coiled

" spring, which can be drawn out to any required extent, and " when not in use fits flat against the bracket, thereby also serv- " ing as a covering or protection to the lens."

[Printed, 3*d.*]

### A.D. 1855, November 14.—N° 2573.

MÖLLER, JOHANNES.—(*Provisional Protection only.*) The title of this invention is, "Producing a transparent photographic pic- " ture on ivory without injuring the nature of the ivory, so as to " be able to finish the picture with colours like other miniatures."

This invention "consists in producing a transparent positive " photographic picture, and bringing or transferring the same " upon ivory by means of a gelatine or glass medium, which " leaves the ivory free from all injury, and will enable the artist " thus to finish the picture so brought or transferred on ivory " with colours like other miniatures prepared on ivory."

[Printed, 3*d.*]

### A.D. 1855, December 12.—N° 2808.

HAY, GEORGE HERON, and HAY, DAVID SYME.—This invention relates to colouring photographic pictures; it consists in applying amber or other clear varnish to the collodion film, and, when this first coating is dry, applying a secondary coat of drying oil and varnish on which the colours are laid whilst the drying varnish is still moist. "The colours may be applied either in the " form of ordinary oil paint, or in the dry condition."

According to another process "the colours may be applied with- " out the previous application of the medium or secondary " coating, in which case of course oil or other moist colours only " are to be used.

"The backgrounds and other portions of the picture may be " finished up in any desired style, and the colouring matter is " then coated with a varnish or protective cover. This system of " treatment is obviously applicable to collodion and other photo- " graphs of various kinds, and it renders such pictures extremely " durable and unliable to injury or decay."

[Printed, 3*d.*]

A.D. 1855, December 13.—N° 2815.

POITEVIN, ALPHONSE LOUIS.—The title of this invention is, "Improved photographic printing."

A mixture is made, consisting of "a concentrated solution of "albumen, fibrine, gum arabic, gelatine, or similar organic sub-"stance, and a concentrated solution of a chromate or bichromate "of potash, or of any base which does not precipitate the organic "matter of the first solution." This mixture is applied "upon "the surface which is to receive the design," so as to form a layer or film; the layer is used dry when the picture is impressed by contact, and moist when used in the camera. When the picture is impressed, if the surface is not moist enough, it must be moistened and a "greasy ink" applied to it; the ink "will be found "to adhere to those parts only which have been affected by the "light," and the print thus obtained "may be retained on the sur-"face on which it is first produced," or impressions may be taken from it in the manner of lithographic printing. Lithographic stone, metal, glass, wood, or other suitable material may be used to receive the photographic image.

A design is produced in colour by mixing a suitable colour with the above-mentioned organic mixture, and, when the photograph is impressed, washing away those portions of the mixture which have not been acted upon by the light. A design in various colours may be produced by suitably applying different colours to different parts of the surface.

[Printed, 3d.]

A.D. 1855, December 13.—N° 2816.

POITEVIN, ALPHONSE LOUIS.—The title of this invention is, "Improved photographic engraving."

A glass or other surface is covered with a layer of "chromatized "gelatine," either by pouring on a uniform layer of gelatine and plunging the whole into a concentrated solution of bichromate of potash, or by using the gelatine and bichromate mixed to pour on to the glass plate. The plate being thus sensitized the image is impressed and the plate plunged into water; "the parts which "have not been acted upon by the light absorb the water and "swell or expand, while the parts which have been acted upon

" by the light only become slightly moistened, and thus form " hollows."

A plaster cast of the plate whose hollows thus correspond to the lights is taken by pouring a solution of protosulphate of iron upon it, then washing it, and pouring the plaster upon it, a suitable frame having been fitted to it; when the plaster has set it is carefully removed. The protosulphate of iron gives consistence and strength to the plaster, so that it may be successfully removed from the film of gelatine.

The surface of gelatine is metallized for yielding an electrotype mould by the following process:—Immediately after exposure to light it is immersed in a solution of iodide of potassium, then in a solution of nitrate of silver. The whole surface is then exposed to light, and, ultimately, the silver is completely reduced by a solution of protosulphate of iron.

The engraved plates that may be produced from these moulds are adapted for printing and embossing, also for moulds for earthenware, &c.

[Printed, 4*d.*]

---

## 1856.

### A.D. 1856, March 18.—N° 646.

MAW, ARTHUR.—(*Provisional Protection refused.*) " Improved " means of ornamenting the surfaces of woven, knitted, or felted " fabrics, such as cloths, stuffs, ribbons, and other fabrics; or of " parchment, vellum, leather, or other animal tissues, and render- " ing such fabrics or tissues applicable to various purposes."

This invention " consists in the employment of certain processes, " closely allied to the process of photography, and whereby the " surfaces of these materials may be rendered sensitive to light, " so that, having obtained a negative picture on glass or paper, or " some other transparent or translucent material, a positive im- " pression or image may be transferred to the prepared surface of " the fabric or tissue. By these means any variety of design or " pattern may be transferred to these materials; but the particu- " lar nature of the design or pattern must depend not only upon

" the taste of the operator, but also upon the purpose for which " the ornamented materials are to be employed.

" It will be clearly understood, from the foregoing, that the " invention admits of an infinite variety of applications ; for in- " stance, it may be successfully applied to ornamenting various " articles of dress, and also for ornamenting books, both exter- " nally and internally, as well as various articles of furniture, and " to other purposes too numerous to mention."

[Printed, 3*d.*]

A.D. 1856, April 12.—N° 875.

SCHULTZ, LUDWIG.—(*Provisional Protection only.*) The title of this invention is, " Improvements in obtaining photographic " pictures upon paper, glass, metal plates, and other fibrous " substances."

The inventor states :—" The invention consists in preparing " my paper or other fibrous substances which I intend to use for " obtaining photographic pictures upon, by first coating them " over with a suitable varnish, preferably of a dark color, so as " to give them a smooth surface, and at the same time to prevent " them from being injuriously acted upon when immersed in the " liquids for subsequent preparation. I apply in like manner the " same varnish to glass or other metal plates, for the purpose of " giving them a proper smooth surface to receive the picture " upon."

[Printed, 3*d.*]

A.D. 1856, April 15.—N° 896.

OLLEY, WILLIAM HENRY.—(*Provisional Protection only.*) The title of this invention is, " Taking photographic impressions or " pictures of microscopic objects by reflection, such reflection " being effected by the combined aid of the microscope and " camera obscura and camera lucida, or other reflectors that may " be employed in place of the latter."

" The four-sided reflecting glass prism, commonly known as the " camera lucida," is fitted " on the eye piece of the microscope, " the cap by which it (the eye piece) is usually surmounted being " previously removed, and the body or tube of the instrument " brought into a horizontal position ;" an inverted camera obscura

" or darkened box" is then placed over the said prism. The camera fits closely to the tube of the microscope, and has a second box open at the lower end sliding within it that carries the plate frame; the image of an object properly placed at the object end of the microscope formed by reflection from the camera lucida can thus be correctly focussed on to a ground glass plate, temporarily substituted in the usual manner for the sensitive plate. The picture is then taken, developed, and fixed in the ordinary way.

"The invention is applicable to the obtaining microscopic im-" pressions, by reflection, by the use of iodized silver plates, and " every modification of the daguerreotype process, albumenized " photographic glass plates, calotype paper, and all other papers " or surfaces impressionable by light."

Plate glass or other reflectors may be used instead of the camera lucida.

[Printed, 3*d.*]

## A.D. 1856, May 8.—Nº 1078.

MAYER, Louis Frédéric.—(*Provisional Protection only.*) This invention "consists in substituting linen and other woven " fabrics for the ordinary materials, on the surface of which pho-" tographic images are deposited or printed.

"The surface of the said woven fabrics is prepared in the " ordinary manner in which the surface of paper and other " materials have been hitherto prepared, so as to render it sus-" ceptible of receiving an image by means of the photographic " process; and the process is carried on in the usual manner, " except that the chemical ingredients in ordinary use for the " purpose are varied in their proportions, so as to adapt the process " to the particular nature or character of the fabric."

[Printed, 3*d.*]

## A.D. 1856, May 13.—Nº 1123.

PARKES, Alexander.—(*Provisional Protection only.*) The title of this invention is, "Improvements in the use of collodion " in photography."

The inventor states:—"Heretofore in taking photographic " pictures by what is known as the collodion process, a sheet of

" glass has been used to support the film of prepared collodion. " Now this invention consists in substituting for this sheet of " glass a sheet of collodion, of sufficient thickness to support " the prepared film; or a thick layer of collodion may be first " formed on the glass, and on this layer the film of prepared " collodion may be produced, and the picture taken thereon and " suitably varnished or protected, afterwards the whole may be " stripped from the glass together. And when making positive " pictures by either of these processes, I coat the back of the " collodion plate which supports the picture with a black flexible " varnish, or the plate of collodion which supports the picture " may be of a black color."

[Printed, 3*d.*]

### A.D. 1856, May 16.—N° 1159.

THISTLETHWAITE, WILLIAM (*a communication from Louis Angamarre*).—" This invention consists in producing and fixing " photographic pictures upon all kinds of fabrics and materials."

A very thick preparation of collodion is made, " consisting of " sulphuric ether, alcohol, gun cotton, and iodide of ammoniac;" " to make the collodion more sensible," certain proportions of alcohol, " iodide of ammoniac, bromide of ammoniac, and chloride " of ammoniac," are mixed with it (the " chloride of ammoniac " is not mentioned in the Complete Specification). This composition is termed " the pedicle," and is poured upon glass and then plunged into a solution of nitrate of silver. When the photographic image is impressed, the negative picture is obtained by introducing the plate " into a bath composed of distilled water, " protosulphate of iron, acetic acid crystallized, and sulphuric " acid;" the positive picture is fixed by plunging the plate into a solution of cyanide of potassium. " The proof is then dried, " and the pedicle may be seperated from the glass with a piece of " blotting paper and placed upon the necessary material, and " passed under a roller or rollers so as to well fix it upon the " material; and after the blotting paper is removed, spirits of " wine (of 60 degrees centigrade) is poured upon the likeness or " image to give it the required tone and brilliancy."

The following is a mode of separating the " pedicle " that does not invert the picture:—The " pedicle " is made to float upon the top of the glass, by plunging it (the glass) into a weak solution of hydrochloric acid; the "pedicle" is then transferred and finished by laying the glass down on a flat surface, placing the material upon it, taking up the whole, so that the glass is uppermost, pressing "the fabric underneath it with the fingers, to force " out all air bubbles," removing the fabric and "pedicle" from the glass, pouring water over it and drying the picture. The surface is then moistened and cleansed from all imperfections by rubbing. The picture may then be coloured in the usual manner. " This last mode of removing the pedicle from the glass" is preferred.

[Printed, 3*d.*]

## A.D. 1856, May 21.—N° 1201.

DUFRESNE, ALEXANDRE HENRI.—" An improved process of " gilding and ornamenting steel and other metals," in which photography is used to produce " reserves."

This invention consists in :—

1st. " The employment of one or several intermediate metals, " deposited either by chemical, electro-chemical, or mechanical " processes, on the metal to be gilded, silvered, or ornamented."

2nd. " The manual or mechanical application of a protecting " matter, such as varnish, bitumen of Judea, printers' ink, &c., " upon the intermediate metals to form the reserves to be gilded, " silvered, or ornamented, such reserves being produced by photo- " graphic means, or by a general coating sensible to light, such " as bitumen of Judea." The ordinary processes of photography may be employed " either directly or by transfer on the surface to " be gilded, silvered, or ornamented."

3rd. " The destruction of the intermediate unreserved metals, " by baths of different kinds," " for the production of flat and " relievo designs."

4th. " The removal of the protecting matters which have served " to preserve the surfaces operated on."

5th. " Gilding or silvering the surfaces by means of mercury, " according to the ordinary processes of gilding and silvering by " amalgamation, and finally the volatilization of the mercury " by heat."

Methods of operating on iron, steel, "platina," and silver, according to the processes comprised in this invention are given at length.

[Printed, 3*d.*]

A.D. 1856, May 24.—N° 1245.

JUNDZILL, ADAM DUNIN.—(*Provisional Protection only.*) The title of this invention is, "An instrument for animating " stereoscopic figures."

The inventor states:—" This invention, for which I request a " patent, is an application of the stereoscope to the apparatus well " known under the name of phanakisticope" [phenakistoscope?] " or sobroscope, invented by Plateau; the last shews the figures in " motion. My apparatus makes the figures which stand out by " the stereoscope to be seen in motion; this is its description:—

" Two round cartoons, with figures of phanakisticope set in " the same manner and symetrically, are put in motion by means " of a crank which gives them a speed perfectly identical. A " small pasteboard wheel, bored on the edge, is put in motion by " the same crank in such a manner that its speed be the same as " that of the aforesaid two cartoons with figures. Two stereo- " scope glasses are set into their frames forward the bored wheel, " and are placed in such a manner that both eyes may see the " two figures reflected in a looking glass which is set before the " apparatus. Therefore, when any one intends to use the appa- " ratus, he will begin by setting it at such a distance from the " looking glass that the figures motionless be confounded, and " stand out as in the stereoscope. Then the crank will be put in " motion. The two cartoons with figures being identical, and " their speed being the same, it is obvious that their effect will be " the same as in the phanakisticope, and therefore the figures " will be seen in natural motion."

This apparatus is called the "*kinimoscope.*"

Any of the ordinary means can be used to put " the cartoons " in motion.

[Printed, 3*d.*]

A.D. 1856, May 31.—N° 1295.

FOWKE, FRANCIS.—(*Provisional Protection only.*) "An im- " proved portable photographic camera."

This "improvement consists in certain mechanical arrange-"ments, by means of which the frame of the photographic "camera can be brought to lie close together, and packed in the "least possible space.

"The bottom which has usually been employed, is partly "dispensed with, by constructing the front part of the camera "in the form of a four sided cone, made of wood, each side "hinged, and differing sufficiently in length to fold down flat "one over the other, so as to be enclosed in a shallow frame. "This peculiar arrangement of the front part has the advantage "of giving it sufficient rigidity to sustain itself."

[Printed, 3*d.*]

A.D. 1856, June 5.—N° 1344.

DALLAS, Duncan Campbell.—(*Provisional Protection only.*) The title of this invention is, "Improvements in chemical pre-"parations applicable to the photographic and photo-galvano-"graphic processes."

The inventor states:—"The peculiarity of my said invention "consists in the employment of chromic acid, either alone or in "combination with photographic or other materials, as hereafter "mentioned.

"I put a solution of chromic acid to a solution of gelatine; or "with the solution of chromic acid I mix a solution of iodide of "potassium, & add them to one of gelatine; or the solution of "chromic acid may be mixed with one of nitrate of silver, and "be added to one of gelatine; or, if required, all the four solu-"tions may be combined."

[Printed, 3*d.*]

A.D. 1856, June 10.—N° 1373.

SKAIFE, Thomas.—This invention is entitled "Spring-folding "camera shutters, for the more speedy and convenient mode of "taking photographic pictures than has been hitherto adopted."

The invention consists:—

1st. In substituting for the ordinary dark slide of the camera, a pair of folding shutters. These shutters are fixed on vertical spindles attached to the plate frame; each spindle carries at its upper end a drum, and an ndia-rubber elastic band and cord are so arranged round the drums that, upon a ring being drawn parallel to the optical axis of the camera, the band stretches and rotates the drums, thence opening the shutters. Upon the ring being let loose the reaction of the India-rubber spring closes the shutters. When the shutters are large, other springs are applied, so as to prevent them opening when out of the camera.

2nd. In adapting a similar arrangement to that of the 1st part of the invention, to the camera itself, immediately behind the lens. The cord from this second pair of shutters may be attached by a hook to the ring of the larger shutters above mentioned, and thus both pair of shutters opened at once, or they may act independently of the frame shutters. "The object of this second pair of "shutters is to keep the lens covered until the first pair is at "least three quarters opened."

When an instantaneous picture is required, a steel plate with a circular aperture is used instead of the second pair of shutters. To allow the aperture to be suddenly and momentarily projected behind the lens, a thread attached to the steel slide is acted on by an India-rubber tube. A bolt of steel or "trigger" retains the opaque part of the slide behind the lens until the picture is required to be taken. When the "trigger" is pressed, the slide is immediately projected behind the lens, thus instantly admitting and shutting off the light in consequence of the India-rubber tube being free to act. A magnet may be placed so as to accelerate the movement of the slide.

[Printed, *7d.*]

### A.D. 1856, July 12.—N° 1651.

AVERY, John (*a communication from Madame Millot*).—"An "improved 'plate-holder' for photographic and other purposes."

The Specification and Drawings describe and show an apparatus possessing the general appearance and many of the characteristics of an ordinary pistol. The part corresponding to the barrel is a "rod of wood or other suitable material" fixed to the stock or handle, over which "rod" a moveable tube or "hollow rod" fits. "Both rods are provided with jaws or clamps" "cut aslant or "hollowed on the parts opposite to each other." By sliding the "hollow rod" upon the other, the jaws can be set at any distance apart to suit plates of various sizes. The jaw upon the "move-

" able rod " is fixed; that upon the stock slides in a groove and is provided with a suitable spring and trigger.

The trigger being fixed to the jaw on the stock, by pressing the trigger, the action of the spring is neutralized, and the plate can either be fixed in the apparatus or removed therefrom. When a plate is required to be coated with a thin layer of collodion or solution, this apparatus enables the operation to be performed easily and readily, owing to the freedom of motion permitted by the position of the handle.

When large plates are to be operated upon, a metal point at the end of the "moveable rod," "stuck either in the wall or a " door," gives the necessary support to the plate.

The spring attached to the moveable jaw may be of caoutchouc, or a magnet may be used instead of a spring.

[Printed, 10*d.*]

A.D. 1856, July 26.—N° 1782.

COOKE, George Colleton. — "Improvements in stereo- " scopes."

This invention consists :—

1st. In "the application to stereoscopes of conical, "or pyra- " midal, or trumpet-mouth tubes for the eye pieces, in place of " the ordinary cylindrical tubes usually employed for this pur- " pose."

2nd. In "the adaptation to the eye pieces of stereoscopes of " additional moveable lenses of various kinds." "These lenses " may be colored or plain," and "are moved into and out of " use by means of small levers projecting through the sides of " the box;" "colored glasses may be adapted to the instrument " in a somewhat similar manner." When it is desired to magnify the picture, the additional lenses should be placed about half way between the picture and the eye pieces. If magnifying lenses of short focal distance are placed immediately under the ordinary lenses, the picture is placed on a shelf about midway between the bottom of the box and the eye pieces.

3rd. In "the adaptation to stereoscopes, immediately above the " space occupied by the picture, of a double 'passe-partout,' or " frame, for the purpose of preserving uniformity of size in the

" pictures, as well as to prevent any light from being reflected " from their margin and confusing the eye."

[Printed, 10*d.*]

## A.D. 1856, August 19.—N° 1935.

SUTTON, Edwin.—(*Provisional Protection only.*) The inventor states :—" My invention of an improved 'construction of stereo- " 'scope,' has for its object, firstly, the constructing and so com- " bining the various parts of which the stereoscope is commonly " composed, that it may be easily compressed into a small case, " & thereby rendered more conveniently portable than the usual " rigid form stereoscope.

" Secondly, it is made to possess all the advantages of the " most improved rigid stereoscope by the construction & appli- " cation of entirely newly formed hinges & spring. These " hinges I fix to the bottom of the folding ends that support the " top that carries the lenses or eye pieces. The centre pivots or " axles of the hinges I make so as to work in holes in the sides " of the case, into which the instrument is compressed; upon " these axles I fix a piece of metal with angular hinges. I also " attach to the inside of the case a flat steel spring, made in such " a manner as to be always kept in tension by the angular edges " of one hinge at each end, and as the flat edge of the spring " presses against the hinges, it will always have a tendency to find " a flat baring, & consequently will firmly keep the parts of " the instrument in position when in use & at rest."

[Printed, 3*d.*]

## A.D. 1856, August 23.—N° 1965.

BENOIST, Philippe.—(*Provisional Protection only.*) " An im- " provement in the construction of sterioscopes."

" This invention is to combine the effects of the 'fenakisticope' " [phenakistoscope ?] " and the sterioscope; the former gives move- " ment to the figures, but no relief; the latter relief, but no move- " ment. The animated sterioscope unites the two results in one " instrument. The 'animated sterioscope' is contained in a box " scarcely larger than the ordinary sterioscope, but in doubling the " pictures of the sterioscope it preserves the relief of that instru- " ment, and adds the movements of the 'fenakisticope.' The "figures in the animated sterioscope are, as in the two instruments "above mentioned, obtained in the desired position, and they are "seen alternately without visible displacement in the substitution. "To obtain this effect they are placed one upon the other, "so that they may be seen alternately, the one without the other, "by the following means:—The two sterioscopic pictures are "placed at right angles, the plain of the one being perpendicular "to the plain of the other. A grooved glass cuts the angle in two, "and consequently places an angle of forty-five degrees between "the glass and each figure. The motion which is given to the "glass by a spring at the side of the instrument displaces it "without altering its inclination, so as not to change the angles "which it makes with each picture. As the glass is only trans- "parent in those parts necessary to obtain the desired effect, "it shews alternately the two figures, the one by reflection, the "other by refraction."

[Printed, 3*d.*]

A.D. 1856, August 26.—N° 1983.

PERRY, John.—This invention relates to the obtainment of photographic positive pictures from negatives previously obtained on glass or other suitable material, the said positive pictures being "absolutely permanent and insensible to the action of solar light."

The paper to be used is floated on a solution of chromate of ammonium or bi-chromate of potash, dried, impressed with a positive picture, washed, dipped in a solution of protosulphate of iron, washed again, and treated with a solution of gallic or pyrogallic acid or of tannin.

In this process other combinations of chromium may be employed, and if chromate of iron is used, the picture is ready for the application of the gallic acid, without the use of any other salt of chromic acid or of iron, thus simplifying the process.

Soluble salts of silver, tin, lead, zinc, cadmium, antimony, nickel, or bismuth, with or without ammonia, may be used, "previous to the application of the chromium," to produce depth of colour.

Instead of using protosulphate of iron to develope the picture, other salts of iron may be used for that purpose, and the salts of silver, copper, tin, lead, zinc, cadmium, antimony, nickel, or bismuth may be employed at the same stage of the process. The salt of iron may be employed, combined with ammonia, without the chromium, but in conjunction with the subsequent use of gallic acid.

To produce colour and improve the general effect, cyanogen and other analogous articles may be applied in the same manner as the gallic acid.

If paper is used for this invention, its surface may be prepared with gelatine prior to other treatment.

[Printed, 4*d.*]

A.D. 1856, September 1.—N° 2029.

NORRIS, RICHARD HILL.—" Certain improvements in photo-" graphy by the use of collodion in a dry condition, and for " a means of transferring photographic films."

Under ordinary circumstances, the sensitiveness and porosity of collodion films are destroyed by dessication, but by the following process pictures may be produced " on perfectly *dry* and hard " collodion films:—Having produced in the film the sensitive " iodide of silver," the film is immersed " in a solution of gum " arabic, or of dextrine, starch, gelatine, albumen, gum traga-" canth, vegetable mucilages, caseine, gluten, or other such like " substances," that will by " occupying the pores of the collodion " film, prevent its condensation on drying, and retain it in a " sensitive and pervious state;" " the films are then dried and are " ready for exposure to light, or may be kept for any convenient " length of time and used as desired."

To transfer photographic films from glass plates to elastic plates of gelatine, the following process is used:—" A strong solution " of gelatine is poured over the film and permitted to dry; when " dry it is coated with plain collodion, and may be readily " stripped from the glass, and then possesses the advantages of " elasticity, compactness, lightness, and freedom from breakage. " This mode of transfer is applicable to photographic films in " general."

[Printed, 3*d.*]

A.D. 1856, September 5.—N° 2064.

DANCER, JOHN BENJAMIN.—" Improvements in photographic " cameras, and in the apparatus connected therewith."

1st. To adjust the focus of binocular stereoscopic cameras, a rack or chain movement is used; one extremity of the rack or chain is fixed to the camera box, the other to the sliding piece carrying the lenses. When a chain is used, a fluted spindle terminated by an exterior milled head, produces the necessary motion; and if a rack is employed, it is worked by means of a pinion.

2nd. The application of a fixed or permanent level "to cameras "in general." The Drawings show an ordinary spirit level mounted on the top of the camera box.

3rd. "The use of revolving or sliding diaphragms or plates "pierced with apertures of various areas, which may at pleasure "be brought into use." The lenses are attached to the front of the camera by flanges, and project into the camera; the flanges carry the perforated discs by means of centre pins. The discs are turned upon the centre pins until the required apertures are opposite those of the flanges, and are held in the required position by means of springs fixed to the camera box, which take into indentations on the discs. To shut off the light from the lenses of binocular cameras, a plate is mounted upon a centre pin situated between the two lenses; by the turning of the plate the two apertures are closed at the same time.

4th. "A method of altering the level of the lenses of binocular "cameras." The lenses are mounted upon a piece capable of sliding in vertical guides; to this piece is attached a central nut, within which is a vertical screw; by turning the milled head of the screw the lenses may be raised or lowered as required.

5th. Apparatus whereby "a number of prepared plates or other "sensitive surfaces" may be successively exposed to light, without the aid of a dark room or other similar contrivance. According to one method, the camera box is supported on the top of a box, within which slides another box containing the sensitive plates, separated by suitable compartments. The plates are mounted in frames, which are (by rack and pinion motion) successively brought under an aperture in the camera box; a rod, situated in or near the focal plane of the lenses, is moveable vertically, so as to raise the plates through the said aperture into the proper position in the camera for receiving the image. When a plate has had the image impressed upon it, it is restored to the dark box by depressing the rod; the rod is then detached from the plate, the box moved so that the next compartment is under the aperture, and another plate impressed and restored to the box in like manner. The lid of the plate box is kept away from the aperture in the camera box by a screw, which is undone, to allow the lid to be closed, when the box of impressed plates is removed from the camera.

The box may be on either side or at the top of the camera, instead of underneath. According to another modification, the box may be stationary, and the vertical rod moveable to the places of the respective plates. The plate box may be detached from the camera, and the dark frame may be adjustable to it, so as to take out or restore a plate without exposing the said plate to light; for this purpose the lid of the box is made more than double the length of the box, with an aperture in the middle, and a hinged shutter prevents the light from entering the dark frame.

[Printed, 11*d.*]

### A.D. 1856, September 5.—N° 2072.

JOHNSTON, John.—" Improvements in photographic plates."

This invention "consists in substituting in lieu of plates of " glass, as commonly employed for taking photographic portraits " upon, thin plates of metal, for example, sheet iron, one side or " surface of which is coated with black japan, such said japanned " surface being intended to receive the collodion coating, in the " manner usually practised of coating plates of glass; the said " japanned metal plates possessing this advantage, that whereas " for positive pictures on glass the back of the picture requires " to be covered with black varnish to throw up the shadows, by " this invention, when the image or picture is developed the japan " background is at once exposed to view; moreover, the aforesaid " metal plates may be afterwards easily cut to any desired shape."

[Printed, 3*d.*]

### A.D. 1856, September 8.—N° 2092.

SABATIER, Boniface.—This invention relates to the instruments used in photographic manipulations.

This invention consists in making the plates used for photographic purposes with "tenons," for holding them in the various operations they undergo. The Drawings show a glass plate to which these "tenons" are adapted, either by being cut originally in that shape by means of a diamond, by casting, or by affixing by means of marine glue; the "tenon" is represented as a square projection in the middle of each of the shortest sides of the rectangular plate. This tenoned plate is fitted for application to the various instruments by a strip of glass, (the width of the tenons, and rather longer than the total width of the plate and tenons,) which is affixed to the back of the plate (by means of silver wedges) so not to touch any part of the plate itself.

To clean the glass plates.—For the rough cleaning a Baron Gros' instrument is used, to the ends of the plate of which are attached "metal tenons" and clamp screws. When manipulating, the plate of this instrument is attached to a table by a clamp and screw. For giving the finishing polish to the glass, an instrument is employed consisting of two cheeks, one of which slides in a dovetailed groove attached to the other, and is made to clamp the piece of glass at the back of the plate by means of a screw. The tops of the cheeks are flat, to serve "as a table for the tenoned "glass plate to rest upon."

A plate frame fitting into an ordinary camera is described and shown, having recesses for the reception of the tenons of the glass plate. The back of this frame has a recess for the reception of the strip of glass at the back of the plate, and is fastened to the frame by means of hooks and eyes.

An instrument for holding the strip of glass whilst the plate is being detached therefrom, consists of a stand having fork-shaped pieces projecting upwards that carry recesses and wires for that purpose.

An instrument for holding the glass plate whilst developing the image, consists of a slide provided with a handle underneath and carrying a fixed and moveable cheek. By suitably placing and fixing the moveable cheek the glass can be securely held.

A glass bath for containing the sensitizing solution is adapted for use in connection with the tenoned plates, by having two projections from the bottom of the bath on which the tenons rest. The solution should not be higher than the top of these projections.

In a pressure frame adapted to the tenoned plates, a frame with recesses for the tenons and an India-rubber strip placed round the edge for the plate to rest on, is placed in the pressure frame and obviates the necessity for using very thick glass.

In an instrument employed in the albumen and chloride of gold processes, a plate is connected with a stand by means of rods proceeding from its opposite sides to a common stem; a spirit lamp can thus be placed under the plate, and the photographic plate is secured to the plate by means of cramps.

In an iodizing frame the tenoned plate is surrounded with suitably shaped pieces of glass, to prevent evaporation by making an air-tight joint.

[Printed, 7*d.*]

### A.D. 1856, September 16.—N° 2254.

LANGLOIS, Claude.—"This invention consists in a method " of arranging photographic cameras so that the operator can " adjust the focus from the front, or that side next the person or " object to be taken, and also so that the moveable frame for " containing the prepared glass or other material for receiving the " image may also be used for ascertaining the correct focus." On one side of the camera folding doors are hinged to the front and back, and close together in the middle. The door hinged to the front of the camera is inclined to the front 45° when closed; it also has an aperture which serves for focussing and for opening the front door of the slide. The plate frame "has a hinged door " at the front and at the back, one for introducing the plate and " the other to be opened for exposing it;" it is mounted on a centre round which it can be turned half round; upon the frame being placed in the camera, the image is focussed upon the door at the back of the slide—made white for that purpose; the slide then being turned half round, the plate is exposed by the opening of the front door, the operator's hand being introduced through a sleeve for that purpose. "Screw feet" enable the camera to be adjusted to any required inclination.

"This invention also consists in backing photographic pictures " with plaster of Paris." A positive collodion picture (printed by the light of a lamp) is backed with plaster of Paris in the following manner:—The plaster is poured on the collodion surface and a piece of muslin is applied over the plaster, this operation being repeated until the required thickness is obtained. If it be required to remove the glass from the picture, the glass plate is first coated with gum, and when the picture has been printed and backed, the gum is dissolved by warm water, thus leaving the picture on the plaster; the resulting shining appearance of the picture is removed by ether. The plaster may be saturated with wax, when coloring is to be applied, and it may be cemented to metal, &c., to strengthen it.

[Printed, 4*d*.]

A.D. 1856, October 23.—N° 2486.

JOHNS, George Edward.—The title of this invention is, "The application and adaptation of an optical or stereoscopic "arrangement in the manufacture of boxes;" it consists in combining an optical or stereoscopic apparatus with ornamental and other boxes, "capable of such an arrangement, whether such "boxes be manufactured of paper, wood, or any other material or "materials."

The stereoscopic tubes and glasses are fixed to the front of the box; "the object or photographic views" are placed against "the "inner back of the box," the box being empty when used for stereoscopic or optical purposes.

"It will be seen, that all that is necessary to turn any box with "this arrangement into a complete stereoscope, will be to adapt "the focuses of the sterescopic lenses to the proper viewable "distances from the views to be placed against" "the inner "back of the box, and the tubes to the necessary width apart, so "that both eyes of a spectator may, at the same time, through "the tubes, see the object in the box."

"The stereoscopic lenses may be applied to any part of the box "to which it may be necessary or desirable to affix the same."

[Printed, 3*d*.]

A.D. 1856, November 3.—N° 2581.

SCOTT, Ebenezer Erskine.—"Improvements in stereo-"scopes."

This invention consists in "the use of eye pieces composed of "two magnifying lenses, either entire or modified," "placed so "that their centres are at a greater distance apart than the eyes "of the observer, the said distance being either fixed or capable

" of adjustment, so as to produce the maximum of pleasant effect " in the use of the instrument."

The double eye piece is so arranged "that the rays from the " two pictures come to the eye as if they radiated from a point of " medium distance, such as that at which we are in the habit of " looking at natural objects, the effect being" "that the eyes of " the observer instantly form the combined picture without the " slightest pain to the eyes, and the reality of the view is conse- " quently much more strongly impressed on the mind, and the " pleasing illusion much increased."

The lenses may be made achromatic with the plain surface next the eye, and in all cases the chromatic and spherical aberrations are greatly reduced.

An instrument is shown in the Drawings in which a diaphragm enables each eye to view only one picture; the picture is adjustable by means of a screw, to the centre of the field of vision, and suitably placed diaphragms shut off all extraneous light, thus permitting only the view to be seen. The box is mounted on a stand, and is capable of adjustment by a sliding rod and pinching screw.

[Printed, 7*d.*]

## A.D. 1856, November 6.—N° 2614.

OLLEY, William Henry.—"Improvements in obtaining " photographic impressions or pictures of microscopic objects."

This invention consists in obtaining the above-mentioned impressions "by the combined use of the microscope, the camera " lucida, or other suitable reflector or reflectors, and the camera " obscura."

" The metallic cap by which the eye piece of the microscope is " usually surmounted" is removed, and "the four-sided reflecting " glass prism known as the camera lucida, or any other suitable " and analogous reflector or reflectors," is or are placed upon or near the eye piece so as to produce an image of the object of which an impression is to be obtained. "The tube or body of " the instrument having been brought into a horizontal position, " an inverted camera obscura, or darkened box, is placed over " the said prism," it (the camera) being so connected with the microscope tube as to exclude light. Upon the object being properly placed before the object glass of the microscope, an image of the said object will be formed by reflection from the prism or other reflector upon a prepared plate or surface previously introduced into the camera. After the prepared surface has thus been exposed to the action of light, it is removed from the camera, and subjected to the usual processes of developing and fixing.

[Printed, 3*d.*]

### A.D. 1856, November 26.—N° 2806.

PALMER, Henry Eastman.—" This invention consists in a " method of arranging photographic cameras and apparatus so " that the plate may be rendered sensitive and the picture de- " veloped and fixed within the chamber of the camera."

The camera chamber contains the following apparatus:—A " bag or well, of flexible waterproof material, for holding a vessel " containing the solution of nitrate of silver," fixed underneath the plate frame. A tank, also under the camera, and nearer to the lens end, containing water heated by a lamp; the lamp is used to heat the chamber in cold weather. The waste pan, under the camera near to the plate end. A cold water reservoir, of flexible waterproof material, at the back and upper part of the camera, with a flexible tube and stopcock; this reservoir supplies water to the interior of the chamber, to the hot-water tank, and to a condenser that is close by the hot-water tank and in which the steam produced in the hot-water tank is condensed. Manipulating sleeves, attached either to the back or sides of the camera. Two eye pieces, and a small window glazed with yellow glass, to enable the photographer to see his operations. The plate frame; this can be slid to and from the lens by means of a rack, and can also be placed in an inclined position. The frame, containing the lens, attached to the camera by means of a flexible tube capable of adjustment by means of a set screw. Vessels containing the developing and fixing solutions, suspended from the rods that support the top of the camera and the cold water reservoir.

The sides of the chamber are made of waterproof fabric, supported by rods set upon the ordinary tripod.

The cap of the lens is a disc with a hinge.

[Printed, 7*d.*]

A.D. 1856, November 29.—N° 2832.

HARMER, Richard.—" Improvements in stereoscopic pictures."

The inventor states: — " My invention relates to colouring " stereoscopic pictures, that is, pictures to be viewed in the " stereoscope produced by the camera or any other means, the " same being printed on paper chemically prepared by the " agency of light or otherwise, as usual.

" In pictures viewed in the stereoscope two similar pictures " are introduced to produce one view, which two pictures hereto- " fore have been exactly similar in every particular in colour and " outline, or as nearly so as may be, whereas according to my " invention, I produce and use two pictures in the ordinary way, " similar in outline but different in colour. For example, one " picture may be coloured emerald green and carmine, and the cor- " responding picture used therewith may be coloured blue and " violet, which colours when viewed in the stereoscope blend with " each other, and greatly enhance the effect produced. Any other " colours may be used, and if selected and used with judgment " produce beautiful effects in such pictures."

The colours are applied by hand.

[Printed, 5*d*.]

A.D. 1856, December 3.—N° 2871.

CHEETHAM, James Kinder.—" Improvements in the appli- " cation of photographic pictures to metal and other surfaces, " and in rendering the same applicable as printing surfaces."

1st. Obtaining designs upon copper which may be used for ornamental purposes, " or upon which the engraver may work by " any of the usual methods." A varnish composed of shellac dissolved in a solution of borax is applied to a collodion photo- graph; " when this is dry, a piece of paper painted with the same " materials is applied wet, and the whole allowed to dry, after " which the plate is placed in cold water, and the film will then " separate from the glass." This film is placed upon a sheet of copper that has been coated with mercury, the reduced silver of the photograph transferred to the mercury by rubbing or pressure, and the film separated from the metallic surface by heat. A variation of this process consists in removing the film with its picture from the glass by means of a solution of gutta percha in benzole, applying it (the film with its picture) "to the mercu-" rialized metallic surface," and removing the film only from the metallic surface by dissolving it in methylated chloroform; "the " mercury may then be driven off by heat if desired." It is important to observe that "the picture is not caused to adhere to the " metallic surface by any glutinous substance," " but is connected " thereto by the mercury, after the ordinary manner of gilding."

2nd. Printing surfaces are obtained from photographs by etching the silver picture, obtained according to the method set forth in the 1st improvement, so as to leave one portion of the surface in relief. The etching fluid may be nitric acid.

3rd. Producing lithographs by transferring the aforesaid silver picture on to a surface of copper electro-deposited upon a stone. The "bare copper" is then dissolved away, the stone run over with an inking roller, and the metal picture removed, "leaving a " clear surface of stone for the light portions." "This operation " may be reversed."

[Printed, 3*d.*]

A.D. 1856, December 5.—N° 2887.

KLOEN, William, and JONES, Daniel.—(*Provisional Protection only.*) "An improvement or improvements in photography."

This invention "consists in substituting the materials, herein-" after enumerated, in place of glass, for taking the photographic " pictures upon, called positive collodion pictures. When glass is " used for the said pictures, the said glass requires to be blackened " at the back if colorless, and when, to avoid this trouble, black " glass is employed, the brittleness, which is one of the greatest " objections to the use of glass is unremedied."

This invention "consists in the use of materials which present a " black surface, which are not brittle, and which have such a " chemical composition as not to act prejudicially on the materials " employed in taking the picture." The material preferred "is " papier mâché, and the various compositions and materials which " are or may be used as substitutes for papier mâché, such as " plates of metal, wood, or other substance covered with the " varnish or japan with which papier mâché is usually covered. " The plates on which the pictures have been taken may be " applied to trays and such other articles as are or may be made " of papier mâché."

[Printed, 3*d.*]

A.D. 1856, December 9.—N° 2914.

BROWNING, John.—(*Provisional Protection only.*) " Improve- " ments in stereoscopes."

This invention " consists in wholly or partially coating the ex- " terior of the instrument with glass, either silvered, painted, " etched, stained, engraved, or otherwise ornamented, and in the " employment of colored screens of glass to intercept the light " transmitted directly through glass or other transparent slides, " or reflected upon paper or other opaque slides, thereby produc- " ing effects upon the pictures, when viewed with the improved " stereoscope, resembling those produced by what are known " as Claude Lorrain glasses." When the improved stereoscope is used with transparent slides, pieces of coloured glass, of any required tint, are placed " in a groove at the bottom of the instru- " ment underneath the slide containing the picture or other " object. When opaque slides are used, the colored glass " screens are slid into a groove in front of the instrument, over " the usual aperture made for that purpose."

" The bottom portion of the front of the instrument being " composed of glass," the door and reflector at present in use may be dispensed with, or coloured reflectors of metal or glass may be employed as may be required. When transparent slides are examined, an opaque glass closes " the aperture in front of the " instrument;" this opaque glass " can be concealed within the " instrument when opaque objects are viewed."

[Printed, 3*d.*]

A.D. 1856, December 19.—N° 3009.

MASSI, Charles.—(*Provisional Protection only.*) " This inven- " tion has for its object improvements in apparatus for mounting " cameras. For these purposes, when two cameras are used to " take two pictures at the same time and at an angle to each other " for stereoscopic purposes, the two cameras are placed on a suit- " able surface or stand, through which are two parallel slits or " openings for the passage of two pairs of studs or upright pro- " jections, which are fixed to two bars, one pair to each bar. The " two studs or uprights of each bar are at a distance apart from " each other, so as to pass through the two parallel slits in the

" surface above. The two cameras are each placed on to the " two stems or uprights of one of the bars, by which means, so " long as the two bars are parallel to each other, the cameras will " be parallel, but when the bars are made to incline to each other, " a similar inclination is given to the two cameras. The two " bars are actuated by means of two screw shafts or axes, each " having a right handed screw at one end, and a left handed screw at " the other end. The two bars have each two screw nuts attached " thereto, through which the two screw shafts or axes pass, and " they actuate the two cameras in such manner, that if both the " screw shafts or axes are simultaneously turned, the cameras " approach to or recede from each other according to the direc- " tions in which the screw shafts or axes are turned. And when " only one of the screw shafts or axes is turned, or when one is " turned in excess of the other, the cameras are made to incline to " each other to the extent desired. By these means two pictures " may be more conveniently and more correctly taken at the " same time than heretofore."

[Printed, 3*d.*]

---

## 1857.

A.D. 1857, January 1.—N° 11.

PHILLIPS, William Henry. — "Improvements on stereo- " scopes."

"The frame or case of a stereoscope is made in two parts, " hinged or connected together at or near the base of each in such " manner that the eye pieces may be brought to or moved from " each other with greater facility than heretofore to accommodate " for different sights. The movement of the two parts may be " accomplished by hand when holding a stereoscope to the eyes, " or by a screw, or other convenient means. The lenses and eye- " pieces, in place of being mounted in sliding tubes, as heretofore, " are each mounted on a tube, produced by winding a plate or " strip of sheet metal or other suitable material spirally, and the " elongating or contracting of such tube may be by a screw or " other convenient means. In order to give a better finish to the " picture, it is arranged to be seen through openings of another

" photographic picture, representing a frame or curtain, or other " suitable device, according to taste, and such second pictures " may be either fixed or moveable in respect to the stereoscope, " and also in respect to the pictures used therewith. When " separate from the pictures different pictures may be viewed " through the openings of the same second picture."

The Drawings show a stereoscope in which the two parts above mentioned that carry the eye pieces are connected to the two ends of the base by a flexible material; a hinge joint connects the two parts together. The distance between the eye pieces is regulated by a right and left handed screw. The elongation or contraction of the eye-piece tubes is accomplished by means of a right and left handed screw. In another form of stereoscope the spiral tube is the body of the instrument, and carries both eye pieces.

[Printed, 8*d.*]

## A.D. 1857, January 20.—N° 168.

QUIN, Richard.—(*Provisional Protection only.*) "This inven- " tion consists of improved methods of arranging stereoscopes. " For this purpose the front and back of the instrument, which " are rigid, are hinged or jointed to the top or portion containing " the eye pieces, and they are connected together at the sides by " flexible gussets, so that they may be folded down parallel to " each other when the instrument is out of use. To the back of " the instrument a stiff partition is hinged, which turns up so as " to keep the front and back at a suitable distance apart when the " instrument is set up for use; or in place of a central partition " the instrument may have two stiff sides turning up in a similar " manner, in place of having flexible side gussets, as previously " mentioned.

" In the front of the instrument is a reflector of the ordinary " description, which, when the instrument is folded, is fastened in " its place by the same clasp as that which holds the front and " back together. This clasp is made in three portions; one fixed " to the back of the instrument, another folding over the edge, " and the last turning down on the front, and catching over a pin " fixed on the outside of the reflector." The instrument, when arranged in this manner, forms a box capable of containing a number of stereoscopic pictures.

Another arrangement consists in using a stiff bottom, to which the front and back of the instrument are connected by a flexible joint; the top of the instrument being connected in a similar way to the front and back, "the parts of the front and of the back " double in and fold one over the other when the instrument is " out of use. A central partition is used to make this instrument " rigid when set up."

Folding stereoscopes may be made with a stiff front, to which the top and bottom are jointed; the back and sides are flexible.

[Printed, 3*d.*]

### A.D. 1857, January 22.—N° 191.

MANDER, Elisha, and MORGAN, William.—(*Provisional Protection only.*) " Improvements in the manufacture of photo- " graphic, jewellers', and other cases having wood or papier mâché " foundations, and where raised, regular, or irregular forms are " required in such cases, and the machinery for carrying out such " improvements, parts of which are applicable to other purposes " where sawing or shaping is required."

The improvements relate to various modes of "mitring" the angles; holding the material together; preventing warping; combining papier mâché, millboard, and wood; forming the tops and bottoms; and "the employment of composition or glue, or marine " glue with sawdust, in hot moulds to form cases."

The saw of the sawing machine is placed at an angle of 45° to the table, and traverses "by suitable contrivances, the material " being held stationary in guides, or the saw may be stationary " and the material moving against it.

"The shaping of the various portions of cases is performed by a " machine, having two lathe heads geared together and arranged " either side by side or facing each other, one having an arrange- " ment for holding the material to be cut, and the other a maundril " corresponding with or the reverse of the form required, the " cutter running in a rocking shaft or spring slide placed on a " slide passing before or between such heads; or a single head " may be used, the maundril being placed on the spindle of the " said head; or the action may be reversed, the head may be made " to rock instead of the cutter."

[Printed, 3*d.*]

A.D. 1857, February 4.—N° 321.

LEWIS, Edward, and BÖHM, Gideon.—(*Provisional Protection only.*) The title of this invention is, "Improvements in "printing in colours, called an improved photo-galvanographic "chromographic process."

The inventors state:—"The object of this invention is to use "plates or impressions from plates prepared by a process called "the photo-galvanographic process, in the art of colour printing or "chromography. The first named process, which forms the "subject of Letters Patent granted to M. Pretsch, consists in the "peculiar adaptation of the photographic process to the produc-"tion of printing plates in copper and other metals. Now, our "invention consists of the use of these plates or impressions as a "basis, outline, or groundwork for pictures printed in colours on "wood, stone, metal, or other substance."

[Printed, 3*d.*]

A.D. 1857, February 9.—N° 374.

TAYLOR, Thomas John.—(*Provisional Protection only.*) The title of this invention is, "An improved construction of stereo-"scope."

The inventor states:—"This invention relates to the application "to stereoscopes of a reflecting surface, which will enable opaque "photographic pictures to be viewed with greater ease, and seen "with greater clearness than where the ordinary construction of "stereoscope is employed. For this purpose, I apply to the under "surface of the hinged door of the stereoscope a plane or concave "mirror, (preferring the use of silvered glass,) and the door thus "provided I set at any required angle with respect to the plane of "the photographic pictures, by means of a quadrant, arc, or other "suitable contrivance, in order that the reflecting surface may "receive the light and transmit it to the surface of the double "picture. Or, I apply to the stereoscope a fixed surface or screen, "instead of a movable one, provided with a plane or concave "reflector.

"By this arrangement not only will the picture be more clearly "seen than heretofore, but the spectator when using the stereo-

" scope will not require to stoop in order to permit the rays of " light to strike directly upon the picture."

[Printed, 3*d.*]

### A.D. 1857, February 18.—N° 478.

MOULE, John.—This invention consists of " improved apparatus to be used for burning pyrotechnic compositions or preparations," and " relates to a means of burning any of the " chemical compounds that are usually employed for producing " various colored lights, such as are required for theatrical performances; some of which chemical compounds will, however, " produce lights of such a quality as will admit of their being " used for photographic purposes."

In one apparatus, a wide shallow vessel containing sand, is provided for the reception of the pyrotechnic composition; this vessel is placed within a glass cylinder or other suitably shaped case with openings at the bottom " to admit air in vertical columns " all round the burning composition," and a moveable cap connected with a flue or chimney. A hinged door enables access to be obtained to the interior of the glass case.

In another apparatus, the pyrotechnic composition is contained in a tube " up which it is pressed or forced by means of a coiled " spring," " somewhat in the same manner in which candles are " pressed up in candle lamps." The charged tube is fixed in a case similar to that of the apparatus first described; it has a knife attached to a lever for the purpose of dividing the charge into several parts.

The effect of either of these apparatus is to carry off the noxious vapours or gases from the burning composition, and thus prevent annoyance to bystanders.

The glass surrounding the burning composition may be coloured so as to render the light " useful for photographic purposes."

[Printed, 9*d.*]

### A.D. 1857, February 20.—N° 501.

GLOVER, Joseph, and BOLD, John, the younger.—The title of this invention is, " Improvements consisting of extended uses " of photography as applied to dials, tablets, and pictures."

Enamelled glass, metals, or other suitable mineral substances are operated upon in the following manner :—The surface of the material is first washed with fluoric acid to destroy the gloss, and then with water. The material thus prepared is ready to receive the collodion or other sensitive substance, and to be subjected to the ordinary photographic operations of receiving the image, and developing and fixing the same. "Translucent or transparent "tablets and dials may be taken on glass or other transparent, "semi-transparent, or opaque substances by the above process. "By thus preparing the enamelled or other surface, pictures or "subjects printed thereon are rendered capable of receiving water "colors, oil colors, dry colors, and varnish colors."

[Printed, 3*d*.]

## A.D. 1857, April 9.—N° 1005.

PURNELL, JOSEPH.—(*Provisional Protection only.*) "Im-"provements in apparatus for taking photographic pictures."

An apparatus is described in which "the photographic surface "is rendered sensitive within the camera." A clamp, suspended by a rod passing through the top of the camera, presses against the two edges of the plate by means of two jaws. The prepared plate having been placed in the clamp, is lowered by means of the rod into an exciting bath "let into the bottom of the camera." When it has remained in the exciting bath a sufficient time, "it is "in the same way raised out of it, and the picture is taken while "it is still held by the clamp; afterwards the hand of the operator "is introduced through a sleeve into the interior of the camera, "and the plate is removed from the clamp and placed in a drawer "at the bottom of the camera, which is then closed by a slide, "and the drawer is removed for the developing operation to be "effected." To develope the picture a cover fits on the top of the drawer; "this cover is furnished with eye-pieces, glazed with yellow glass," "the bottom of the drawer is also made of yellow "glass;" the picture is developed by pouring the requisite solution over the plate, then water is poured over the plate "through "a trapped hole in the side of the cover, and it is allowed to "run out by a valve at the bottom of the drawer. The picture "is fixed in the ordinary manner."

For taking stereoscopic pictures the lens of the camera is

mounted on a slide, and in taking the two pictures is pushed first to one side and then to the other of the camera. A moveable partition prevents the light falling on one side of the plate while the picture is being taken on the other side. To move the camera suitably in order to take the second picture, it has a rectilinear motion along a board, and an angular motion by turning the board on its centre.

[Printed, 3*d.*]

A.D. 1857, May 4.—N° 1253.

MOSELEY, THOMAS BEEBY.—(*Provisional Protection only.*) The inventor states:—"This invention relates to a particular "description of instrument called a pneumatic holder, used very "generally in photographic manipulations, and the improvement "which constitutes this invention consists in constructing the "same in the following manner, that is to say:—Instead of "simply affixing a bell-shaped piece of india-rubber to an up-"right handle and exerting pressure direct thereon, I affix an "annular piece of wood or other light substance to one end of "a piece of wood or other material which serves as a handle. "Into the aforesaid annular piece of wood I place a bell-shaped "piece of india-rubber, and connect it by a wire to one end of a "horizontal lever, the fulcrum of which is a pin passing through "the handle of the instrument, so that in using this instrument "the operator has simply to place the india-rubber surface upon "the article to be held thereby, and by forcing down the aforesaid "lever, I raise the bell-shaped piece of india-rubber within the "before-mentioned annular piece of wood, thereby producing a "partial vacuum therein and causing the india-rubber to adhere "closely to the surface, which it is desired to hold. The aforesaid "lever may be securely held in this position by a sliding link or "other suitable contrivance."

[Printed, 3*d.*]

A.D. 1857, May 27.—N° 1511.

NEWTON, WILLIAM EDWARD (*a communication*).—"An im-"proved method of applying photography to the use of engravers."

This invention has for its object "to produce a photographic "picture upon a surface of wood that shall entirely answer the "requirements of the engraver;" and it "consists in producing "a surface upon the block by rubbing into it a volatile varnish "made so limpid as easily to soak into and fill the pores of the "wood, and produce a smooth and polished surface without "leaving a pellicle thereon of appreciable thickness to obstruct "the operations of the engraver."

"The surface of an ordinary engraver's block is smoothed in "the customary manner to prepare it for the engraver's use; a "mixture is then made of alsphaltum varnish, ether, and lamp-"black. This mixture is rubbed into the surface of the block "with a piece of buckskin or cloth, two or three thin coatings "being applied so that the pores may be thoroughly filled, but "no pellicle of varnish of appreciable thickness left upon the "surface. An even, smooth, and polished surface is thus obtained, "upon which to take the photographic picture; upon this surface "the collodion is poured in the ordinary manner of taking pho-"tographic pictures; the silver bath is then applied. The block "is then exposed in the camera a sufficient length of time to take "the picture," and it is then subjected to a developing solution compounded of sulphate of iron, acetic acid, water, and alcohol. The picture is then fixed by means of a solution of "cyanuret "of potassium," "and the block is to be washed in clear water "and dried, when it is fit for the engraver."

[Printed, 3*d.*]

## A.D. 1857, June 2.—N° 1550.

SHAW, Charles.—(*Provisional Protection only.*) This invention is entitled, "A new or improved manufacture of matts for "photographic and other pictures."

The inventor states:—"My said invention consists in manufac-"turing the said matts of card or paper, plain or coloured, and "gilded or partially gilded, or covered with a metallic coating. "The said matts may be cut out and embossed at a press, or "manufactured by rolling, or by pressure otherwise applied. "Any desired ornamental form may be given to the said matts "by suitably engraving the dies or tools by which the said matts "are made. Matts made of gilded or plated paper or card "according to my invention cannot be distinguished when con-

" nected with a photographic or other miniature from matts made
" of metal. Matts made according to my invention are less costly
" than the ordinary metallic matts. Although I prefer to use
" paper or card in carrying my invention into effect, yet papier
" mâché, or other composition or mixture consisting mainly of
" vegetable fibre, may be employed with the same or nearly the
" same effect."

[Printed, 3*d.*]

A.D. 1857, June 3.—N° 1558.

CHAPPUIS, Paul Emile.—(*Provisional Protection only.*) "Improvements in stereoscopes."

"This invention consists of the use of metallic or other
" reflectors in or on stereoscopes, for the purpose of increasing
" the intensity of light brought to bear on the objects viewed
" through that apparatus."

[Printed, 3*d.*]

A.D. 1857, June 6.—N° 1595.

NOÉ, Henri Joseph.—(*Provisional Protection only.*) The title of this invention is, "Improvements in portable stereoscopes."

The inventor states :—" In constructing portable stereoscopes, I
" form each end piece in two parts, connected by a folding joint,
" and over the eye glasses in the upper piece I place a folding
" piece or jointed lid. The bottom of the instrument, or the piece
" which supports the pictures to be viewed, is connected to the
" upper portion of the instrument by an expanding connecting
" piece, or bellows joint, so that the distance between the pictures
" and the eye glasses may be varied by means of adjusting screws,
" or otherwise, to suit the vision of the person using the instru-
" ment. I also colour the reflecting surface of the instrument
" variously, to produce different effects, as may be desirable. I
" sometimes affix the bottom of the instrument to a case or port-
" folio, which, when the stereoscope is not in use, may be folded
" over it, and fastened by an elastic strap, or otherwise."

[Printed, 3*d.*]

A.D. 1857, July 1.—N° 1835.

NEWTON, William Edward (*a communication from Charles Nègre*).—The object of this invention " is to reproduce copies of "photographic images, drawings, and prints, by obtaining from "them by the action of light in conjunction with the employ- "ment of chemical and galvanic operations engraved plates, "either in intaglio or in relief, upon iron, steel, copper and its "alloys, zinc, silver, tin, aluminium, gold, and other metals; also "to produce damaskened designs formed by metals of various "colors; also designs formed of different colors, by means of "several engraved plates, registering one with the other; also "designs of different colors, gold and platinum upon porcelain, "earthenware, stoneware, &c.; also enamel or 'nielle' work and "inlaid work upon metals, marble, stone, &c."

The photographs employed are "negative reversed proofs, and "ordinary positive and negative proofs."

A layer of organic matter is spread upon a plate of polished metal. It is essential that this layer is sensitive to light, and can at the same time act as a varnish, to protect the metallic surface underneath from the action of acids. The sensitive layer is then acted upon by light, in a manner suitable to the effect ultimately to be produced, and the parts of the layer that have not been exposed to light are removed by a suitable solvent. All parts of the metal that have been laid bare by the solvent are then coated with a less oxidizable metal than that of the underneath plate, by means of electric force; the protecting layer is removed; the result is a "heliographic incrustation" of a different metal to that forming the ground of the picture. The plate may now either be oxidized or sulphurized, to obtain a colored damaskene, or it may be engraved by immersion in acids, or in electro-etching solutions; after the first biting in, the parts in relief are inked, and the biting in finished. To obtain an inlaid design from an etched plate, the parts in relief are protected by varnish, and the sunken parts are filled up by electro-deposited metal.

"'Nielle' work is obtained by filling up the hollow parts with "a substance which will melt into an enamel by means of fire."

Instead of etching the design upon the plate, an electro-cast may be taken from it.

If a layer of gelatine, added to bichromate of potash, has had a design impressed upon it by means of a "reversed negative," and has been acted upon by a solvent, "the polished metal will be laid "bare at those parts which correspond to the light parts of the "design;" these polished metal portions having been covered with printing ink, and the gelatine (and metal under it) having been acted upon by acid, an engraving in intaglio or in relief is obtained.

Engraved plates in intaglio and in relief may also be obtained from photographs, &c., by reducing the metal "in the parts of the " sensitive layer acted upon by the light, when this layer consists of " a salt of silver or any other metallic salt sensitive to light," and depositing copper upon the metallized design. "When the " deposit is considered sufficient it may be detached."

[Printed, 5*d.*]

## A.D. 1857, July 2.—N° 1843.

McCRAW, WILLIAM.—(*Provisional Protection only.*) "This " invention relates to certain improved processes of producing " positive photographic images or pictures on white or light-" tinted substances, either vitreous, animal, or vegetable."

According to one process, a slab of unglazed porcelain is coated with collodion, excited and exposed in the camera in the usual way; it is then developed by a weak developing solution, and " immediately washed before any visible or appreciable effect is " produced." After a momentary exposure to light, it is again treated with a developing fluid; the result is a positive photograph, "with the right and left reversed."

According to another process, a glass negative photograph is "placed in front of the camera at a suitable distance off, with a " mirror or reflector placed behind the image at an angle of about " 45 degrees, to act upon the principle of the microscopic re-" flector." The prepared porcelain slab is exposed in the camera, and the image developed, fixed, and washed in the usual manner. A clear and well-defined positive and direct photograph is the result; this may be treated with chloride of gold solution to heighten the effect, dried, and varnished, also coloured, if required.

By either of these processes positive photographs may be produced "on porcelain, china, ivory, bone, mother-o'-pearl, white " enamel, white or opal glass, and other surfaces, which, by " reason of the inequalities of their surfaces, or the rigid nature " of the materials, are rendered unfit for receiving photographic " images from negative photographs in the ordinary way."

Stereoscopic slides on white or opal glass can be produced by either of the above-mentioned processes.

[Printed, 3*d.*]

A.D. 1857, July 7.—N° 1883.

BÉRARD, PETER HIPPOLYTE GUSTAVE.—"Improvements in "manufacturing azotic cotton or pyroxile for photographic and "other purposes."

The cotton used for this process "is that obtained from the "shearing of swan skins (tontisse de molleton)." Two glass bottles are taken; a certain proportion of concentrated sulphuric acid and "powdered azotate of potash" is placed in one bottle, the bottle closed at once with its stopper, and shaken; the liquid is then poured into the other bottle containing the shearings of cotton, and the bottle closed directly. "The mixture forms a "thick paste, which is shaken several times," and the operation is completed in from six to ten minutes. The product is then washed by pouring filtered water into the bottle, pouring the whole into a large glazed earthen vessel containing filtered water, draining off the superincumbent liquid by means of a stopcock, and pouring in fresh water and so on until the liquid no longer acts on litmus paper. The cotton is then dried "by means of an "apparatus rotating quickly (une essorense);" "the drying is "then completed in a chamber heated by steam pipes."

"In order to obtain a greater solubility, ensure its preservation, "and avoid the causes of spontaneous alteration and the danger "attending removals, the azotic cotton is dissolved either in ether "or in the usual mixture of ether and alcohol; it is then left to "settle, and the solution being drawn off is evaporated by means "of a still so as to collect the ether which distils over."

The collodion thus obtained "is dry, diaphanous, entirely "soluble in ether and in alcoholic mixtures, and is ready for use "either for surgical or for photographic and other operations."

[Printed, 3*d.*]

A.D. 1857, July 7.—N° 1884.

BÉRARD, PETER HIPPOLYTE GUSTAVE.—"Improvements in "manufacturing and applying concentrated collodion," applicable to the collodion used for photographic purposes.

"Concentrated collodion" is economically prepared, "either by "dissolving the azotic cotton in a warm state, or by concentrating "the solution, in which case sixty per cent. at least of the ether "used can be recovered. The modus operandi consists in dis-"solving the azotic cotton with the assistance of heat in a "common distilling apparatus, composed either of a metallic, glass, "or earthen vessel, heated by means of a water bath, and pro-"vided with a worm as usual, which worm is cooled by cold "water." The distillation is continued "until the collodion is "concentrated to the required degree for the preparation to "which it is intended to be applied." The quantity of ether distilled and condensed, as shown in a graduated vessel, is an indication of the amount of concentration.

The collodion, thus manufactured, can be made to yield sheets as thin as the thinnest paper, as well as plates as thick as stout leather. In this process, instead of the cotton remaining several days in cold ether, to effect its solution, the solution of the cotton is almost instantaneous.

To prepare a coloured collodion, the colouring material (ground with castor oil) is added to the distilling vessel, and the distillation carried on as far as necessary.

[Printed, 3*d*.]

## A.D. 1857, July 28.—N° 2058.

BAXTER, Edward William.—"An improved mode of pre-"paring glass labels, advertizing tablets, and ornamental devices "upon glass," in which photographic processes are used.

To copy the device in gold or silver.—That part of the surface of the glass which is to receive the device is coated with weak isinglass size, and overlaid with gold or silver leaf. The metal leaf is then backed with a varnish composed of "asphaltum" dissolved in turpentine, which will become insoluble when exposed to the action of light. A negative design, made upon a transparent medium, either by the aid of photography or by painting or printing, is then applied to the sensitive surface and exposed to the action of light, thus fixing the parts left uncovered by the negative. The negative is then removed, the unfixed portions of the varnish dissolved off by means of turpentine, and the superfluous gold or silver wiped away by means of a damp piece of cotton wool; "there will remain on the glass a perfect counter-" part of the device contained on the negative plate."

"When using a pigment in place of metal leaf, it is to be mixed " with resinous varnish or gold size, and laid thinly upon the " glass surface, and when dry it is to be subjected to the above-" named treatment."

To produce devices in frosted silver.—A positive design is painted or printed upon a sheet of transparent paper. The glass is coated with a sensitive varnish as above described; the sensitive surface is then overlaid with the positive and exposed to light. The soluble parts of the varnish are then removed; the exposed portion of the glass is acted upon by fluoric acid, and the ornamented surface of the glass is silvered in the usual way.

[Printed, 4*d.*]

A.D. 1857, July 30.—N° 2078.

BAUERRICHTER, Henry, and GOTTGETREU, Gustavus.—(*Provisional Protection only.*) "The improvements relate to " adapting stereoscopic apparatus so as to obtain simplicity of " parts, and in combining therewith a box or case, which, whilst " it is suitable to hold such stereoscopic apparatus, is at the same " time adapted as a receptacle for handkerchiefs, gloves, stereo-" scopic slides, or other articles.

"The stereoscopic lenses are supported from one end of a plat-" form or frame, the other end of which is provided with a rest " for the pictures, and the apparatus thus formed fits into a box " or case in such manner as not to interfere (or only slightly so) " with the internal form of the box, which is also of form to " receive the stereoscope without interfering with its other uses; " the front or other side of the box or case being cut to admit of " the lenses projecting or being seen even when the box is " closed."

The stereoscope and box is also formed "so that the one fits " into and for the time when not in use forms or appears to form " part of the other without the lenses appearing externally. For " this purpose the part of the stereoscope carrying the lenses and " that forming the support for the picture (either or both of " them) fold to form part of the hollow of such box or case "without otherwise interfering with the internal form of the box "or case, and its adaptation as a receptacle for handkerchiefs, "gloves, stereoscopic slides, or other articles."

[Printed, 3*d.*]

A.D. 1857, September 1.—N° 2295.

ELLIOTT, Robinson.—"Improvements in photography, by "which the lensular defects of the present processes of taking "photographic prints are avoided, and impressions are obtained "of any size." "The lensular defects alluded to are the different "ways in which the camera represents certain colours; blue, for "instance, however powerful or deep, is represented as white, and "red as black."

To attain the above-mentioned objects the following method of copying a picture is pursued:—The surface of a piece of good glass is brushed over with gum water, and the glass is fixed in a frame. The picture is placed underneath the glass, and its outline traced on the coated surface of the glass by means of "a fine colored point of chalk." The picture is then removed, a black cloth placed in lieu thereof, and the outline on the glass is painted in the requisite gradations of light and shade by means of white lead, the dark background forming the shades, and the white lead the lights of the picture or painting. The painting has a piece of sensitive paper placed on the unpainted side, and the whole is submitted to the action of light in the ordinary photographic printing frame. The photographic copy thus obtained is fixed in the ordinary manner.

If the copy is required of a different size to the original, the glass painting is taken from an outline on paper of the desired size. The picture may be an original one on the glass itself.

In taking impressions from lace, &c., it may be fixed on to the glass by means of gelatine; the sensitive paper is placed next the lace or other fabric.

[Printed, 4*d.*]

A.D. 1857, September 4.—N° 2312.

GODET, Prosper Bernard.—(*Provisional Protection only.*) The title of this invention is, "Improvements in stereoscopes."

The inventor states:—"Hitherto the stereoscope and the pic-"tures to be inserted therein for view formed separate parts. My

" invention consists in collecting the pictures in the manner as " the folios of an album or keepsake, the cover of which latter " may be folded up so as to form a box open at both ends. In " one part of this cover are fixed the lenses, so that each picture " may be placed at its turn in the visual axis of the lenses, in order " to be seen in the proper stereoscopic view."

[Printed, 3*d.*]

A.D. 1857, September 4.—N° 2315.

FERRIER, Jacques Alexander.—(*Provisional Protection only.*) This invention relates:—

1st. " To a new means of reproducing transparent photographic " representations or pictures on other substances than glass."

2nd. " To the application of such pictures or representations to " stereoscopes."

The substances alluded to in the 1st portion of the invention are " gelatine, gutta percha dissolved in chloroform, in benzine or its " solvents, pharmaceutical collodion and india-rubber dissolved, " siccative oils, mucilagenous substances, starch, or other amy- " laceous substances, such as arrow root, tapioca, &c., soluble or " fusible resins, galipot, pounce, white or yellow wax, either in " solution or in fusion."

The process is as follows:—" A light coating of neuter collo- " dion" is spread over " a smooth and transparent surface," allowed to dry, and coated with albumen. The surface is sensitized by means of " aceto-nitrate of silver," washed, dried, and applied on to a glass negative. Exposure to light then takes place, the picture is developed by means of gallic acid and the nitrate of silver, fixed with hyposulphite of soda, and coloured. A solution of gelatine is then poured over the picture, allowed to dry, and the whole detached from the smooth surface on which it has been placed. The picture is thus protected on one side by a coat of collodion, and on the other by a coat of gelatine, which may be coloured. Any of the above-mentioned materials, either singly or in combination, may be employed instead of gelatine.

A stereoscopic picture may be made by detaching from glass a picture on photographic collodion by means of gelatine or gutta percha.

[Printed, 3*d.*]

A.D. 1857, September 7.—N° 2332.

LEWIS, William, and LEWIS, William Henry.—The title of this invention is, "Improvements in plateholders or frames for "photographic cameras."

These improvements "relate to the frames that receive the glass "or other plate on which photographic and similar pictures are "taken in cameras," and consist:—

1st. In a glass receptacle inserted in the bottom of the frame, so as to retain "any chemicals that may drip from the plate." This receptacle is in the form of a horizontal bottle, with a neck and cork or stopper at one end, "and a dishing or concave upper "surface, catching drippings, and passing the same into the "receptacle through the mouth or opening;" small projections may be provided on this receptacle, against which the slide shuts down.

2nd. In forming the corners or supports, that hold the photographic plate, of solid glass; these fit into the frame by means of ribs that project into corresponding grooves formed on the inside of the frame; they are fixed by means of putty.

3rd. In a "cut off," made of a bar of black glass. This arrangement prevents the entry of the light into the frame during the withdrawal of the slide. When the slide is withdrawn, the "cut off" is forced by a spring into a slight groove, "so as "perfectly to exclude the light."

[Printed, 6*d*.]

A.D. 1857, September 16.—N° 2396.

GODET, Prosper Bernard.—(*Provisional Protection only.*) The title of this invention is, "A new mode of illustrating literary "productions."

The inventor states:—"The invention consists in illustrating "literary productions by means of photographic stereoscopic pic- "tures taken from groups or scenes formed by living figures or "laymen, dressed arranged and combined with the required ac- "cessories according to the narratives described in the book or "other literary production to be illustrated, and from which nar- "ratives or sceneries the said photographic stereoscopic pictures "may thus offer a perfect natural illustration, corresponding with "the ideas of the author of the book or other literary production;

" thus for instance, I intend to give illustrations of the most " conspicuous scenes of living or bygone authors, such as Shaks- " peare, Dickins, Lamartine, historical or other works, and even " of sacred books, such as the Holy Bible, prayer books, or any " other suitable literary productions in general."

[Printed, 3*d.*]

A.D. 1857, September 22.—N° 2459.

NEWTON, ALFRED VINCENT (*a communication from David A. Woodward*).—" Improvements in obtaining photographic pictures."

These improvements consist in " adapting to the photographing " camera a lens and reflector in rear of the object glass, in such " manner that the instrument may be mode to answer the twofold " purpose of a camera obscura and a camera lucida."

By means of this apparatus a negative is taken in the usual way. For this purpose the ordinary photographic lenses have a rack and pinion movement for focussing the object on to the place afterwards occupied by the sensitive plate. The negative thus produced is then fixed in an open " sheath " or frame, " and placed " in the instrument in the same position as before so that the light " may pass through it." The reflector is made to pass through an aperture in the shutter of a darkened room, and by means of this reflector and a lens at the back of the camera " the sun's rays " are reflected with great power through the negative, which is " focussed on the sensitized paper or canvas ;" the open frame containing the negative is mounted in a box which slides by means of rack and pinion movement, the front lenses have a similar mode of adjustment, and by these motions combined the image is focussed of the required size.

" To reflect a positive picture for tracing, &c.," a positive picture is placed in the open " sheath " or frame, and its image is thrown on to the canvas, in a similar way to that used in throwing the image of the negative picture on to the sensitive surface, as described above.

[Printed, 6*d.*]

A.D. 1857, September 28.—N° 2494.

QUIN, RICHARD.—(*Provisional Protection only.*) " Improve- " ments in the construction of cases suitable for containing pho- " tographic and other pictures."

These cases satisfy the three following requirements:—The picture may be enclosed for protection; the picture may, when placed on a table, be supported by the case in an inclined position; the picture may be hung so as to exhibit the picture in an ordinary picture frame. The picture is mounted in an inner frame, which is backed with leather and serves as the back of the complete case; an exterior frame, capable of turning on pins attached to the inner frame, has a lid of the ordinary description hinged to it.

By turning the picture half round on its axes within the exterior frame, and using the lid as a strut, the picture may be placed on a table at the desired inclination.

To hang the picture, the inner frame is turned half round and the lid is closed down on the outer frame.

Another way of constructing cases consists of mounting the picture in a frame enclosed at the back and placing this within a similar frame. "There is a strut or foot hinged to the outer case " to support it in an inclined position."

If the lid of the case be hinged by a double hinge to the centre of the top of the case, it will serve either to enclose the picture or to form a strut. To hang the picture the lid is turned over against the back of the case.

If the picture is mounted in a frame hinged to the front of the ordinary case, a spring may throw it up into an inclined position on the opening of the case.

[Printed, 3*d*.]

## A.D. 1857, October 5.—N° 2551.

**BECKERS**, Louis (*a communication*).—(*Provisional Protection only.*) The title of this invention is, "Improvements in appara- " tuses for exhibiting daguerreotype, photographic, and other " stereoscopic views and pictures."

The inventor states:—"My improvements consist in arranging " inside a box, case, or chamber an endless belt or apron having a " series of frames or slides to receive views and pictures attached " to it, in such manner as to stand perpendicular to its face; by " moving the belt through a handle outside the box, the pictures " are presented successively in a vertical or other suitable position, " opposite to magnifying glasses or other transparent media, or " to suitable openings. By arranging the pictures at right angles " to the face of the endless belt or chain, a larger number of pic-

" tures may be contained and exhibited in a box or case of a given " size than by the arrangement of the same parallel with the face " of the belt, as is usual in moving panoramas. I secure the " pictures in the slides by means of elastic bands and notches " formed in the ends of the grooved portions of the slides."

[Printed, 3*d*.]

A.D. 1857, October 6.—Nº 2560.

BROOMAN, RICHARD ARCHIBALD (*a communication from M. Garella*).—" Improvements in apparatuses for taking photographic " pictures."

"This invention consists in a method of taking photographic " views or pictures subtending any angle up to 360° on a plane " surface."

"To effect this, the object glass, together with the whole appa- " ratus, is made to turn on an axis, so as to present the glass, " paper, plate, or other sensitive surface, as the case may be, to " the object to be taken. The vertical slide or sash slides hori- " zontally into the dark chamber through the side of the camera, " and the object glass, instead of being opposite to the centre of " the dark chamber, is fitted at the side of the same. The camera " is fixed to a board which rests on a bed, and this bed is caused " to turn, and with it the rest of the apparatus, by means of " suitable gearing at the will of the operator. The moveable sash " which receives the sensitive surface is composed of two distinct " parts, and is provided with a grooved shutter hinged at one end, " so that the shutter can be raised when the sash is fixed in the " frame, and returned when the sensitive surface has received the " image. The whole is caused to move along the bed by means " of grooves and rollers suitably arranged. In the dark chamber, " between the sash frame and the object glass, there is a partition " extending the whole length of the dark chamber, provided with " an opening, which is covered by a diaphragm or card.

" A guide curve is used for the purpose of imparting to the sen- " sitized glass, paper, or plate such a motion that every point of " the picture shall remain on the same point of the sensitized sur- " face as near as may be during the exposure of the surface; this " guide curve is formed of hard wood or other suitable material " to ensure accuracy. In some cases a mirror or prism is used " before the object glass;" positive pictures are thus produced.

When negative proofs are required, and when the object glass alone is used, the axis of rotation is placed under the object glass; when positive pictures are required and a mirror is therefore used, the axis is placed at a distance behind the plate equal to the focal distance. The relative motion of the plate, in the former case, is therefore in the same direction as that of the rotation of the instrument, but in the latter case is in the opposite direction.

The Drawings show an apparatus for "obtaining pictures sub- " tending an angle of 100°." The modification of the apparatus necessary to produce positive pictures by the aid of a mirror is also shown.

As the panoramic views produced upon a plane surface are all distorted, the following apparatus is devised in order to view them free from distortion or deformity :—The picture is placed in a case so as "to assume a position resembling that in which it was " taken." The case is semi-cylindrical, and "formed with two " horizontal semicircular ends connected by a curved vertical side, " its radius being equal to the focal distance of the object glass " with which the picture to be examined was produced." The picture is viewed by means of a moveable lens fitted to a vertical pivot in the axis of the semi-cylindrical case. The images are thus seen "as they appeared when transferred to the sensitized " surface, and with a certain increase of size and a relation be- " tween their parts somewhat analagous to the illusion produced " by the stereoscope."

[Printed, 1*s.* 4*d.*]

## A.D. 1857, October 8.—N° 2574.

GRUBB, THOMAS.—" An improved photographic lens."

This lens may be either used alone and as a " view lens," or in combination with other lenses of similar or ordinary construction. The distinctive characteristics of the construction of this lens are :—" That the two kinds of glass of which it is formed " occupy inverted places in the compound, and that the internal " or separating curve is also inverted," and much deeper in the improved lens than in the ordinary or existing photographic lenses.

To construct the improved photographic lens :—

Firstly, a crown lens of suitable focus is first formed; this lens is " of a form approximating to that known as plano-convex ; one

" surface is generally, by preference, made slightly concave, and " the lense is in such case a 'meniscus.'"

Secondly, a flint lens is formed; one side or surface of this lens is concave and of the same radius of curvature as the deeper side of the crown lens; the other side of this lens is formed " of " that curvature which will cause the lens to form with the crown " lens previously described a nearly achromatized compound, or " more strictly speaking a compound corrected for actinic dis- " persion, and which second surface of the flint lens will neces- " sarily be a convex surface."

Lastly, the two lenses are placed together " with or without a " cement or fluid interposed, so that the surfaces of similar radii " shall be adjacent." The " crown lens is next the object."

[Printed, 5*d.*]

## A.D. 1857, November 3.—N° 2792.

SWEET, Henry Kinsman (*a communication*).—(*Provisional Protection only.*) " Improvements in photographic portraits and " pictures."

This invention " consists in taking such portraits or pictures on " concave or convex plates of glass or other material in place of " taking them on flat plates as heretofore, by which means effects " in some respects superior to those hitherto obtained may be " produced."

[Printed, 3*d.*]

## A.D. 1857, November 9.—N° 2827.

HARDIE, Walter.—(*Provisional Protection only.*) " An im- " proved stereoscope."

" The optical part of the instrument consists of two plane " mirrors," " placed side by side, with their reflecting surfaces at " right angles to a vertical plane midway between them, but in- " clined to each other at an angle." " The pictures are viewed " by reflection from these mirrors, the eyes being placed so that " the line joining them is perpendicular to and bisected by the " above-mentioned vertical plane."

" The pictures must be placed one above the other, head to " foot, inverted with respect to the observer, and with a lateral " separation of the corresponding vanishing points amounting to

" about two inches and a half (the distance between the eyes); " those points in the picture for the right eye being to the right " of the corresponding points in the other picture." Both pictures are not in the same plane, but are inclined to each other; the angle of inclination depends upon their distance from the mirrors and the angle of the mirrors.

" There is no necessary limit to the breadth of the pictures; " panoramic views and pictures broader than the field of view " may be exhibited by being drawn across the field of view." If necessary, prisms and lenses may be used with this stereoscope in conjunction with the above-mentioned mirrors.

A stereoscope is shown in the Drawings to exhibit pictures up to ten inches in height. Hinges enable the instrument to be compactly folded up.

" When not in use the pictures may be folded together face to " face;" they may be bound together as the leaves of a book, and exhibited in succession by turning over the leaves without removing them from the stereoscope."

[Printed, 9*d.*]

## A.D. 1857, November 19.—N° 2903.

GILL, SETH, and NEWTON, HENRY.—(*Provisional Protection only.*) " Improvements in obtaining stereoscopic pictures."

" This invention consists in obtaining the double image to " form the stereoscopic picture by a double reflector, and then " taking the pictures by a camera from the reflectors, which may " consist of two silvered plates placed at a convenient angle " to each other.'

[Printed, 3*d.*]

## A.D. 1857, November 24.—N° 2940.

SANDS, CHARLES.—" Improvements in stereoscopes."

This invention relates " to the actuating, adjusting, and sus- " taining of the flap or door of stereoscopes, by which the light " is admitted to the picture, such door having a reflector on the " inside which reflects and concentrates the light upon the pic- " ture."

The door is placed upon a stiff hinge, " so that it will remain in " any position in which it is placed," and " the joint pin or rod "(which is affixed to the door)" is extended beyond the sides of the instrument and has thumb knobs applied at either end. "By "taking hold of the thumb knob the position of the door and "reflector can be readily adjusted and regulated while viewing the "picture, and it will remain of itself in the position adjusted.

"Instead of extending the joint pin to carry the thumb knobs, "a separate rod or axis may be applied for the purpose just above "the joint, and a short arm extended therefrom with which the "door and reflector is connected. This additional rod with "thumb knobs is fitted in its bearings to rotate therein with "sufficient friction to sustain the door and reflector as explained, "or the door may move with sufficient friction on its hinges for "the purpose required."

[Printed, 9*d.*]

A.D. 1857, December 8.—N° 3034.

PERSHOUSE, HENRY.—(*Provisional Protection only.*) The title of the invention is, "An improvement or improvements in "stereoscopes."

The inventor states:—"In stereoscopes as ordinarily con-"structed the lenses are removed for cleaning their inner surfaces "by either pulling the end of the instrument from out the body, "or by lifting up the said end upon the hinge, by which it is "jointed to the body. My invention consists in making the end "of the instrument in which the lenses are inserted engage with "the body of the instrument by a lateral sliding motion, the said "end being drawn from off the body by a motion similar to that "of the sliding lid of a colour box. I prefer to insert a spring "in one of the grooves in which the lid slides for the purpose of "producing such an amount of friction as will prevent the end "from sliding out by its own weight."

[Printed, 3*d.*]

A.D. 1857, December 12.—N° 3066.

COWPER, CHARLES (*a communication from Testud de Beauregard*).—This invention consists in "producing photographic "proofs or pictures by means of carbon or other colouring "matter," applied by superposition to a coating sensitive to the action of light.

Paper is immersed in or floated on a warm solution of "bichro-" "mate of potash or ammonia," mixed with gelatine; it "is" "then dried, and its surface covered with the pigment." The pigment may be rubbed over the dry surface with a pad of leather or other suitable material; or it may be "ground up very fine" "with nut oil or other oil, and rubbed over the surface, which is" "subsequently immersed in a bath of ether to which a little col-" "lodion may be added; or the paper may be immersed in a bath" "of indian ink or other pigment ground up very fine with" "water and mixed with gelatine and a little gum or dextrine, and" "used hot; or rollers or presses or other apparatus may be em-" "ployed to apply the pigment or assist the operation."

The paper, having been prepared in the dark, is exposed to the action of light and washed in hot water. This "dissolves the" "gelatine which has not been acted upon by the light, but does" "not dissolve that which has been rendered insoluble by the" "action of the light, and which insoluble gelatine retains the" "pigment and thus produces the image."

Glass or other substances may be substituted for paper.

By employing carbon, pigments, or finely divided gold or silver "photographs or pictures of the greatest permanence and dura-" "bility may be obtained."

[Printed, 4*d*.]

### A.D. 1857, December 17.—N° 3101.

HIGHTON, Edward.—" Improvements in electric telegraphs."

1st. The substitution of an electro-magnet for a permanent magnet in "the gold leaf telegraph."

2nd. "Recording signs or indications made by telegraphic" "instruments by means of photography." A strip of paper, rendered as sensitive as possible to the action of light, is moved "uniformly forward by suitable apparatus, as is well understood," "& in front of this travelling strip of sensitive paper a mag-" "netic needle or bar is mounted, and this is caused by a suitable" "arrangement of coils to move in one or other direction, when a" "current is passing in the line wire. In the magnetic needle or" "bar a hole is formed, and by suitable optical apparatus light is" "concentrated at and around this hole, and a portion of it passes" "through on to the paper, and marks it, so that the position of

" the needle is constantly registered on the paper. In place of " perforating the needle or bar itself, it may, if preferred, be " caused to actuate a perforated screen."

3rd. " A method of protecting telegraphic wires when buried " in the ground."

4th. " A peculiar form of code table."

[Printed, 5*d.*]

A.D. 1857, December 23.—N° 3148.

NUNN, WILLIAM.—" Improvements in stereoscopic apparatus."

The improvements relate :—

1st. " To the application of glass or other reflectors to stereo- " scopes, in such manner that more than one person may be " capable of looking through suitable glasses or eye-pieces at the " same picture or the reflection of it at the same time." " The " reflecting surface employed is placed in such position that, " whilst the direct view of the picture is obtained as heretofore " through one pair of glasses, the reflection thereof may be seen " at the same time through other glasses." In the Drawings, one instrument is shown with a reflector, complete in itself, and another consisting of duplicate apparatus, each of which is capable of being used as an ordinary stereoscope; prismatic glasses are also shown near the eye pieces, " to counteract the reversing effect " produced by the reflector."

2nd. " To the application of glass or other reflectors to the " sides of stereoscopes, in such manner as to facilitate the intro- " duction of reflected light on to the picture or pictures side- " ways." The Drawings show reflectors on hinges, capable of folding down to the side of the stereoscope.

3rd. To the table or other support to receive the above-men- tioned stereoscope. The stereoscope is mounted in a stem sliding freely in a hollow pillar, and provided with a spring stop. The pillar, stem, and stereoscope, when out of use, may be lowered within the central table leg and enclosed by a flap, so that the table may be used for other purposes.

4th. To " the application of india rubber or other elastic means " as connections between the eye-pieces and the bodies of stereo- " scopes, so as to admit of elasticity, and thereby prevent injury " to the eyes whilst viewing objects."

[Printed, 9*d.*]

A.D. 1857, December 24.—N° 3164.

BURLEIGH, Benjamin, and DANCHELL, Frederick Ludwig.—This invention has for its object the formation of "receptacles and other articles for photographists" from carbonaceous matters solidified by percussive force, "and afterwards submitted "to the action of heat in closed vessels."

The carbonaceous substances, in a pulverized form, are mixed with "either moist, bituminous, resinous, gummy, oleaginous, "saccharine or glutinous matters," and forced "into moulds, "suitable for the objects to be constructed, by stamping or beating "with suitable tools by means of machine or other power to the "required degree of consistency or solidity, after which the "objects are baked or burned in closed vessels."

"To render crucibles and other objects made of solidified "carbon proof against the action of oxygen when exposed to "fire, the said articles" are "coated outside with silicious glaze "or other substance not liable to be acted upon by oxygen or "atmospheric air."

"To render vessels and other objects impermeable to fluids and "gases," they are glazed or coated "either on the outside or "inside, or on both sides when required, with a varnish or lacquer "insoluble in the particular fluid or gas intended to be contained "in the vessel."

Many other applications of this invention are set forth.

[Printed, 4*d*.]

---

## 1858.

A.D. 1858, January 21.—N° 115.

HERMAGIS, Hyacinthe. — (*Provisional Protection only.*) "Improvements in stereoscopes."

This invention "relates to improvements in the optical parts of "stereoscopes, which improvements produce four important "results,—

"1st, perfect rectitude of the lines of photographic pictures.

"2ndly, more zones or fringes of prismatic colours.

"3rdly, greater increase of the size of the pictures than can "be obtained with the prismatic lenticular system heretofore "employed.

"4thly, less fatigue to the eyes as regards their position in "the axes of parallel spherical lenses.

"These improvements consist in adapting to the stereoscope, "without the aid of any prism, spherical lenses, either simple or "achromatic, having parallel spherical surfaces compelling the "eyes to be placed at a reasonable distance from the optical lenses, "perpendicular to the axes of the said lenses, which allows of "the view being more easily taken in by reason of its position "towards the converging point of the luminous zones proper to "produce the effect of relief in pictures placed in the stereo- "scope."

[Printed, 3*d.*]

A.D. 1858, February 8.—N° 228.

MATHIEU, François.—"Improvements in stereoscopes."

"The lenses are fixed in a short box with an opening or open- "ings at the back. A piece of wood or other suitable material "is hinged to the back, and another piece of wood is hinged to "the first piece and carries the picture holder. These pieces fold "up on the box, and are secured by a catch when the instrument "is not in use. The picture holder is either permanently fixed, "or jointed, or otherwise attached to its base, with or without "means of adjustment in either direction, and it lies in a suitable "space or recess when the instrument is shut up. The picture "holder may be a simple upright piece or pieces, with a spring or "springs to secure the picture, or grooved upright pieces or other "similar means may be employed to support the picture at the "proper distance from the lenses, which may be of any of the "known forms. A large amount of light can thus be allowed to "fall on the picture, and the instrument is very portable."

"The materials of which the apparatus is constructed may be "varied."

[Printed, 6*d.*]

A.D. 1858, February 9.—N° 329.

THOMSON, William.—The title of this invention is, "Im- "provements in testing and working electric telegraphs;

some of the improvements relate to the application of photography to the above-named purposes.

The description of the invention is divided into the following parts:—

1st. Testing the insulation of a telegraphic conductor by means of certain electro-static instruments.

2nd. Testing "the insulation of a telegraphic wire by comparing its resistance to the flow of electricity from a constant source with the resistance of a standard wire to the flow of electricity from the same, or another constant source."

3rd. Testing "a submarine telegraph wire during the operation of laying it by measuring from time to time the strength of currents produced in it by the electro-motive force of a constant battery or batteries." Certain galvanometers adapted for use at sea are employed for this purpose.

4th. The use of "a double bifilar suspension" for the "indicator" of electric telegraph instruments.

5th. The use of a conducting suspension wire for the above-named indicators; also the use of a conductor connected with the said indicator and dipping into a conducting liquid, in order to maintain electric communication.

6th. The use of "electro-motive forces, of several different positive and negative strengths, to give different signals in telegraphing." Certain arrangements of electro-dynamic coils and magnets are used to determine the direction and amount of the electro-motive forces.

7th. A "galvanometric relay or receiving instrument."

8th. The use of the electrometers, described in the 1st part of the invention as receiving instruments for telegraphic signals made by electricity other than frictional.

9th. The use of the thermal or thermo-electric effects of rays reflected from the indicator of a receiving instrument for the purpose of recording its indications.

10th. The use of *photography* "for recording electric signals as indicated, by motions of an indicator or indicators, either by light reflected from a mirror attached to the indicator, or transmitted by an aperture or transparent part in the indicator, or by influences of a shadow or image of the indicator, or of the shadow or image of any part of it." "A band or piece of "photographic paper or other substance sensitive to photographic "action" is, "by any ordinary mechanism," caused "to travel "at a steady rate across the line of motion of the indicating "beam whose movements are recorded by the marks made on "the paper or other sensitive substance."

11th. A method of "transmitting telegraph messages between "one or more conductors and a different number of conducters." Also a method "by which two or more messages can be sent at "the same time from the same station along one telegraph wire."

12th. The use of electric sparks *in connection with photographic action* for receiving or recording telegraphic signals. "Electric "sparks excited by a Ruhmkorff coil or any other convenient instru- "ment," are passed "between any part of the indicator and a sheet "of matter fixed or carried along by a regular motion near it;" these sparks record the motions of the indicator by photographic action, by impressions, by chemical changes, or by perforations. "In using electric sparks to record the indications of a moveable "body in a galvanometer or galvanometric relay employed for "telegraph purposes," a regular succession of sparks is employed; these sparks follow one another at very short intervals of time, and each make a mark "on the paper or other sheet receiving the "impression or perforation." In using this part of the invention it is preferred "to cause a strip of paper sensitive to photographic "action, or otherwise, to travel with a steady motion across the "path of the end of the indicator over a metal plate, above "which plate the end of a conductor projecting from, or forming "part of the indicator, moves to and fro in consequence of the "action of the current transmitting signals. The series of "sparks excited by the Ruhmkorff coil or other instrument takes "place between the before-mentioned end of a conductor and the "metal plate, and makes a series of marks on the paper either by "photographic action or perforation, or both. The result is an "undulating curve of dots, the character and succession of the "undulations of which give the different signals desired to be "sent."

13th. "An apparatus by means of which electric signals may "be read off" by observing the amount of motion of an indicator. The greater or less amount of motion of the indicator is produced by currents of greater or less strength. It is preferred to use a light mirror attached to the indicator; a beam of light is thus produced, which beam produces a luminous spot on a graduated screen.

14th Various apparatus "for rapidly producing and for maintaining electric currents of stated strengths in telegraphic wires, "and for discharging currents from such wires."

15th. Certain "arrangements for throwing the receiving instrument at the transmitting end out of circuit during part of the "time during which a signal is being transmitted from that end."

16th. Means and apparatus for "compensating the effects produced on the receiving instrument at the transmitting end by "operations performed at that station."

17th. A method and apparatus for transmitting signals in opposite directions at the same time.

18th. "The use, for receiving and recording signals, of two or "more instruments at or near the same station, thrown into "operation during regulated intervals of time."

19th. A means of checking the motion of any moveable indicator in a receiving instrument after its indication has been read off. This portion of the invention is mentioned in the Provisional Specification, but is not proceeded with.

20th. A method and apparatus for increasing, diminishing, or regulating the stability of the indicator of a telegraphic instrument.

21st. Certain means of compensating the effect of previous signals on the telegraph wire. This portion of the invention is mentioned in the Provisional Specification, but it is not proceeded with, except so far as is comprehended under the 16th part of the invention.

22nd. A method and apparatus "for producing at a receiving "station, lines, figures, letters, or symbols of given shapes." Two mirrors are attached "to the indicators of two galvanometers, at "the receiving end of, and receiving currents from one transmit-"ting station by two separate telegraph lines;" a ray of light is caused "to be reflected from these two mirrors in succession, "and to be thrown on a screen, so as to give a visible effect, *or by "photographic or thermal influence*, a recorded effect, by which "any lines, figures, letters or symbols, drawn or written at the "transmitting station upon an instrument adapted to give dif-"ferent degrees of electro-motive force to the two lines, according "to the two independent variables thus dealt with may be written,

" drawn, or made visible at the receiving station." " For trans-" mitting a line, drawing, or writing of any kind by telegraph," a third telegraphic wire is added, " which is so connected and " arranged, that when the tracing point at the transmitting station " is lifted from the surface on which the curve is drawn, the " indicating or marking beam of light is cut off at the receiving " station."

[Printed, 1*s.* 6*d.*]

A.D. 1858, February 23.—N° 357.

NEWTON, William Edward (*a communication from A. Cutting and Lodowick H. Bradford*).—" An improved process for " producing photographic pictures or designs on the surface of " stone or metals so that impressions may be taken therefrom by " the process of lithographic printing."

" The surface of the stone after being prepared either by " polishing or graining," is covered with a solution containing water, gum arabic, sugar, and bichromate of potash, exposed to the action of light, washed with a solution of soap, thoroughly washed with water and dried; it is then ready to be inked by means of a roller.

The action of this process upon the stone is as follows :—The sugar deprives the gum arabic solution of its power to adhere firmly to the surface of the stone until it has been subjected to the subsequent processes. The exposure to light, by its action upon the bichromate of potash, causes the gum to adhere to the stone upon those parts of the picture where lights appear. The application of the solution of soap removes the gum from those parts which have been protected from light, and leaves an insoluble soap in place of the coating removed; the gum still adheres to those parts which have been fixed by the access of light. The gum is entirely removed by the thorough washing with water, and the insoluble soap forms " the groundwork from which the " picture is printed, as in the ordinary lithographic process."

[Printed, 4*d.*]

A.D. 1858, February 27.—N° 396.

CLARK, William (*a communication from M. Niepce de St. Victor*).—(*Provisional Protection only.*) " Improvements in pre-

" paring paper for and in obtaining photographic proofs or " impressions."

The method of preparing the paper " is based on the property " that all bodies have of absorbing a greater or less quantity of " light." This new photographic process is called " photography " by absorption of light." To produce a positive picture, a sheet of paper is kept in the dark for a certain time, immersed in a solution of "azotate of oxide of uranium" (or other salt of oxide of uranium), dried, and impressed with the photographic image. The picture is then developed by means of a solution of " azotate of silver," and fixed by immersion in pure water, " which dissolves all that part of a salt of oxyd of uranium " which by reason of the dark parts of the negative proof have " not received the action of light."

When it is desired to give a black tint to a photograph of the above description, either chloride of gold solution may be applied to it after the process just described, or bichloride of mercury solution may be applied between the exposure to light and immersion in nitrate of silver solution. Chloride of gold may be used to develope the image instead of nitrate of silver; in this case the proof has a very dark blue tint.

Negative impressions are produced by placing " a sheet of " paper impregnated with salt of uranium " in the dark.

This process is adapted for use on glass by means of a gelatinous or gummy solution of " azotate of uranium," which is insoluble when exposed to light.

This process is simple and rapid, and gives more stable photographs than the ordinary process.

[Printed, 3*d.*]

## A.D. 1858, March 22.—N° 590.

BROOMAN, Richard Archibald (*a communication*).—(*Provisional Protection only.*) The title of this invention is, " Improve- " ments in apparatuses for exhibiting daguerreotype, photographic, " and other stereoscopic views and pictures."

" These improvements consist in arranging inside a box, case, " or chamber an endless belt or apron having a series of frames " or slides to receive views and pictures attached to it, in such " manner as to stand perpendicular to its face. By moving the

" belt through a handle outside the box the pictures are pre-" sented successively in a vertical or other suitable position oppo-" site to magnifying glasses, or other transparent media, or to " suitable openings. By arranging the pictures at right angles " to the face of the endless belt or chain, a larger number of " pictures may be contained and exhibited in a box or case of a " given size than by the arrangement of the same parallel with " the face of the belt, as is usual in moving panoramas." The pictures are secured " in the slides by means of elastic bands and " notches formed in the ends of the grooved portions of the " slides."

[Printed, 3*d.*]

## A.D. 1858, March 22.—N° 591.

MANWARING, EDWARD JOHN.—" Improvements applicabl " to stereoscopic apparatus."

This invention " consists in the application to or combination " with the stereoscope, or apparatus for exhibiting stereoscopic, " photographic, or other similar views, of a system of rollers, " whereby views produced or placed upon flexible material may be " wound, unwound, and rewound upon or from such rollers, so " as to be brought up to and taken away from sight by acting " upon the said rollers, without the trouble of taking away each " separate view and putting another in its place, as heretofore " practised."

The Drawings show a box or case containing the apparatus to be attached to the bottom of a stereoscope. Two rollers, worked by a pulley and driving band, receive the band of views; two other rollers keep the views flat for observation. The driving pulley is actuated by a knob or handle outside the case, and the rollers are suffered to turn one way only in consequence of a catch and spring on each roller. Each roller can be unshipped by means of " a " screw, which forms the axis of one end of roller, working in a " socket in that end of said roller."

Transparent views may be exhibited by this apparatus, transparent textile material being used.

[Printed, 6*d.*]

A.D. 1858, March 25.—N° 636.

CHEVALLIER, François Auguste. — " Improvements in " photographic apparatus."

The apparatus which is the subject of this invention enables the exact reproduction to be made of the entire horizon surrounding it. For this purpose it is made to revolve slowly upon a vertical axis placed under the sensitive plate, whilst the sensitive plate has a corresponding motion, so as to expose a certain portion of its surface to a corresponding point of the horizon. The sensitive plate is circular, and revolves in a vertical plane by means of toothed gearing connected with the vertical axis carrying the camera. By means of " eccentric " or radial shutters and a divided arc, a sector of any number of degrees can be photographed at once, and then another similar sector until the whole horizon is completed, or such portion of it as may be required; when, however, the whole horizon has to be taken, it is preferred " to operate with sectors of 2 or 3 degrees, imparting to the " apparatus a continuous rotary motion slow enough for the " image to be reproduced." Auxiliary fittings contribute to the perfect working of the instrument; a divided circle upon the vertical axis enables the camera to be moved exactly through a given arc; " a magnetic or compass table " fixes the geographical position of the various parts of the photographs; and a special stereoscope, having a transparent band of silk (tinted to represent various atmospheric effects and mounted on moveable rollers) and also having internal mirrors, enables the views to be seen under various apparent circumstances.

The rollers and silk band may be adapted to ordinary stereoscopes.

The pictures obtained by means of this apparatus may be applied to fans, as they are circular and radiate towards the centre.

[Printed, 10*d*.]

A.D. 1858, April 5.—N° 725.

SARONY, Oliver.—" Improvements in producing photographic " portraits."

This invention consists in producing a positive portrait by means of two or more negatives. A negative portrait is first taken, "in which every portion of the figure, excepting one, is " sacrificed in order to obtain an accurate representation of that " one portion;" a second negative is taken, in which another portion of the figure is accurately taken, all the rest being sacrificed; and so on, until accurate portraits of the whole figure are taken. The accurate portions are then printed in their proper relative positions upon a positive plate, those portions sacrificed being stopped off by masks; a positive portrait is thus produced, consisting of all the accurate portions of the negatives. To prevent the stunted appearance that photographic portraits usually have, in the positive picture the head is placed "somewhat further " from the waist than it appears in either of the negatives."

These improvements "may also be effected by taking up the " different portions of the collodion film from the glass of one or " more negatives and laying them down on a glass or in the " printing frame in the positions above described, and then " printing from them without masks."

When all the portions of the portrait have been thus printed in, "the portion of the figure between them is drawn in by an " artist."

[Printed, 3*d.*]

### A.D. 1858, April 10.—N° 780.

POUNCY, John.—(*Provisional Protection only.*) The title of this invention is, " Improvements in the production of photographic pictures."

The inventor states:—" According to my invention I prepare " the paper or other surface for having the picture produced on " it, by applying over its whole surface the coloring matter which " is to form the picture, and together with this coloring matter " is applied a substance which is acted on by the light. The " following is the manner in which I proceed when printing positive pictures on paper from negative pictures:—I coat the paper " or surface which is to receive the picture with a composition of " vegetable carbon, gum arabic, and bichromate of potash, and " on to this prepared surface I place the negative picture, and " expose it to the light in the usual way, afterwards the surface " is washed with water, which dissolves the composition at the " parts on which the light has not acted, but fails to affect those "parts of the surface on which the light has acted; consequently, "on those parts of the surface the coloring matter remains in the "state in which it was applied, having experienced no chemical "change. Sometimes for the vegetable carbon I substitute "bitumen, or other coloring matter may be employed. By this "process pictures are obtained which are not liable to fade like "ordinary photographs."

[Printed, 3*d.*]

A.D. 1858, April 13.—Nº 794.

DEAN, George Alfred Henry.—(*Provisional Protection only.*) "An improvement in stereoscope's slides," "whereby the "figures on the slide may have movement and motion imparted "to them. This is managed by first taking your stereoscopic "background, and then taking your figures or groupings. The "figures required to move are cut out and affixed in their position "on the slide by a shaft running at the back of slide with shafts "or arms passing through passages cut through the cardboard "of slide, to which arms or shafts the figures required to move "are affixed by either wire, thread, gum, paste, or any other "adhesive or sticking substance; a mechanical motion is thus "obtained and given to the figures or groups of figures by the "shaft being pulled at will up or down, backwards or forwards."

[Printed, 3*d.*]

A.D. 1858, April 20.—Nº 860.

DEROGY, Eugene.—"Improvements in instruments and ap-"paratus applicable to photographic purposes."

This invention consists of a new system and combination of lenses and diaphragms in the usual lens holders or mountings of photographic cameras.

By means of certain adjustments, arrangements, and re-arrangements of the lenses and diaphragms the same results are obtained "which now require the powers and use of at least 3 "separate instruments or sets of apparatus." Also representations may be taken "on a larger or smaller scale by lengthening "or shortening the focus;" they may also be taken instantaneously.

The peculiarities of the instrument may be set forth as follows:—

The tube forming the body of the instrument is made so that the usual cell and diaphragm holder is removeable therefrom so as to be placed at either end of the said tube; certain lenses and diaphragms are thus placed in the centre of the tube from either end of the said tube. The cap covers the exterior end of the tube when the above-mentioned cell and diaphragm holder with the sun shade is placed in the end of the tube that is inserted in the camera. To fix the cell and diaphragm holder, as well as a lens cell used for portraits, a bayonet joint is used. Two extra lens cells may be placed in the usual cell and diaphragm holder; one when the representations are required smaller, the other when they are required larger than the above-mentioned combinations can give. The combination of ten foci in one apparatus.

[Printed, 7*d.*]

A.D. 1858, April 21.—N° 875.

TALBOT, William Henry Fox.—" Improvements in the art " of engraving," in which photographic processes are used.

In the first place a mixture of gelatine with bichromate of potash is poured upon a metal plate prepared for engraving; this is left to dry, placed in the photographic copying frame in contact with the object to be copied, and exposed to the action of light so as to produce a photographic impression on the gelatine; thus far the process is identical to that described in N° 565 (A.D. 1852).

The novelty of the present invention consists in the mode of etching the image obtained as above; this is done as follows:—The picture is not washed, but as soon as it is removed from the copying frame it is covered with pulverised copal or other resin. This powder is melted by heating it strongly over a lamp. When the plate has cooled it is etched by means of a nearly saturated solution of perchloride of iron in water; the etching being accomplished in consequence of the perchloride solution penetrating the gelatine wherever the light has not acted upon it, but refusing to penetrate those parts upon which the light has sufficiently acted. " When the etching is effected the etching liquid is washed " off with a rapid current of cold water " and the plate is cleaned.

Another mode of etching the image consists in washing the image (when taken out of the copying frame) with a mixture of water and spirits of wine, and then with spirits of wine alone. The plate is dried, dipped into warm water, and dried again; the powdered resin is then put on and the rest of the process conducted as described above, but with etching liquid more diluted with water. This mode of etching the image is only mentioned in the Provisional Specification.

The perchloride of iron may be employed for ordinary etching.

Another part of this invention, only detailed in the Provisional Specification, is as follows :—A metal plate is engraved or etched all over, so that it will print a dark and uniform shade upon paper; it is then covered with a mixture of gelatine and bichromate of potash, impressed with a photographic image, and washed as above described, so as to expose the etched surface of the metal at those parts on which the light has not acted. An impression of the plate is then take on a slab of warm gutta percha, this impression is made conductive of electricity, and an electrotype etching is obtained therefrom. " By a nearly similar process a " photographic etching may be made upon the surface of the " stones usually employed for lithography."

The processes here described are named " photoglyphic en- " graving."

[Printed, 5*d.*]

## A.D. 1858, May 7.—N° 1027.

COGGAN, George Bartlett.—(*Provisional Protection only.*)

The title of this invention is, " A new portable apparatus to be " called a 'Stereoscopia,' for exhibiting stereoscopic pictures."

The inventor states :—" This invention consists of a square " platform mounted on wheels for transit from one place to " another, having a box or case about two and a half feet square " and about seven feet high securely fixed thereto, the aforesaid " box or case having a glass top either ground or otherwise. At " the front and back of said box or case I propose inserting two " or more sets of lenses, and in the interior I place a cylinder of " wood or other suitable material extending from side to side, " having an axis at each end, upon one of which I place a winch " handle for the purpose of changing the subjects. The lower " portion of this apparatus I propose using to contain and for " conveying photographic apparatus."

[Printed, 3*d.*]

A.D. 1858, May 21.—N° 1136.

BRYER, Stephen.—" Improved instruments to be used in the " sensitizing and developing of photographic plates."

" For lifting the plate off the dipper (by which the plate is im- " mersed in the bath) and transferring it to the plate frame," two pairs of forceps are used; these have two curved limbs jointed together and provided with elastic cushions at their ends, the plate is thus held firmly " by simply gripping its edges." " In " order to take up the plate after it has been removed from the " camera and hold it firmly while the developing and washing " operations are proceeding," " the developing forceps " is used. The limbs of this forceps have attached to their ends, by loose joints, clips of L shaped metal coated inside with gutta percha. A screw and nut " after the manner of compasses " are also fitted to the forceps for retaining the instrument in a closed position. The construction of these forceps enables them to accommodate themselves to various sizes of plates and to firmly grip them.

The transferring forceps above described may also have jointed ends or clips, the said clips being discs of metal covered on their inner face with India-rubber.

" When operating upon plates of a size that could not be con- " veniently spanned by forceps," an instrument is used consisting of the common elastic suction holder attached to the face of a blade of wood; the said blade forms a handle for the operator to grasp at one end, and at the other end its continuation facilitates the manipulation of the plates.

[Printed, *7d.*]

A.D. 1858, June 1.—N° 1230.

GRANT, Alonzo Gaylord. — (*Provisional Protection only.*) The title of this invention is, " A method of preparing paper, in " order to render it waterproof and adapted for the reception of " photographic pictures."

" Paper with or without size, giving preference, however, to " that in which little or no size has been used," is dipped in or impregnated with a " composition of the following materials:— " Asphalte, oil, turpentine, Canada balsam, caoutchouc, gum lac, " and white wax, with or without the addition of chloroform."

" The paper, coated or impregnated, may or may not be passed " through rolls, and whether passed through rolls or not it is " next placed in an oven or drying chamber, wherein the heat is " raised to a point just below that at which the material will " ignite, where it remains until it is perfectly dry ; after drying, " the paper or material prepared as above described is fit for use. " When to be employed for photographic purposes, it receives " any of the usual preparations employed in the art to render it " sensitive."

[Printed, 3*d*.]

A.D. 1858, June 1.—N° 1231.

GRANT, Alonzo Gaylord. — (*Provisional Protection only.*) " An improved stand or rest for" [photographic ?] " cameras, " theodolites, guns, and other articles."

" This stand consists of three or more legs which, when brought " together, unite in the form of a cylinder. The upper ends of " these legs are hinged or jointed to a metal boss, with a hole " through the centre thereof. In the hollow of the boss there is " a collar in the shape of a split ring, and through this collar " passes a shaft carrying at top a table or other holder, according " to the nature of the article to be held or supported. To the " inside of each of the legs one end of a rod is connected, while " the other end of each rod is connected to a collar through " which the central shaft passes. When folded up, the rods lie " in a groove made for their reception inside the legs. In order " to fix the table or other holder at any required height, the split " ring is made to grasp and nip the central rod by a screw pass- " ing through the side of and worked from the outside of the " boss."

[Printed, 3*d*.]

A.D. 1858, July 3.—N° 1501.

SARONY, Oliver.—" Improvements in treating and coloring " photographic pictures."

The object of the process which is the subject of this invention is to imitate paintings upon ivory by means of paper photographs, to which a medium is applied, " which penetrates the paper, " destroys its opacity, and allows the artist's work to appear

" floating within the substance of the paper. This appearance is " obtained by the following method :—The paper on which the " photograph has been taken is stretched on a frame having a " moveable panel, so that the back of the photograph picture " may be uncovered when required without unstretching. The " photograph is then colored in the ordinary way with water " colors on the front side until the picture is worked up to the " required finish; the panel at the back is removed, the photo- " graph picture is then placed face downwards upon a hot metal " plate, the back is then, by a brush, covered all over with melted " bees'-wax until the paper is perfectly saturated and the wax " appears at the other side."

" The photograph picture may now be mounted like any piece " of ivory upon a light or cream-coloured ground of paper, " enamel, or other substance."

A similar effect may be produced "by using oil, varnish, or " gum, to give a semi-transparency to the paper."

[Printed, 3*d.*]

### A.D. 1858, August 16.—N° 1866.

CHAPPUIS, Paul Emile.—" Improvements in stereoscopes and " stereoscopic apparatus."

1st. " Making a folding stereoscope in the form of a book." The back of the instrument is made with a double hinge in the centre, so as to form, when shut up, the back and sides of the book. " The lenses and divider or partition fold inwards," and the frame supporting the picture is made to slide in a groove in the back of the instrument so as to lengthen or shorten the focus. To open the stereoscope a spring in front of the book is pressed, the sides are expanded, the lenses raised to a vertical position, and the partition raised until the top corner fits into a slit in the upper part of the lens frame; the inverse operations close the instrument.

2nd. " Another form of portable folding stereoscopes consists " of an oblong box resembling a color box with a sliding lid. " When the lid is removed it is turned over and slid into grooves " made in the sides of the box, and the divider which is laid flat " is raised up; the focus may be adjusted by shifting the lid on " which the picture rests."

3rd. The "improvements in stereoscopic apparatus consist of a "box or case for presenting the slides or pictures for exhibition "through the lenses." The slides are retained in slips of wood glued to an endless band; the band is stretched over rollers which are caused to revolve by means of a thumb screw. "When the "slides have been presented to the glasses they fall into a bag or "other suitable receptacle, and are to be replaced in the slips. "This apparatus may be used with any ordinary stereoscope, or "with the folding sort herein-before described."

[Printed, 7*d.*]

### A.D. 1858, October 11.—N° 2262.

ENGLAND, John.—(*Provisional Protection only.*) "Improvements in apparatus for cleaning the plates used in photography."

This apparatus "is for the purpose of providing greater "facilities for holding the glass plates used by photographers." The glass plate is fixed on the ordinary pneumatic sucker. The apparatus consists of a frame supported on a suitable table, which also carries a pair of clips worked by a right and left handed screw. The frame carries and supports the glass at the same time that the clips hold the pneumatic sucker firmly. The top surface of the frame "is provided with india-rubber," "on which "the glass rests in a horizontal position, and is favorably placed "for being operated upon."

The parts forming the table are provided with hinges, so as to be capable of being folded up.

[Printed, 6*d.*]

### A.D. 1858, October 14.—N° 2295.

BAXTER, George.—"Improvements in coloring photographic "pictures."

This invention "consists in combining with photographic "pictures the processes of intaglio, plate, and surface printing, "and also the processes of lithographic and zincographic printing "as a means of coloring such photographic pictures."

"To color photographic pictures by means of wood blocks."—As many positives as it is desired to use blocks in coloring are taken from a negative on to transfer paper, the said negative having two dots for each block. The surfaces of the wood blocks are coated with flake white and gold size, then with transfer varnish. A positive, marked with a composition of vermilion, mastic, varnish, and turpentine on the parts to be coloured by one of the blocks, is laid on one of the prepared blocks, and the marking transferred to the surface of the block by pressure. When the transfer varnish has hardened, the paper is washed off by means of dilute acid, "the marking or drawing remaining on "the block." "The marked blocks obtained in this manner are "engraved by hand in the usual manner." Register is obtained by puncturing the sheets to be printed with holes exactly corresponding with the dots on the negative, the said holes being placed over two register points on the tympan of the press. Stereotype plates may be used instead of wood blocks.

"To color photographic pictures by means of intaglio engraved "plates."—Tracings, one for each colour, are transferred to a series of copper plates, and each plate is engraved in accordance with the tracing. Register is obtained in a similar manner to that above detailed, the dots being traced on to the copper plate and points let in where the dots are; the sheets to be printed have corresponding holes, which are placed over the points.

"To color photographic pictures by means of lithographic "stones or zinc plates."—The parts to be coloured with each tint are drawn in transfer ink on a series of photographs printed on lithographic transfer paper. The pictures are printed in succession, until they are completely colored, with the stones or plates to which the pictures are transferred. The register is obtained in the usual manner or by means of dots. Another method of operating consists in transferring tracings to the stones or plates, and making the drawing for the different tints on the stones or plates themselves.

[Printed, 4*d.*]

## A.D. 1858, October 20.—N° 2342.

STORTZ, PHILIP CHRISTIAN.—(*Provisional Protection only.*

"Improvements in the materials of photographic plates."

This invention "consists in substituting prepared copper or "other metal plates, pannels of wood, millboard, canvass, or "other materials having oiled or varnished surfaces, such as

" usually used for painting upon in oil, for the ordinary materials " upon which photographic images are deposited or printed, and " for producing photographic images or pictures direct from the " object, with or without printing from a negative.

" The prepared surface of the said plates, pannels, millboard, " canvass or other materials is first cleared of all oil or varnish " glaze. The edges of the said plates, pannels, millboard, can- " vass, or other materials are then hermetically sealed, and the " surface prepared in the ordinary manner so as to render it sus- " ceptible of receiving photographic images by means of the " photographic process, and the process is carried on in the usual " manner, except that the chemical ingredients in ordinary use " for the purpose are varied in their proportions so as to adapt " the process to the particular nature or character of the said " plates, pannels, millboard, canvass or other materials."

[Printed, 3*d*.]

A.D. 1858, October 23.—N° 2367.

STORTZ, PHILIP CHRISTIAN. — (*Provisional Protection only.*) " An improvement for taking life-size pictures from smaller " pictures, either with or without the aid of photography."

This invention " consists in taking and tracing life-size pictures " from smaller negative or positive pictures in the most simple " manner.

" Place a double or single lens with camera so that the lens " is turned in a dark room; place a transparent positive or " negative image against the daylight. In the dark room is " placed on an easel a cardboard, canvas, glass, or paper, on " which will be seen the image on a large scale. If the image is " negative take Indian ink or any color, pencil, or chalk, and fill " up all white lines and designs, which, when seen in daylight, " will produce an exact image of the smaller one. If it appear " positive on the canvas, take the same ink or color, but do only " the outlines, which will produce the same effect; or in rendering " the surface of the said cardboard, canvas, glass, or paper " sensitive to light, an image by photographic means may be " produced."

[Printed, 3*d*.]

A.D. 1858, October 27.—N° 2391.

BEAU, Adolphe Paul Auguste.—" A pocket stereoscope."

The lens frame is attached to two covers or flaps, "so as to " resemble a small pocket book or note book." The flaps have pockets on the inside to carry small articles, or " even the " stereoscopic views." When the instrument is in use, it is held by means of the flaps or covers, which are, for this purpose, thrown backwards; a moveable diaphragm is also then let into a groove in the lens frame, so as to project at right angles from it, and thus separate the field of view of each lens. The eye pieces may be formed of a "spiral" [helical?] spring, covered with leather, so as to shut off the external light from the eyes. The diaphragm may slide within a sheath, by which means it may be lengthened or shortened. "The stereoscopic views may be held " at the end of the diaphragm by any suitable arrangement."

Instead of the above-described arrangement, the lens frame may simply slide within a sheath, " which serves for a handle when the " latter is pulled out and the instrument in use, or a handle may " be attached to the frame itself."

[Printed, 6*d.*]

A.D. 1858, October 29.—N° 2425.

JOHNSON, John Henry (*a communication from Gaspard Felix Tournachon, called Nadar*).—(*Provisional Protection only.*) " Improvements in photography, and in the apparatus connected " therewith."

"This invention relates to an entirely new system of photo- " graphy, whereby photographic bird's-eye views of portions of " the earth's surface may be taken, and consists in fitting a " photographic camera in a vertical position, with the lens down- " wards, into the bottom or side of the car of a balloon, such car " being formed, when requisite, into a 'dark chamber,' by " surrounding or enclosing it at the upper part with black silk or " other fabric, and covering or closing the space or chamber so " formed by a moveable roof or ceiling of yellow silk, or other " fabric for the purpose of obtaining the required yellow or " negative light. The lens is covered or uncovered in a ready "manner by means either of a lever acatthed" [attached?] "to "the car, and worked by a cord, so as to bring or remove a cap or "cover over or from the lens, or if preferred, a horizontal disc "turning on a vertical axis, may be employed for that purpose.

"Photographic views may thus be readily taken of the sub-"jacent plane either during the passage of the balloon, or when "anchored and maintained stationary at an elevation in the air, "for which latter purpose three ropes are employed, extending "from the balloon to suitable windlasses below."

[Printed, 3*d*.]

### A.D. 1858, November 4.—N° 2468.

BAGGS, Isham.—(*Provisional Protection only.*) "Improvements "in telegraphing by electricity;" in one of these improvements photography is employed.

This invention "consists in the employment of a fixed micro-"scope, telescope, or other optical power or instrument in combi-"nation with an instrument or apparatus for indicating currents "or effects produced in telegraphing by electricity, whether "needle instrument, or an instrument in which movements or "signs are capable of being produced or exerted by electricity, "howsoever obtained or applied, and whether the electricity be "frictional, voltaic, or otherwise."

In carrying out this invention, "the chemical agency of electri-"city" is preferred to be used, "as acting upon particles of "matter, and not upon masses, and being thereby practically "independent of the opposing forces of inertia and friction." Various illustrations of this mode of operating are given.

Another part of this invention relates to the generation of gas by electricity for telegraph purposes, "under a vacuum, or partial "vacuum."

Another part of this invention "consists *in the application of* "*photography* conducted by any suitable means whatever, to "impress upon paper, glass, or other material the required tele-"graphic communications."

[Printed, 3*d*.]

A.D. 1858, November 20.—N° 2644.

SWAN, Henry.—(*Provisional Protection only.*) "Improvements in stereoscopes and other optical instruments, and in "stands or supports for stereoscopes."

The "standard" connecting the "stage" of the stereoscope with the lens frame "is for the purpose of adjustment divided transversely into two parts, and in one of these parts two pro- 'jecting wires or prongs are fixed, and these enter corresponding "holes in the other part. The holes are lined interiorly with "cloth, so that sufficient friction may be offered to the sliding of "the prongs in the holes, to retain them in any position in which 'they may be placed. When the instrument is in use it is held "by a button attached to the lower part of the standard, and the "adjustment of the distance between the picture and the glasses "is effected by placing the end of the thumb in a recess formed "in the upper portion of the standard, and moving it up or down "till the proper distance is found." To render the instrument more portable, the "standard" may be attached to the "stage" and to the frame of the glasses "by hinges furnished with springs "similar to the springs of a pocket knife."

A method of arranging opera glasses is set forth.

In arranging stereoscopes, the partition is formed of ribbed glass or other transparent or translucent material.

In the ordinary stereoscope, instead of the single reflector usually employed, a series of narrow reflectors like a Venetian blind is used.

"In arranging stands or supports for stereoscopes two (or other "number of) series of levers, parallel the one to the other, and "arranged on what is known as the 'lazy-tongs principle,' are "employed." "The height of the stand or support is adjusted "by altering the angle, which the levers of each pair make with "one another."

[Printed, 3*d.*]

A.D. 1858, December 20.—N° 2909.

MACLEHOSE, Robert.—(*Provisional Protection only.*) "Im- 'provements in stereoscopes."

The stereoscope described "is in the form of a deep rectangular

" box, fitted at one end, near the top, with the usual stereoscopic " glasses."

An open drum or roller is placed " in the line of sight, and " between the glasses and the extreme back of the box or case." This drum works " upon a horizontal spindle, to which is attached " a projecting handle for turning it by. Behind this drum or " cylinder and in the end of the case there is a hinged door, fold- " ing downwards, and fitted with a mirror for throwing light " into the interior of the case. The pictures are disposed in the " chain form over the drum or roller, and so arranged that one " end of the chain can fall down into the bottom of the case at " the front and the other at the back portion of the case. With " this arrangement, as the drum is turned, the pictures are suc- " cessively drawn up from the back section of the bottom of the " case, passed over the faces of the drum, on the vertical side of " which for the time being, they are successively displayed, and " then passed down in a zig-zag or regularly folded pile in the " bottom of the front portion of the case." If necessary, two persons can see pictures at once by means of stereoscopic glasses fitted " on the top of the back part of the case, to look in the " reverse direction of the fixed glasses. The sight is, in this in- " stance, directed upon a mirror," which receives " the image from " the pictures upon the drum beneath it." There are also mirrors and side doors to improve the effect of the light.

" After a picture has been seen and passed on, it can at once be " brought back, if necessary, and any picture seen by one viewer " can be passed on to be seen by the other."

[Printed, 3*d*.]

## A.D. 1858, December 27.—N° 2961.

MARION, Claude Mamès Augustin.—" An improved box or " case for containing and preserving sensitive photographic paper, " called 'Marion's box.' "

This invention " consists of a box with a double bottom her- " metically closed, and containing in the one part chloride of " calcium, and in the other the paper to be preserved."

The first arrangement consists of a rectangular box, " divided " into two parts by wire gauze, the upper part of the box con- " taining the paper to be preserved, and the lower part containing

" a drawer in which is placed chloride of calcium;" the lid of the upper part of this box is hermetically closed by means of a strip of India-rubber.

The second arrangement consists of two rectangular boxes placed one on the top of the other; the upper one contains the paper to be preserved, the lower one the chloride of calcium. The upper box is made in a similar manner to the upper compartment of the first arrangement, it has, however, a perforated bottom which fits hermetically on to the lower box.

The third arrangement consists of a " cylindrical box for con-" taining a roll of the nitrated paper, the bottom of which is " pierced with holes, and forms the lid of the lower box, which is " filled with the chloride of calcium." The lid of the top box, and the upper and lower boxes are respectively joined together hermetically.

A " reservoir " of chloride of calcium, to be used separately, consists of a frame strengthened by means of divisions, and having its bottom formed of a zinc plate and the upper surface of muslin and wire gauze.

" In cylindrical boxes, the chloride of calcium may be contained " in an open cylinder which would serve as a roller for the paper " placed at the centre of the box."

[Printed, 8*d.*]

---

## 1859.

### A.D. 1859, January 7.—N° 51.

SPENCE, William (*a communication from Charles B. Boyle*).— " Improvements in the mode of taking photographic pictures on " wood."

This invention " consists in a process of preparing the block " and applying the silver solution thereto, by which an unchange-" able picture may be produced directly upon the surface of the " wood without injury to its fibre."

The white of an egg, mixed with water, is beaten into a white froth; the face of the block is moistened with this fluid, which is allowed " to sink in and dry by natural evaporation." An aqueous and warm solution of Russian isinglass and chloride of sodium is then rubbed over the face of the block and allowed to sink in and dry. The underlying albumen is coagulated by heat, another coat of gelatine applied, and the excess of gelatine scraped off, sufficient friction being used to enable the silver solution to be brought into direct contact with the surface of the wood itself. The photographic impression is then made in the ordinary manner, except that it is much darker than usual. The solvent or fixing and toning solution is then applied in a heated state; the solution removes the gelatine and effaces all parts of the picture except those on the surface of the wood.

The following process may be applied to blocks which have been previously pumiced and whitened:—The albumen of an egg, mixed with salt and water, is beaten up; the mixture is then poured upon the horizontal block, and is coagulated by heat; this process is repeated until the pores are filled. The surface is then ready for the nitrate solution, and the picture is "taken on sub-" stantially the same principle as a photograph on paper."

[Printed, 4*d.*]

## A.D. 1859, February 2.—N° 296.

ALLEN, Edward Ellis.—"Improvements in stereoscopic " apparatuses." The complete apparatus is called "a compound " collapsible stereoscope."

"The object of these improvements is to render stereoscopic " apparatuses portable."

The views are attached to one side or to both sides of a flexible band which passes round rollers. Each roller rotates in a case of its own, the cases of the two rollers being capable of sliding one inside the other; by this means, when the instrument is out of use, the cases are brought close together, but "when the apparatus " is in use the cases are forcibly kept apart by flaps, rods, or " otherwise," "so as to ensure the straining of the views."

The Drawings show another arrangement, in which the cases are cylindrical and have slits in them for the passage of the views. These cases are connected together by means of a "lazy-tongs;" they can either be thus shut together or kept apart as desired.

The stereoscope moves in slits in the case, "and is made " collapsible, being kept rigid when in use by a flap" hinged on

" the piece containing the eye glasses." The following modes of fitting the stereoscope to the cylindrical cases are described and shown :—The sides and eye glasses are made to fold down close to the cylinders. The eye pieces are divided, each being attached to a "mouse-trap" spring. The eye-glass frame "is attached at " either end to rods hinged on the cylinders." "A system of " bars" connect the eye-glass frame and the "lazy-tongs."

Cases of glass or card slides are rendered portable by arranging the views in two piles, one on each side of the stereoscope; those from the top of one pile are successively transferred to the top of the other, and the bottom slides at the same time undergo an inverse transfer. A system of rods or levers are used for this purpose.

[Printed, 11*d.*]

## A.D. 1859, February 7.—N° 344.

SIMS, THOMAS.—(*Provisional Protection only.*) "Improvements " in the application of photography to engraving and printing."

A collodion photograph is produced "either in the wet or dry " state in the ordinary manner," and is submitted "to the action " of bichloride of mercury and ammonio-chloride of gold or " other chemicals capable of causing the lines of the picture to " swell or rise and become absorbent." "This operation is " however not always necessary." The film is then covered " with a thin coat of varnish, which speedily dries on the porous " raised lines or parts, leaving the other parts wet or damp." A quantity of powdered resin or other suitable powder is then dusted over the picture; this "adheres to the wet parts, and does " not adhere to the dry parts." The picture can now be printed or transferred, "or otherwise converted into a printing surface or " plate."

The invention further states :—"I also make electrotypes from " the films prepared as above-mentioned after metallizing the " surface, or rendering it a conductor of electricity. In some " cases I apply plumbago or other suitable powder to the film " without varnish, and it then adheres chiefly to the raised lines " or parts, and can be transferred or electrotyped. Or, I thoroughly " dry the film on the glass, and coat it or not with oil of turpen- " tine or other essential oil, and then apply resin or other suitable

" powder, and fix it by heat, and I then etch the glass by hydro-" fluoric acid. The glass so etched is then used for printing, or " the pictures may be preserved on the glass itself."

[Printed, 3*d.*]

### A.D. 1859, February 14.—N° 410.

SANDERS, CHARLES.—" Certain improvements in ornamenting " English passe-partouts for photographic pictures, also the glasses " used with the same, and which said modes of ornamenting are " also applicable to ornamenting photographic frames generally."

This invention relates to " the application to the passe-partouts " and front glasses of photographic portraits or pictures (either " singly or combined) of ornamental designs."

The invention consists in " applying to photographic mounts " the well-known process or processes of gilding, painting, " staining, etching, or printing, singly or combined, the inner " side of the front glass around that portion that does not " intercept the picture or the moulding that surrounds it, or to " ornament in like manner by any suitable means, such as by " gilding, painting, staining, printing, or embossing, or either " one or more of these processes combined, that portion of the " front of the passe-partout that surrounds the picture before " referred to."

In many cases it is preferred to produce a part of the design upon the glass and the remainder of the said design upon the passe-partout; when these come together they form the whole design and improve " the brilliancy and general effect."

[Printed, 4*d.*]

### A.D. 1859, February 17.—N° 444.

SAILLARD, BENOIT.—(*Provisional Protection only.*) " An " improved mode of obtaining printing plates from collodion " pictures."

" This invention relates to the obtaining, by the aid of the " electrotype process, metal printing plates from collodion pic-" tures." A collodion photograph is developed by means of a solution containing water, pyrogallic or gallic acid, and nitrate of silver; it is then fixed by hyposulphite of soda or cyanide of potassium solution, washed, and dried. The resulting image is treated with solution of bichloride of mercury, washed with bichromate of potash solution, and then with distilled water; "when dry, the design will appear on the glass plate in relief."

The inventor further states:—"As a protection thereto I find "it convenient to coat the design with photographic varnish, but "this will in a measure detract from the sharpness of the plate "ultimately obtained. I next submit the design, drawing, or "picture to the electrotype process for the purpose of throwing "down copper or other metal thereon by electro-deposition, and I "thus obtain a reproduction in metal suitable for yielding im-"pressions in ink of the raised picture or design produced upon "the glass."

[Printed, 3*d.*]

## A.D. 1859, March 15.—N° 653.

CLARK, William (*a communication from Victor Louis Marie Serrin*).—"Improvements in the apparatus of electric lamps or "lights," applicable (amongst other purposes) to photography.

This invention relates to an improved self-acting regulator. The approach of the carbon points is "effected according to the "relative wear of the carbons, so as to maintain the light in one "uniform position, whereby the circuit can also be broken or re-"established as often as it is wished, and the apparatus be still "always in working order."

To compensate for the loss of weight of the carbon points in the proportion in which it occurs, the positive electrode actuates a train of wheelwork by its gravity combined with that of a weight; whenever the wheelwork is allowed to act (by reason of the electric current ceasing to permit the full action of an electro-magnet in the circuit), the positive electrode descends and actuates a chain passing round a drum, which causes the negative electrode to ascend in a proportion to the movement of the positive electrode which is defined by the relative leverage at which their chains are attached to the said wheelwork.

The other apparatus in the lamp, combined with that above described and compensating the variation in electric force, consists of an "oscillating frame," which is acted upon by the electro-magnet and sustained by a "suspension spring;" by the joint and alternate action of these contrary forces the wheelwork is fixed or liberated.

"As a permanent light this regulator can be applied to light-
"houses, fixed lights, optical experiments, physical and photo-
"graphical purposes."

[Printed, 10*d.*]

## A.D. 1859, April 21.—N° 1011.

PEPPER, John Henry.—(*Provisional Protection only.*) "Improvements in apparatus for showing stereoscopic pictures."

This invention consists "in so combining apparatus, that the "two stereoscopic pictures shall be thrown side by side on to a "screen or surface by means of lanterns having suitable lenses, "in such manner that when seen by several persons through "different lenses (having focal powers according to their several "distances from the screen or surface) each person shall see the "two projected pictures as one largely magnified stereoscopic "picture. For these purposes, a theatre or room is fitted at one "part or end with a suitable screen or surface to receive the two "pictures thrown thereon, by two suitable lanterns, and in front "of the seat for each observer is stationed a suitable standard or "holder of lenses of focal powers coinciding with their distance "from the two projected pictures on the distant screen or surface, "by which, when looked at through such distant lenses, the two "pictures will be resolved into one largely magnified stereoscopic "picture."

[Printed, 3*d.*]

## A.D. 1859, May 6.—N° 1139.

HART, Frederic William. — (*Provisional Protection only.*) "Improvements in photographic apparatus."

1st. "Constructing a photographic printing frame in such a "manner as to admit of adjusting thereby the negative stencil or "mask on the prepared paper or surface, and securing the same "thereon as required." A frame, capable of being moved in any direction over the bed on which the prepared surface is fixed, is secured in the required position by screws, "so as to hold the "stencil mask or negative on the prepared surface ready for "printing." This frame is hinged at one end, so that the reverse end may be raised when required in order to inspect the picture as a whole, and afterwards secured in its position "without disturb-

" ing the original arrangement of the stencil mask or negative on " the prepared surface."

2nd. " Constructing the said moveable frame so as to adapt it " for holding tablets or surfaces to be printed from of various " sizes. For this purpose a curved or other slide is adapted and " applied thereto so as to be capable of being moved along the " frame to the position required in order to grasp or hold the said " printing tablet or surface."

3rd. " Applying vulcanized india-rubber, or other similar " elastic material, to the bed or backboard for the prepared sur- " face to rest upon, in order to facilitate the bringing of the " printing and the prepared surfaces into the required close con- " tact."

[Printed, 3*d.*]

### A.D. 1859, May 7.—N° 1156.

JEFFERY, WALTER.—" Rendering more convenient out-door " manipulations in photography, by means of an improved " portable photographic tent and tent camera."

In one form of the apparatus the base " folds in two towards " the under side." The sides " are formed of folding frames, and " covered first with stout cartridge paper, and then with an " American cloth," " to prevent the light from penetrating into " the chamber." " These sides, when shut up, fall towards the " base, thus the whole folds into a space of half the surface of " the base." The sides of the chamber are kept apart by means of a double framework, which is hinged at either end to the said sides and fastened by means of hooks and eyes.

When the manipulating chamber is " required to be used as a " camera," the following apparatus are added :—A camera front with an opening for the lens, a frame to carry the sensitive plates, and a stereoscopic camera. These are respectively made to fold down upon the base of the instrument, when not in use.

" When the tent is fixed for use, the whole of the upper part " is covered by a double thickness of calico ; " an apron is fixed to the front, which apron reaches to the ground and entirely covers the manipulator. The lenses are covered by means of bags with elastic rings sewn in them ; these bags are fastened to the front. The lenses can be opened and shut from the interior.

When the apparatus is in use it is fixed on a tripod stand by means of battens carrying angle pieces; the battens are fixed to the base of the chamber by thumb screws.

Ventilators "made at right angles" are placed in the tent; by this means no light is allowed to enter.

Another form of framework is described and shown, which is hinged to one of the sides, and folds up against the said side.

[Printed, 1*s.*]

A.D. 1859, May 26.—N° 1300.

PATRICK, Hugh William.—"A new substance or material "to be used in lieu of ivory and other like substances;" amongst the applications of this invention "photographic pur-"poses" is mentioned.

In preparing the new material the following substances are employed, either separately or combined in suitable proportions:—"Amber, Canada balsam, the Australian gum kowrie, potato "flour or fecula," "meerschaum, paper pulp, calcined bones, "fluorate of silicia, sulphide or sulphurets of mercury (vermilion) "or of other metals, chlorides of zinc or other metals, alkaline pre-"parations, asbustos, fluxed or fritted colors, or finely powdered "pumicestone, sulphur, India-rubber, or similar gums."

"The combinations may be effected in various ways, such as by "reducing the gums to solution or 'hard bodies' or precipitates, "or by the application of heat." Where solutions are adopted, the gums employed are dissolved in "naptha, mithilated spirits, "chloroform or essential oils, or other suitable solvent." In this state of solution, or "in a state of precipitate," the gums are added to the above-mentioned substances or to such of them as may be necessary; the latter substances being in a state of powder, "or they may also be mixed with the gums whilst they are in a "state of fusion." When thoroughly mixed, the whole is evaporated "to a thick paste, and when at a proper consistency" it is rolled, cut, or moulded. "Where shaped or moulded, the "new material is hardened by the application of heat, and will "bear a very high polish."

This invention is (amongst other purposes) applicable to "photographic purposes, such as the working into sheets or "tablets."

[Printed, 3*d.*]

A.D. 1859, May 31.—N° 1343.

WANSBROUGH, James (*a communication from Egbert Moxham*).—" Improvements in the construction of stereoscopes."

This invention "relates to a mode of rectifying in the stereo-" scope the distortion which is common to photographs of " architectural subjects, and which arises from the camera used in " taking such pictures being set at an incline, in order to take in " the upper portion of the view." In the photographic representation of a street, for example, made by this method, the rows of buildings appear falling forwards, "the vertical lines being " at the same time elongated or thrown out of proportion." "To " remedy this defect when viewing such pictures stereoscopically," the plane of the pictures may be adjusted at a suitable inclination to the axis of vision. "The optical laws on which this adjustment " is based being that the apparent size of an object diminishes " in proportion as it recedes from the eye, or as its distance from " the eye of the observer is increased," "the want of parallelism " in the vertical lines" is thereby neutralized, and the truthfulness of the representation restored.

The Drawings show a stereoscope whose bottom plate, that carries the views, is hinged to the front of the instrument, and is capable of being fixed at the required inclination by means of a spring catch bearing against the back of the case. The said back is provided with a stop, which prevents "the frame from " swinging loose on its hinges."

[Printed, 5*d*.]

A.D. 1859, June 16.—N° 1450.

JONES, Thomas Wharton.—The title of this invention is, " Stereoscopic glasses for single pictures," and it relates to "a " new form and arrangement of optical glasses or lenses which " give a stereoscopic effect" "to single pictures of any kind," and to real objects "viewed through them with both eyes, that is " one glass before each eye."

The fundamental form of the glasses is "plain on one side and " concave on the other;" other glasses or curved surfaces, suitable for near or long sight, may be superadded to the above-mentioned glasses, if required. The concave curve of the horizontal section in its fundamental form "is a somewhat cycloidoidal curve;" that of the vertical section "is parabolical."

These glasses are mounted either in spectacle frames or in double eye-glass frames, and give a dissimilar perspective of picture on each retina, "and the result is in accordance with the "conditions for stereoscopic vision, the perception by the observer "of a stereoscopic effect or appearance of relief in the picture."

Other forms of glasses besides that above described are capable of producing a stereoscopic effect. Glasses of the fundamental forms above described may be fitted to a binocular, opera, or field glass arrangement of the ordinary kind, and thus enable the relative position of real objects at some distance to be determined with greater exactness than is possible under ordinary circumstances.

[Printed, 6*d.*]

### A.D. 1859, June 17.—N° 1463.

VASSEROT, Charles Frederic (*a communication from Henry Adolphe Corbin*).—"Improvements in the construction of "stereoscopes."

"The object of this invention is to give the power of seeing in "relief pictures of any size (according to the proportions of the "apparatus) much more enlarged or amplified than can be "effected by the present system of stereoscopes."

In this invention the stereoscopic pictures are separate and are inclined to one another in such a manner that one of them is viewed directly by the eye, and the other is viewed by means of a mirror placed at the bisection of the angle that the views respectively make with each other. The two views are thus made to coincide, and are viewed by means of a pair of lenses. It will be immediately perceived that by this means an angle of vision is obtained "more considerable than it is possible with the ordinary "system."

Although it is preferred to place the reflected picture at right angles to the one viewed by direct vision, and to see one picture direct and the other reflected, both pictures may be seen by reflection from a mirror. When this is the case the mirrors make equal angles with the axis of vision, and the pictures are placed at a suitable angle, the angle of each picture with the mirror reflecting it being the same.

The Drawings show a box made according to the above principles, which is covered by rough glass or a removeable opaque lid so as to regulate the light. By cutting out the partitions at the back of the pictures and closing the lid, transparent pictures may be shown.

The picture viewed in the instrument by reflection must be taken "inverted," or with its unprepared surface towards the object.

[Printed, 9*d.*]

## A.D. 1859, June 25.—N° 1527.

NEWTON, William Edward (*a communication from J. Stuart Perry*).—"Apparatus for exhibiting stereoscopic pictures."

This apparatus consists of "a movable framework for holding "a series of stereoscopic pictures, from which the pictures are "brought to be inspected, and then returned to it again by a "mechanism operated by the user."

The instrument shown in the Drawings consists of a cylindrical drum mounted on a centre. The views are placed in radial compartments of the drum, and each picture or pair of pictures is brought in succession underneath a vertical frame into which it is projected from its compartment to be exhibited.

The action of the apparatus is as follows:—The rotation of the driving spindle, by means of a hand winch, moves one picture underneath the vertical frame for every revolution of the spindle; a tooth for this purpose gearing into a cog wheel fixed on the drum axle. The picture is then risen into the frame by the half-revolution of a crank on the driving axis, and depressed into the box by the other half-revolution of the crank; the tooth then brings another picture under the vertical frame, which is in its turn elevated and depressed, and so on. The centre on which the drum revolves is fixed by means of a friction brake during the movement of the views into and out of the vertical frame.

The carrying frame that catches each picture in succession is moved by the action of the crank in a horizontal slot.

One or two lens frames are placed opposite to the picture, to produce the stereoscopic effect; when two frames are used, one is placed on each side of the pair of pictures.

[Printed, 7*d.*]

A.D. 1859, July 12.—N° 1653.

PROAL, CAMILLE JOSEPH (*a communication from Jules Bernard*).—" The application of photographic impressions or pictures " upon fabrics or tissues for rendering such fabrics or tissues " applicable to various useful purposes."

This invention consists " in applying to oilcloth, glazed cotton, " leather, and other fabrics or tissues, one or more photographic " impressions, and employing the fabrics thus prepared for " making up pads for writing blotters, chimney fronts, table " covers of any form, lamp and other stands, portfolios, purses " or money holders, tea trays, covers for books and journals, and " other similar articles, and more especially such articles as boxes, " and objects of cardboard. The fabrics having received photo- " graphic impressions may be ornamented with gilt thread, or gilt " decorations of any kind, and enriched by designs of various " colours appropriate to the photographic subjects.

" The photographic pictures may represent views, portraits, " monuments, battles, or other photographic reproduction; they " may be colored or not colored, ornamented or not ornamented " with decorations. A varnish may at option be applied over the " photographic impressions to preserve their duration and bril- " liancy."

[Printed, 3*d*.]

A.D. 1859, September 3.—N° 2020.

SWAN, HENRY.—(*Provisional Protection only.*) " Improve- " ments in stereoscopes and stereoscopic pictures."

The inventor states :—" According to my invention I construct " stereoscopes with prisms or lenses of different magnifying " powers, that is to say, the prism or lens opposite one eye " exceeds in magnifying power the prism or lens opposite to the " other eye, and I use such instruments in conjunction with pic- " tures in which the two views forming the stereoscopic combi- " nation are of different sizes, corresponding with the different " powers of the prisms or lenses opposite the two eyes, and these " pictures according to my invention, I connect together by " attaching them both in suitable positions to the same mounting. " In this manner, by the use of the large picture only, I am

" enabled to obtain stereoscopic effect combined with minute-
" ness of detail, such as a large picture alone can give; and am
" also enabled to arrange the instrument in a form which is much
" more convenient than that of the instruments heretofore em-
" ployed for viewing large pictures. In some cases I construct
" instruments having one prism or lens only, the large picture
" being then seen without the aid of a prism or lens.

" Magnifying mirrors may be employed in place of prisms and
" lenses, but not so advantageously."

[Printed, 3*d.*]

### A.D. 1859, September 8.—N° 2050.

SMALL, THOMAS OSWALD.—The title of this invention is,
" Improvements in the stereoscope by means of tinted media, and
" by the application of the pulley and lever or spring in changing
" the lights."

The inventor states:—" My invention consists in the first place
" in the substitution of coloured or tinted glasses or other tinted
" media in place of the uncolored glasses or other substances
" heretofore used for the purpose of reflecting light in stereoscopic
" slides; and, secondly, in the application of a pulley or lever
" for the raising or depression of such media, by the use of
" which coloured or tinted glasses, or other tinted media, assisted
" by the said pulley or lever, I am enabled to throw upon stereo-
" scopic slides every degree and variety of light which may be
" required."

[Printed, 3*d.*]

### A.D. 1859, September 12.—N° 2081.

COLLINS, HENRY GEORGE.—1st. Certain improvements in producing printing surfaces and in transfer inks.

2nd. " The application of india-rubber in obtaining printing
" surfaces from enlarged or reduced photographic images." A mixture, containing water, gum arabic, bichromate of potash, and sugar is applied to the surface of a sheet of vulcanized India-rubber, and a photographic image is produced thereon in the ordinary manner. " In order to bring the photographic image to
" the size desired," the sheet of India-rubber is either extended or allowed to contract, and the sensitive coating is transferred " on

"to a stone, metal, or other surface." The coating is then washed with a mixture of gum arabic, yellow soap, and water, "which removes the portions of the coating unchanged by the "light." The surface is then inked and printed from "as is "usual in lithographic printing." Another method consists in producing the photographic image directly upon a stone or other printing surface covered with the above described sensitive coating, washing the surface with mucilage, charging it with ink, and transferring the same to a sheet of India-rubber which is allowed to extend or contract as may be desired; the design is then transferred on to the printing surface, "from which the copies re-"quired may be taken." The method of operating with the India-rubber is set forth in N° 439 (A.D. 1858).

3rd. Methods of producing printing surfaces.

[Printed, *3d.*]

A.D. 1859, September 14.—N° 2095.

BESLAY, Charles.—The title of this invention is, "Improve-"ments in preparing and obtaining printing surfaces with designs "sunk, as also in relief," and some of the processes set forth involve the employment of photography.

The general outline of the process constituting the improvements is as follows:—A design is drawn on a varnish-coated sheet of glass "by removing the varnish from the glass, so as to pro-"duce the design or drawing required." Increased thicknesses of the varnish are applied "at places where large blanks or whites "are to appear in the print." An electrotype is then taken of the glass plate thus prepared, and backed with lead or other suitable material; a printing surface, representing the design in relief, is thus produced. When it is required to produce a sunken design, it is drawn on the glass in varnish, so as to form a raised surface; a sunken design is then produced in the electro-deposited metal.

In order to test the effect of the design made by removing the varnish from the glass, "a sheet of sensitive photographic paper "is placed on the design, the plate is then reversed, exposed to "the light, & a positive proof thus obtained."

To reproduce a photograph.—The photograph is fixed on the sheet of glass, which is then coated, on the reverse side, with transparent varnish; the photographic proof is then copied

" through the transparency of the glass." " This method may " be varied by covering the photographic proof on the glass " with a transparent varnish. The varnish may further be first " prepared, and then covered with a layer of sensitive varnish, " and so obtain a proof on the glass thus prepared." A design also may be obtained by photographing a proof fixed on the glass, also by means of a paper or other proof.

Other details not photographic are also fully set forth.

[Printed, 4*d.*]

A.D. 1859, September 16.—N° 2112.

BECK, Joseph.—" Improvements in stereoscopes."

To admit light to the picture " the stereoscope is left entirely " open in front and at the two sides," and a reflector of silvered glass is fixed to the back of the instrument; by this arrangement " the texture of the paper will not be prominently seen." The admission of light at the sides of the stereoscope is not an essential feature of this invention, although it is preferred.

The Drawings show a stereoscope in which the back or body of the instrument is made in two parts, one of which supports the lens frame, and the other the platform carrying the picture. The pictures are held on the under side of the platform by means of springs, hinged so that they may be turned back, and thus allow the instrument to stand flat when not in use. A screen of ground glass is attached to the middle of the platform, at right angles to the plane of the picture, so as " to limit the sight of each eye of " the person using the instrument to one only of of the pictures " forming the stereoscopic combination;" this is an important feature of the invention. The distance between the lenses and the picture is alterable by means of a rack and pinion movement which connects the two parts of the instrument.

[Printed, 6*d.*]

A.D. 1859, September 28.—N° 2193.

SUTTON, Thomas.—" Improvements in the construction of " apparatus for taking photographic pictures, consisting of and " entitled ' an improved panoramic lens for taking photographic " ' pictures.' "

"The compound lens is composed of two single thick concavo-" convex lenses made of glass, the curved surfaces of which are " portions of concentric spheres. They are secured to a suitable " mount in such a manner and position that the curved surfaces " of both of them (that is to say, all the four curved surfaces) are " concentric, their common centre being a point in the axis of the " compound lens, and the lenses having their concave surfaces " opposite to each other. In the space or cavity between the con-" cavo-convex lenses is contained a transparent fluid of lower " refractive and dispersive power than the glass of which the " lenses are made." "The two glass lenses may be made of the " same kind of glass, and equal in all respects;" or they may be made of different kinds of glass, having suitable radii in order to correct the compound lens for colour and for spherical aberration. To give sharp definition when objects at different distances from the lens are included in the view, a diaphragm of suitable size is placed between the lenses, within the fluid, at the centre of the compound lens.

The photographic pictures taken with this lens "should be pro-" duced upon tablets, forming either a segment of a sphere or a " segment of a cylinder so placed in the camera as to have the " same centre as the curved surfaces of the lenses."

The Drawings show the two lenses mounted in a zone of brass, which slides in the boss of the usual mounting.

In the Provisional Specification it is proposed to use a conical box, open at each end, fitted with plane partitions, to supply the place of the above-described central diaphragm.

[Printed, 6*d.*]

A.D. 1859, October 3.—Nº 2238.

EARLE, William Richardson, and BARNES, Edwin John. —(*Provisional Protection only.*) The title of this invention is, "Improvements in photographic and other portraits."

The inventors state:—"Our invention of improvements in " photographic and other portraits relates to the application of " some of the hair of the person, of whom the portrait is a repre-" sentation, to the portrait in the position and in the natural " manner of the hair of that person or otherwise, or it may be of " the hair of any other person or kind.

" In its application to photographic portraits, I cement the " hair to the front or back of the material on which the portrait is " taken, if that material is transparent it must however be in " front in cases where the material is opaque.

" We also apply hair in like manner to oil and other painted " portraits to enhance the effect and produce greater re- " semblance.

[Printed, 3*d.*]

### A.D. 1859, October 5.—N° 2258.

FISHER, Robert, and ASPRAY, Charles.—(*Provisional Protection only.*) The title of this invention is, " Improvements in " photographic stereoscope slides, and in the stereoscopes em- " ployed in viewing the same."

The inventors state:—" We take, with a single camera, two " photographs of the same object, but in different attitudes or " positions, and in the stereoscope employed in viewing them we " provide a sliding shade by means of which the right or left " hand picture is viewed alternately, by this arrangement the " effect of motion or change of position is produced."

[Printed, 3*d.*]

### A.D. 1859, October 19.—N° 2393.

COWPER, Charles (*a communication from John Henry Pein*). —" Improvements in photographing on uneven surfaces, and in " apparatus for that purpose."

" By the present invention, the focus is obtained in a direct " manner on the uneven surface itself by means of the apparatus " herein-after described, and the uneven surface is then removed " and covered with a coating of collodion, or other suitable " material, and rendered sensitive to light, and it is then replaced " in the camera in the same position as when the picture " was focussed, and it is exposed a sufficient time, and developed " and fixed in the usual manner."

The apparatus, as shown in the Drawings, is as follows:—A table at the bottom of the camera carries a plate, capable of sliding horizontally in " lateral grooves;" on the plate is fixed a vertical rod that supports a " pan holder," by means of a clamp screw, in a similar way to that of supporting a retort, by an ordinary retort stand; a picture frame is connected with the "pan holder" by a similar supporting apparatus; the vase or other object is placed in a plastic material, with which the pan is filled. A photographic picture is taken by placing the vase in the pan with the surface to be sensitized vertical, and fixing it upon the clay, then focussing the image upon it (having adjusted the vase properly by means of the sliding plate, rod, and clamp, also the picture frame upon it by means of its rod and clamp), removing it to be sensitized, replacing it in exactly the same position, exposing, developing, and fixing it. The camera lens is fixed in a sliding box to facilitate focussing, and the picture or object which serves as the original is placed in front of the lens, at a distance from it depending upon the size of the picture desired.

[Printed, 7*d.*]

A.D. 1859, November 10.—N° 2557.

MELHUISH, Arthur James.—" Improvements in the con-" struction of cameras for obtaining photographic pictures."

The body of the camera is made " of metal (usually of brass), " the bottom, top, and sides of the body are soldered, or otherwise " fixed together, and the front which carries the lens or lenses " " slides in guides fixed to the sides of the body," so that the position of the image can be thereby adjusted " before taking the " picture," and the front " may be entirely removed from the " body of the camera for convenience of packing the slides " therein." The lens may be mounted on a tube fixed to the front, thus reducing the depth of the camera as much as possible. The "dark slide" is introduced into the camera in the usual manner, but the slide is drawn down through a slit in the bottom of the camera in order to expose the sensitive surface when taking a picture; the said "dark slide" is made entirely of metal, and the front slide thereof, when closed, is secured by bolts; spring handles are fitted to the "dark slide" for the convenience of packing. The "dark slide" may be made double, "to receive " two sensitive plates or surfaces back to back;" "for this pur-" pose both sides of the case slide out to expose the sensitive " plates or surfaces, and the top of the case or slide also slides " out to allow the plates or surfaces to be put in. The ground

" glass used in focussing is mounted in a metal frame, that it " may frack" [pack?] "into as small a space as possible."

The Drawings represent a camera of the above description with two lenses, in order to take stereoscopic pictures. In this case there is a partition that separates the images, which partition takes out "in order that the dark slides and focussing glass may " be packed away in the interior of the camera."

[Printed, 1*s.* 1*d.*]

A.D. 1859, December 5.—N° 2755.

NEGRETTI, Enrico Angelo Ludovico, and ZAMBRA, Joseph Warren.—(*Provisional Protection only.*) " Improve- " ments in stereoscopes," to render them portable.

" The instrument is composed of a shallow rectangular box, " capable of containing from six to eight or more stereoscopic " pictures, together with the eye-piece of the instrument, which, " with the pictures, is covered in the box by a moveable lid, pro- " vided with flexible bands or other convenient attachments, " whereby the lid or cover may be temporarily secured to the " box. When the instrument is required for use, the moveable " lid or cover is to be fixed in a vertical position across the centre " of the box, in notches made therein for the purpose. The eye- " piece has a notch or groove cut across the middle of it on the " under side, and is fixed on the top edge of the vertical lid which " fits into the notch on the under side of the eye-piece, the vertical " lid thereby forming a support to the eye-piece, and also forming " a central vertical division between the line of sight of the two " eyes. The pictures are made much smaller than usual, and " when the two views which form a stereoscopic picture are placed " side by side, their edges are made to touch;" the junction line is hid beneath the edge of the vertical lid. " By making the " pictures without any margin round them, the instrument may " be reduced in size without materially or injudiciously increasing " the minuteness of detail in stereoscopic views or pictures."

[Printed, 3*d.*]

A.D. 1859, December 19.—N° 2891.

SMITH, John.—" Improvements in the manufacture of compo- " sition jewellery and ornaments, and in cases for jewellery, pho- " tographs, and for other similar purposes."

This invention consists in the production of the above-mentioned articles "from a composition or plastic material not "hitherto used for these purposes, and in the process of pre-"paring the said composition."

To make a black composition, certain proportions of shellac, ebony dust, "black asphaltum," and ivory black or charcoal powder, or lamp black, are used. The melted shellac is mixed with the ebony dust, and the other materials added as colouring matter.

To make a chocolate-coloured composition, "brown asphaltum" is used in place of the black asphaltum, and ordinary rouge is added "according to the tint required."

"To produce greens, blues, or other dark colours," any suitable colouring matter is employed in proper proportions.

For light colours, dust of boxwood is used as a principal colouring agent, "to which, if still lighter tint be required, a more "intense white in the shape of pigment may be added."

These ingredients may either "be thoroughly amalgamated" so as to produce an even colour, or masses of differently coloured compositions may be twisted or rolled together so as to produce "imitation marbles, onyx, malachite, or other stones."

To manufacture the article required, a lump of heated composition of sufficient size is placed in suitable dies, and "the article "is completed by pressure."

[Printed, 3*d.*]

## A.D. 1859, December 28.—N° 2962.

**ROSTAING**, Charles Sylvester.—"Improvements in com-"bining and mixing gutta percha with mineral and vegetable "substances capable of altering its quality in such a manner as "to produce hard, resistent, unalterable, and imputrescible com-"pounds diversly colored." These compounds are applicable, amongst other purposes, to "photographic and mirror plate "frames," and "to replace wood, ivory, wax, cloth, and other "substances used by photographers upon which to adopt photo-"graphs."

The colours used in this invention are obtained by baking in a muffle or crucible a suitable metallic oxide mixed with water or a mineral acid. The baked oxide after cooling is reduced to impalpable powder and then dried and sieved.

The preparation of the various compositions is divided "into "eight different operations," as follows:—

1st. The partial purification of the gutta percha by boiling it in soft water, flatting it, and again boiling it "in a decoction of "'radix saponaria alba' or soap wort."

2nd. The complete purification of the gutta percha by boiling it in a caustic lye and flatting it.

3rd. The preparation of "artificial silicate of zinc."

4th. The preparation of "silicate of alumina and of zinc."

5th. The preparation of talc "rose colored by oxyde of cobalt."

6th. The preparation of a combination of zinc blende "with "kaolin or with felspar."

7th. The preparation of a combination "of tannin, gutta percha, "chloroform, benzoin, balsam of tolu, or essential oil."

8th. The preparation of a combination of gutta percha with catechu, to which camphine or benzin may be added.

The Drawings show a machine for mixing the above-mentioned compositions, consisting of a vessel heated by a steam jacket and having suitable rollers and scrapers.

[Printed, 1*s*. 5*d*.]

### A.D. 1859, December 28.—N° 2965.

MELHUISH, Arthur James.—(*Provisional Protection only.*) "Improvements in cameras," relating to the use of thin sheet metal in the construction of cameras, "so as at the same time to "obtain compactness and lightness with the requisite degree of "strength and stiffness."

The body of the camera and the dark slides are constructed of thin sheet metal properly strengthened by flattened wire or other suitable means. In folding cameras where the joints butt or come together, a strip of metal is soldered to one of the two parts, so as to overlap and cover the joint when in use. "Where the front "and other parts of sheet metal cameras slide together," guiding ribs are used; "and in some cases the sheet metal front of a "camera is hinged to and is caused to fold into the interior of a "camera, and velvet or cloth (by preference black) is applied to "parts where they come or butt together, in order to shut out the "light from the interior of a camera." A recess in the frame of the camera receives the dark slide which is pressed into the recess by means of springs. Lap joints connect the sides of the camera. The body of the camera may be formed of an open frame of metal covered with an opaque material. The frame to receive the dark slide may be made " distinct from the body of the " camera, but hinged to the bottom of it, so that it will fold down " on to the bottom of the body of the camera to allow of it, the " said body, folding down upon it." This frame serves as a support for the body of the camera when it (the frame) is lifted up for use. Spring handles are applied to the shutters of the dark slides, and the edge of the shutter is bent " over the edge of " the door when only one shutter is used, thus keeping out the " light."

[Printed, 3*d*.]

---

# INDEX OF SUBJECT MATTER.

[The numbers refer to the page in which the paragraph containing the subject commences.]

PATENTS FOR INVENTIONS.

# ABRIDGMENTS

OF

# Specifications

RELATING TO

# PHOTOGRAPHY.

PART II.—A.D. 1860–1866.

PRINTED BY ORDER OF THE COMMISSIONERS OF PATENTS.

*SECOND EDITION.*

LONDON:
PRINTED BY GEORGE E. EYRE AND WILLIAM SPOTTISWOODE,
PRINTERS TO THE QUEEN'S MOST EXCELLENT MAJESTY.
PUBLISHED AT THE
OFFICE OF THE COMMISSIONERS OF PATENTS FOR INVENTIONS,
25, SOUTHAMPTON BUILDINGS, HOLBORN.

1872.

# PREFACE.

The Indexes to Patents are now so numerous and costly as to render their purchase inconvenient to a large number of inventors and others, to whom they have become indispensable.

To obviate this difficulty, short abstracts or abridgments of the Specifications of Patents under each head of invention have been prepared for publication separately, and so arranged as to form at once a Chronological, Alphabetical, Subject-matter, and Reference Index to the class to which they relate. As these publications do not supersede the necessity for consulting the Specifications, the prices at which the printed copies of the latter are sold have been added.

The number of Specifications from the earliest period to the end of the year 1866 amounts to 59,222. A large proportion of the Specifications enrolled under the old law, previous to 1852, embrace several distinct inventions, and many of those filed under the new law of 1852 indicate various applications of the single invention to which the Patent is limited. Considering, therefore, the large number of inventions and applications of inventions to be separately dealt with, it cannot be doubted that several properly belonging to the group which forms the subject of this volume have been overlooked. In the progress of the whole work such omissions, as may from time to time become apparent, will be supplied in future editions.

This volume is a continuation of the "Abridgments of "Specifications relating to Photography" already published, and brings the Abridgments to the end of the year 1866. From that date the Abridgments will be found in chronological order in the "Chronological and Descriptive "Index" (*see* List of Works at the end of this book). It is intended, however, to publish these Abridgments in classes as soon as the Abridgments of all the Specifications from the earliest period to the end of 1866 have appeared in a classified form. Until that takes place, the reader (by the aid of the Subject-matter Index for each year) can continue his examination of the Abridgments relating to the subject of his search in the Chronological and Descriptive Index.

The following quotation from the Preface to the first part also applies to the present volume.

"The Specifications included in this series of inventions "relate :—

"1st. To the processes of Photography.

"2nd. To the chemical, optical, and mechanical apparatus used in the practice of the art, or its applications.

"3rd. To the various applications of Photography.

"4th. To the camera obscura, so far as its use for photographic purposes is concerned.

"5th. To the stereoscope; this instrument depends upon "the results of Photography for its efficient action; it "gives to photographs the nearest possible approach to "reality; therefore, all Specifications relating to stereoscopes "are included in this series."

"The definition of 'Photography' for the purpose of "these abridgments is, 'the art of copying designs or "'images, however they may be produced, by the chemical "'action of light upon surfaces prepared to receive that "'action.'"

In the Subject-matter Index, the most recent chemical names of substances are placed (in *Italic*) after the names that have been obtained from the respective Specifications; this addition is rendered necessary by the general adoption of the new chemical nomenclature.

The Abridgments marked thus (* *) in the following pages were prepared for another series or class, and have been transferred therefrom to this volume.

B. WOODCROFT.

*February*, 1872.

# INDEX OF NAMES.

[The names printed in *Italic* are those of the persons by whom the inventions have been communicated to the Applicants for Letters Patent. The Roman numerals refer to the Introduction.]

# INTRODUCTION.

THE advancement of Photography and of its applications, during the seven years embraced by this volume (1860 to 1866) has been considerable. In this Introduction, every paper on Photography, or involving the chemical action of light, that has been read to the Royal Society, within the above-named seven years, is alluded to.

The following notice should have been inserted in its proper place (chronologically speaking) in the Introduction to Part I. of this series of Abridgments.

A.D.

1853. J. B. DANCER, "early in 1853," gave to JOSEPH SIDEBOTHAM two microscopic photographs; "they are about "one sixteenth of an inch in diameter." GEORGE SHADBOLT, in accordance with this announcement, gave up his claim to being the prior producer of "micro-photographs." (See *The Photographic Journal*, edited by George Shadbolt, of April 15, 1859, p. 91; of May 1, 1859, p. 104; and of May 15, 1859, pp. 118, 125, and 126).

The paragraphs that follow are in continuation of the Introduction to Part I., accordingly they resume the chronological arrangement from the beginning of the year 1860.

1860. Professor WILLIAM B. ROGERS, in June or July 1860, read a paper to the British Association, entitled "Experi-"ments and conclusions on binocular vision." By means of "two slightly inclined luminous lines" "com-"bined into a perspective line," he arrives at the conclusion that the visible direction of an object is not always "in the normal to the point of the retina impressed," "but is felt to be in the middle line" between the two eyes, "that is, in the binocular direction." (See *British Association Report* of the meeting in 1860, p. 17.)

A.D.

1860. A. CLAUDET, in June or July 1860, read a paper to the British Association, "On the means of increasing the "angle of binocular instruments, in order to obtain a "stereoscopic effect in proportion to their magnifying "power." This increase of the angle is accomplished "by adapting to the object glasses two sets of reflecting "prisms." (See *British Association Report* of the meeting in 1860, p. 61.)

1860. A. CLAUDET, in June or July 1860, read a paper to the British Association, "On the principles of the solar "camera." By the use of "the solar camera, invented "by WOODWARD," "small negatives may produce pic-"tures magnified to any extent." (See *British Association Report* of the meeting in 1860, p. 62.)

1860. Dr. HENRY DRAPER, in June or July 1860, read a paper to the British Association, "On a reflecting telescope for "celestial photography, erecting at Hastings, near New "York." A Newtonian reflector is used, so that whatever is the position of the object, the photographic table is always horizontal. Protochloride of palladium is used to darken collodion negatives. (See *British Association Report* of the meeting in 1860, p. 63.)

1860. Messrs. PETSCHLER and MANN, in August 1860, discovered a modification of the collodio-albumen process. The albumen is prepared with a chloride, and the plate is coated with it in the usual manner. The plate is sensitised by washing in water. (See *The Photographic News* for August 24, 1860, p. 195.)

1860. Major-General EDWARD SABINE, on November 15, 1860, read a paper to the Royal Society, upon the "Laws of the "phenomena of the larger disturbances of the magnetic "declination in the Kew Observatory." For the purpose of evolving these laws, a "first approximation" was taken, that was "derived from the photographs in the "years 1858 and 1859, of the self-recording declinometer "of the observatory of the British Association at Kew." "The photographs furnish a continuous record of the "variations which take place in the direction of the

A.D.

"declination-magnet, and admit of exact measurement in "the two relations of time, and of the amount of departure from a zero line. From this automatic record the "direction of the magnet is measured at twenty-four "equal intervals of time in every solar day, which thus "become the equivalents of the 'hourly observations' of "the magnetometers in use at the colonial observatories." (See *Proceedings of the Royal Society*, vol. X., No. 41, p. 624.)

W. T. SHAW sent a communication to the Royal Society,
1861. which was read on the 10th January 1861; in this communication a new optical instrument called the "Stereotrope" was described. In the "stereotrope" the principle of the stereoscope is applied to the phenakistoscope and other similar instruments, so as to give the effect of solidity to the resulting moving image. Either the refracting stereoscope or Professor WHEATSTONE's reflecting stereoscope may be employed to produce the above-mentioned results; the latter is "better adapted for the "exhibition of movements that are not only local but "progressive in space." (See *Proceedings of the Royal Society*, vol. XI., No. 43, p. 70.)

1861. Major-General EDWARD SABINE, on January 10, 1861, read a communication to the Royal Society, "On the "lunar-diurnal variation of the magnetic declination "obtained from the Kew photograms in the years 1858, "1859, and 1860." "The directions of the declination "magnet at the instant of the commencement of every "solar-hour" were "tabulated from the photograms, and "the final normals for each month and hour computed." (See *Proceedings of the Royal Society*, vol. XI., No. 43, p. 73.)

1861. Sir DAVID BREWSTER, in September 1861, read a paper to the British Association, "On photographic micrometers." These micrometers were made by reducing a system of sharply defined lines "to the minutest size," and impressing them "upon collodion or albumen photographically." (See *British Association Report* of the meeting in 1861, p. 28.)

A.D.

1861. Sir DAVID BREWSTER, in September 1861, read a paper to the British Association, "On binocular lustre." He believes that the lustre is due "to the effect of the eyes in "combining the two stereoscopic figures, and to the "dazzle occasioned by the alternating intensities of the "two combined tints." (See *British Association Report* of the meeting in 1861, p. 29.)

1861. THOMAS SUTTON, in September 1861, read a paper to the British Association, "On the panoramic lens." This lens has concentric surfaces, and is capable of producing a perfect optical image, "which includes an angular field "of 100°." (See *British Association Report* of the meeting in 1861, p. 33.)

1861. J. T. GODDARD, in September 1861, read a paper to the British Association, "On the cloud mirror and sunshine "recorder." In the sunshine recorder, the image of a hole in a box carrying photographic paper is imprinted on the paper during sunshine, but not otherwise. (See *British Association Report* of the meeting in 1861, p. 61.)

1861. WARREN DE LA RUE, in September 1861, presented his "Report on the progress of celestial photography" to the British Association. Heliography, as well as lunar and stellar photography, is alluded to. To render the images of the fixed stars visible, it is necessary to put them out of focus. (See *British Association Report* of the meeting in 1861, p. 94.)

1861. Colonel Sir HENRY JAMES, in September 1861, read a paper to the British Association, "On photozincography." An inked gelatine and bichromate of potash positive is transferred to stone or zinc. This process is used to multiply copies of the Ordnance Maps. (See *British Association Report* of the meeting in 1861, p. 263.)

BALFOUR STEWART sent a communication to the Royal
1861. Society, which was read November 21, 1861, "On the "great magnetic disturbance of August 28 to September "7, 1859, as recorded by photography at the Kew "Observatory." The recording was accomplished by means of the "self-recording magnetographs which are

A.D.

"in constant operation." (See *Phil. Trans.* vol. 151, p. 423; also *Proceedings of the Royal Society*, vol. XI., No. 46, p. 407.)

1862. Major-General Edward Sabine, on Feb. 27, 1862, read to the Royal Society, "Notices of some conclusions "derived from the photographic records of the Kew "declinometer in the years 1858, 1859, 1860, and 1861." The methods and processes employed in the elicitation of the results were fully described in a paper communicated by the author to the Society in 1860. (See *Proceedings of the Royal Society*, vol. XI., No. 48, p. 585.)

1862. F. Joubert, on March 4, 1862, read a paper to the Photographic Society "On enamel photographs." A positive picture, made by means of bichromate of ammonia, honey, and albumen, is gently rubbed over with an enamel colour. After the picture is fixed and washed, it is dried and fired. (See *The Photographic News* of March 14, 1862, p. 125.)

1862. Warren de la Rue, on the 10th April 1862, delivered the Bakerian Lecture to the Royal Society, "On the total "solar eclipse of July 18th, 1860, observed at Riva-"bellosa, near Miranda de Ebro, in Spain." About 35 effective photographs of the eclipse were taken by means of the "Kew heliograph;" two totality pictures were obtained (negatives), these showed the rose-coloured prominences with great distinctness. An accurate map of the prominences was made by means of a measuring instrument. The usual collodion process wrs employed for the negatives; the collodion was iodized in London; hyposulphite of soda was the developing medium. (See *Phil. Trans.*, vol. 152, p. 333.)

1862. Dr. W. A. Miller, on June 19, 1862, read a paper to the Royal Society "On the photographic transparency of "various bodies, and on the photographic effects of "metallic and other spectra obtained by means of the "electric spark." The following results were arrived at:—1. Colourless bodies, which possess equal powers of transmitting luminous rays, vary greatly in diactinic

A.D.

power. 2. Diactinic solids preserve their diactinic power, both when liquefied and when converted into vapour. 3. Colourless translucent solids, which absorb chemical rays, "preserve their absorptive power with greater or "less intensity both in the liquid aad the gaseous "state." 4. When polished surfaces are employed to reflect the chemical rays, an amount and character of absorption is exercised by each metallic and other surface, differing with each surface used. 5. The electric spectra of the principal metals, taken in air, exhibit characteristic points, especially the following:—Silver, thallium, bismuth, antimony, tellurium, manganese, iron, cadmium, zinc, and *magnesium*. 6. In alloys, the presence of each metal is for the most part indicated in the spectrum. 7. When the metals are ignited by the electric spark in various gases, "each gas tinges the spark of a characteristic colour; but no judgment can be formed from this "colour of the kind of spectrum which the gas will "furnish." 8. Lines characteristic of the gas are in most cases produced. 9. "The lines due to the gaseous "medium are continuous." The apparatus used in these researches had a prism of rock crystal, and a quartz lens was attached to the camera next to the prism. (See *Phil. Trans.*, vol. 152, p. 861.)

1862. Professor SELWYN, in October 1862, read a paper to the British Association "On autographs of the sun." These "autographs" were taken with the "heliautograph," which consists of a camera and instantaneous slide attached to a refractor. (See *British Association Report* of the meeting in 1862, p. 17.

1862. CHARLES HEISCH, in October 1862, read a paper to the British Association "On a simple method of taking "stereomicro-photographs." A microscope, at its eye end, is fitted to a single lens stereoscopic camera; this apparatus is used in connection with a sliding tube that carries a stop which cuts off half the light and can be turned half round. (See *British Association Report* of the meeting in 1862, p. 46.)

A.D.

1862. T. SUTTON, in October 1862, read a "Description of a "rapid dry-collodion process" to the British Association. Bromo-iodized collodion is used in conjunction with gum arabic. (See *British Association Report* of the meeting in 1862, p. 54.)

1862. Messrs. BUNSEN and ROSCOE, on December 11, 1862, read a paper to the Royal Society entitled "Photo-"chemical researches.—Part V. On the measurement "of the chemical action of direct and diffuse sunlight." These gentlemen proved, within the limits of their experiments, that "equal shades of blackness correspond to "equal products of the intensities of the acting light "into the times of exposure." A pendulum, in connection with the withdrawal of a sheet of darkened mica from a strip of paper prepared with chloride of silver, and with a millimetre scale, is used to expose photographic sensitized paper for times which can be very accurately measured. To compare a given photographic tint with the point of equal shade on a strip, a sliding board is employed in conjunction with the light of a bright soda flame. In the preparation of a standard sensitive paper, nitrate of silver and chloride of sodium baths are employed; the fixing agent is hypo-sulphite of soda. (See *Phil. Trans.*, vol. 153, p. 139; also *Proceedings of the Royal Society*, vol. XII., No. 53, p. 306.)

1863. G. B. AIRY, Astronomer Royal, on the 23rd of April 1863, read a paper to the Royal Society "On the diurnal "inequalities of terrestrial magnetism, as deduced from "observations made at the Royal Observatory, Green-"wich, from 1841 to 1857." During the last ten years of this time "the magnetic indications have been "automically recorded by photographic self-registration," Mr. BROOKE's photographic apparatus being employed. (See *Phil. Trans.*, vol. 153, p. 309; also *Proceedings of the Royal Society*, vol. XII., No. 56, p. 529.)

1863. Major-General EDWARD SABINE, on June 18, 1863, read two papers (one a continuation of the other) to the Royal Society, on the "Results of the magnetic observations at

A.D.

the Kew "Observatory, from 1857 and 1858 to 1862 "inclusive.—Nos. I. and II." The laws of the disturbances of the magnetic declination at Kew are derived from photographic records. (See *Phil. Trans.*, vol. 153, p. 273; also *Proceedings of the Royal Society*, vol. XII., No. 56, p. 263.)

1863. Dr. H. E. Roscoe, on June 18, 1863, read a paper to the Royal Society "On the measurement of the chemical "brightness of various portions of the sun's disc." The result obtained (by the use of "standard photographic "paper," in conjunction with "a graduated photographic "strip insolated in the pendulum photometer") was "that the intensity of the chemically active rays at the "centre is from three to five times as great as that at the "edge of the disc." (See *Proceedings of the Royal Society*, vol. XII., No. 46, p. 649.)

1863. Balfour Stewart, on the 18th June 1863, read a paper to the Royal Society "On the magnetic disturbance "which took place on the 14th December 1862." This disturbance "was registered by means of the Kew "magnetographs." (See *Proceedings of the Royal Society* vol. XII., No. 56, p. 663.)

1863. Professor C. Piazzi Smyth, in August or September 1863, read a paper to the British Association "On a proof of "the dioptric and actinic quality of the atmosphere at a "high elevation." Photographs of distant hills taken at a high level were distinct, and taken at or near the sea level were wanting in details. (See *British Association Report* of the meeting in 1863, p. 25.)

1863. Dr. T. L. Phipson, in August or September 1863, read a paper to the British Association "On a new method of "measuring the chemical action of the sun's rays." A solution of sulphate of molybdic acid is used for this purpose; this solution becomes bluish-green on exposure to sunlight, but is colourless at night. (See *British Association Report* of the meeting in 1863, p. 50.)

1863. G. B. Airy, Astronomer Royal, on December 17, 1863, read a communication to the Royal Society, on the "First

A.D.

" analysis of 177 magnetic storms, registered by the " magnetic instruments in the Royal Observatory, " Greenwich, from 1841 to 1857." "The photographic " ordinates" are converted into numbers. (See *Phil. Trans.*, vol. 153, p. 617; also *Proceedings of the Royal Society*, vol. XIII., No. 59, p. 48.)

1863. BALFOUR STEWART, on December 17, 1863, read a paper to the Royal Society "On the sudden squalls of 30th " October and 21st November 1863." "The gas lights " in the room which contained the barograph" [worked by photographic agency ?] "went out;" the barograph curve, although "incomplete," gives a very rapid rise. (See *Proceedings of the Royal Society*, vol. XIII., No. 59, p. 51.)

Senhor CAPELLO and BALFOUR STEWART, on Jan. 28,
1864. 1864, read a paper to the Royal Society on the "Results " of a comparison of certain traces produced simulta- " neously by the self-recording magnetographs at Kew " and at Lisbon." These instruments record the results by photographic means. (See *Proceedings of the Royal Society*, vol. XIII., No. 60, p. 111.)

1864. BALFOUR STEWART, on March 17, 1864, read a paper to the Royal Society, entitled "Remarks on sun spots," in which he says, "In examining the pictures taken with the " Kew heliograph under the superintendence of Mr. " DE LA RUE, it appears to be a nearly universal law " that the faculæ belonging to a spot appear to the left " of that spot, the motion due to the sun's rotation " being across the picture from left to right." (See *Proceedings of the Royal Society*, vol. XIII., No. 62, p. 168.)

1864. Dr. ROSCOE, in March 1864, used the magnesium light for photographic purposes with success. (See *The Photographic News* of March 18th, 1864, p. 137.)

1864. A. CLAUDET, in September 1864, read a paper to the British Association, on "Photo-sculpture," the invention of M. WILLÈME. Single profile photographs, in con-

A.D.

junction with a pantagraph, were used to work upon the clay, a number of profiles taken at the same moment being used in succession. (See *British Association Report* of the meeting in 1864, p. 10.)

1864. ARTHUR M'DOUGALL, in a paper read to the Chemical Society, and dated October 1864, "On a mode of "measuring the relative sensitiveness of photographic "papers," arrived at the result "that the sensitiveness "of papers containing the same quantity of chlorine "and bromine combined with sodium, potassium, am- "monium, and barium is constant." (See *The Journal of the Chemical Society*, July 1865, p. 183.)

1864. WARREN DE LA RUE, in the latter part of 1864, presented a communication to the Royal Society, entitled "Com- "parison of Mr. De la Rue's and Padre Secchi's eclipse "photographs." By means of an enlargement from a positive print of Padre SECCHI's negative, as it was originally, the two photographs (allowing for difference of position) were found to "accord in their most minute "details." (See *Proceedings of the Royal Society*, vol. XIII., No. 67, p. 442.)

1864. DR. H. E. ROSCOE, on December 22, 1864, read the Bakerian lecture to the Royal Society, "On a method of "meteorological registration of the chemical action of "total daylight." This method is founded upon that described in Phil. Trans., vol. 153, p. 139. "The re- "ciprocals of the times during which the points on the "strip have to be exposed in order to attain the normal "tint, give the intensities of the acting light" in terms of the unit of photo-chemical intensity. (See *Phil. Trans.*, vol. 155, p. 605; also *Proceedings of the Royal Society*, vol. XIII., No. 71, p. 37.)

1865. WARREN DE LA RUE, BALFOUR STEWART, and BEN-JAMIN LOEWY, on January 26, 1865, contributed to the Royal Society a paper entitled "Researches on solar "physics.—Series I. On the nature of solar spots." These researches were aided by pictures taken by the Kew heliograph, and by certain stereoscopic pictures. (See *Proceedings of the Royal Society*, vol. XIV., No. 71, p. 37.)

A.D.

1865. WARREN DE LA RUE, BALFOUR STEWART, and BENJAMIN LOEWY, on February 2nd, 1865, contributed to the Royal Society the second series of their researches on solar physics, "On the behaviour of sun spots with "regard to increase and diminution." The Kew photograms, in connection with Carrington's observations, were used for these investigations. (See *Proceedings of the Royal Society*, vol. XIV., No. 72, p. 59.)

1865. W. WILLIS, on April 11, 1865, read a paper to the Photographic Society on "The aniline process of photographic "printing." The sensitising liquid is an acid bichromate. After exposure to light under a positive photograph, the sensitised paper is placed in the vapour of aniline, the print is then washed. (See *The Photographic News* of April 21, 1865, p. 186; and of April 28, 1865, p. 196.)

1865. A. CLAUDET, in September 1865, read a paper to the British Association "On moving photographic figures." The instrument described in this paper is a combination of the stereoscope with the phenakistoscope of PLATEAU by means of photography. (See *British Association Report* of the meeting in 1865, p. 9.)

1865. Mr. CAREY LEA, in November 1865, published his new form of developer, "in which the decomposition pro- "ducts of gelatine by sulphuric acid take the place of "acetic acid in controlling the development." Iron and acetate of soda are employed in conjunction with the above decomposition products. (See *The Photographic News* of November 24, 1865, p. 560.)

1866. Mr. JABEZ HUGHES, on January 9, 1866, read a paper to the Photographic Society, entitled, "Further observations "on the preparation of the iron developer, so as to pro- "duce dense negatives." To produce the result easily and successfully, a small quantity of solution of "sul- "phuro-gelatine," or of gelatine, is added to the ordinary iron developer. (See *The Photographic News* of January 19, 1866, p. 26.)

1866. Mr. NELSON K. CHERRILL, on January 9, 1866, read a paper to the Photographic Society of London on "The

A.D.

"new organic developer," in which he proposes to add glycocine to the ordinary iron developer. (See *The Photographic News* of January 19, 1866, p. 28.)

1866. Dr. Roscoe and Mr. Baxendell, on February 22, 1866, read a "Note on the relative chemical intensities of direct "sunlight and diffuse daylight at different altitudes of "the sun." The results arrived at were:—1. That the effect of the atmosphere upon chemical rays is not accounted for by reflection by means of hollow vesicles of water. 2. That the ratio of the chemical intensity of direct to diffuse sunlight for a given altitude of the sun at different localities is not constant. 3. That this ratio of "chemical" intensity does not in the least correspond to the ratio of "visible" intensity as estimated by the eye. (See *Proceedings of the Royal Society*, vol. XV., p. 20.)

1866. Lieut.-General Sabine, on March 8, 1866, read to the Royal Society a "Note on a correspondence between Her "Majesty's Government and the President and Council "of the Royal Society regarding meteorological observa- "tions to be made by sea and land." Self-recording instruments (by photographic registration) are recommended; these comprise instruments for recording (1) the variations of the atmospheric pressure; (2) those of the dry and wet thermometers; (3) those of the force and direction of the wind; and (4) those of the atmospheric electricity. (See *Proceedings of the Royal Society*, vol. XV., p. 29.)

1866. Lieut.-General Sabine, on March 23, 1866, published a Postscript to his "Note," &c. of March 8, 1866, in which a Committee of the Board of Trade report favourably on the advantages of self-recording meteorological instruments. (*See Proceedings of the Royal Society*, vol. XV., p. 37.)

Mr. Charles Chambers, in a letter read to the Royal
1866. Society, on April 26, 1866, mentions that, in his Report to the Government, he has, amongst other requirements, requested a set of Kew Magnetographs for the Bombay Magnetic Observatory. (See *Proceedings of the Royal Society*, vol. XV., p. 111.)

A.D.

Dr. Angus Smith, F.R.S., made a communication to the Photographic Society of Scotland, which was read to the
1866. Society on May 8, 1866, entitled, "On a method of "removing the last traces of hyposulphites from positive "prints." He proposes peroxide of hydrogen for oxidising the hyposulphites. (See *The Photographic News* of May 18, 1866, p. 233.)

1866. Mr. Arthur Taylor, on June 1, 1866, published his shellac printing process. A solution of bleached shellac in phosphate of soda is used to prepare the paper, instead of albumen. Photographs prepared by this process are remarkable for their permanency. (See *The Photographic News* of June 1, 1866, p. 256.)

1866. Mr. John Spiller, F.C.S., on June 6, 1866, read a paper before the North London Photographic Society on "Pho-"tography in its chemical aspects." In this paper he points out the use of carbonate of ammonia in combination with hyposulphite of soda to extract the silver from the albumenised coating of an ordinary finished print. (See *The Photographic News* of June 15, 1866, p. 280.)

1866. Mr. F. W. Hart, on June 14, 1866, read a paper before the South London Photographic Society "On the elim-"ination of the double hyposulphites of soda and silver "from photographic prints." He proposes to use the hypochlorite of soda to oxidise the hyposulphite with. (See *The Photographic News* of June 22, 1866, p. 290.)

1866. Lieut.-General Edward Sabine, on June 21, 1866, read a paper to the Royal Society on "Results of the "magnetic observations at the Kew Observatory.— "No. III." "The instruments employed for the deter-"mination of the lunar-diurnal variation furnish a "continuous photographic registry of the changes in the "direction of a magnet whose motion is limited to a "horizontal plane, and in the amounts of the horizontal "and vertical components of the force acting on a freely "suspended magnet." (See *Phil. Trans.*, vol. 156, p. 441.)

A.D.

1866. Mr. WILLIAM ENGLAND, on June 29, 1866, published his modification of the method of preparing dry resinised plates recommended by the Abbé DESPRATZ. To ordinary bromo-iodised collodion, bromide of cadmium and ordinary resin are added. The sensitising bath requires rectification; ammonia or carbonate of soda and cyanide of potassium are used for this purpose. (See *The Photographic News* of June 29, 1866, p. 306.)

1866. Mr. A. CLAUDET, F.R.S., at the end of August 1866, read a paper to the British Association entitled "Optics "of Photography.—On a new process for equalizing the "definition of all the planes of a solid figure represented "in a photographic picture." This is done by changing the focus during the time of sitting, by altering the distance between the two lenses in a double combination. (See *Phil. Mag.*, vol. 32, p. 212.)

# PHOTOGRAPHY.

## 1860.

A.D. 1860, January 9.—N° 58.

CZUGAJEWICZ, Pierre.—" Certain improvements in stereo-" scopes."

" This invention consists of a compound stereoscope, in which " general or panoramic views of boulevards, streets, the banks of " rivers, and coast lines, monuments, sea views, et cetera, may " be displayed by means of the gradual unrolling of one or more " endless slides or bands carrying pictures."

The distinctive feature of this invention is " the adaptation to " stereoscopes of one or more symmetrical independent movable " endless bands, on which are right and left-hand halves or " corresponding parts of a stereoscopic panorama or succession of " pictures. The following is the construction of the aforesaid " improved stereoscope :—The top thereof consists, as usual, of " two lenses or eye glasses, and the bottom thereof is mounted " on a box containing rollers, on which are wound the before-" mentioned endless slides or bands, on which are printed, pasted, " or otherwise appropriately attached the views or pictures in " panoramic succession, also a train of wheelwork for setting the " aforesaid bands in motion. The aforesaid bands and corre-" sponding parts of the pictures thereon are brought under their " respective eye glasses upon a flat stage or platform over which " the bands pass, so that when set in motion a panoramic stereo-" scopic view or picture is thus obtained." Two methods of imparting motion to the bands are described and shown.

The Final Specification sets forth the following method of " inclining the upper part of the stereoscope as may be desired :" —The case is hinged to a bottom frame or base plate, and to the opposite side, external to the spectator, a " flexible piece " is affixed; a winch handle, working a rack and pinion movement, enables the upper part of the stereoscope to be brought to the desired inclination.

[Printed, 8*d*. Drawing.]

A.D. 1860, January 20.—N° 149.

JOUBERT DE LA FERTÉ, Ferdinand Jean.—(*Partly a communication from Henry Garnier.*)—"Improvements in repro-"ducing photographic and other pictures, engravings, prints, "devices, and designs on the surfaces of glass, ceramic and other "substances requiring to be fired, to fix the same thereon."

This invention is carried out by coating the above-mentioned surfaces "with a stickey composition acted on by light, then "printing on the surface in the manner of photographic printing, "thus destroying or changing the stickey composition where the light acts upon it and consequently causing a suitable coloring matter to adhere to such portions of the surface as remain stickey."

The glass is first coated with a solution which, when dry, furnishes the above-mentioned "stickey composition," it is then heated, so as to dry the solution perfectly, and receives the photographic negative image by "printing" in the usual way. The whole surface of the glass, on the image side, is then rubbed over with suitable vitreous colour by means of a camel-hair brush, and the film on the glass is well soaked with an acid solution of spirits of wine; the glass is then dried. When completely dry, the glass is washed in a weak alkaline solution, and when it is again dried, it is "ready to be placed into the muffle and to be fired."

The photographic image can also be produced in a similar manner on well glazed china, or other ceramic substance, and, by successive firings, or even in one firing, polychromes may be produced.

The above-mentioned solution contains bichromate of ammonium, honey, albumen, and distilled water.

[Printed, 4*d*. No Drawings.]

A.D. 1860, February 23.—N° 493.

BROOMAN, Richard Archibald.—(*A communication from Leopold Eidlitz.*)—"An improved method of producing docu-"ments suitable for bank notes and other instruments of value, "in order to prevent forgery thereof."

This invention "consists in the employment of paper having "an irregular surface as ornamental pressed paper, or of paper "manufactured with a movable or variable water-mark, and upon "these marked portions of the paper, some part of the design, as "the vignettes, words, or denominations, is to be produced by "the photographic process, either directly or indirectly."

To make "hand-made paper with a variable or movable water- "mark," "before the pump is spread upon the web or dipper, "a metal water-mark is dropped upon the web or dipper at "various points at random."

The words, letters, or figures to compose the bank note, may be printed or engraved, by the ordinary process, upon the photographic impression of some changing object.

When the photographic process is to be employed alone, a photographic negative is taken of a screen, on which the letters, &c., composing the note are printed or painted "in connection with "the subject selected for the vignette."

"The photographic devices may appear either on the surface of "the paper or below the surface and within the body of the paper, "and either of them may be produced by means already known. "Upon the surface they may also be made by printing from "stones or plates prepared by the photographic process."

The inventor relies upon the impossibility of taking a photograph of these bank notes, without photographing the water-mark also, as the preventive of forgery.

[Printed, 4*d*. No Drawings.]

## A.D. 1860, February 27.—N° 537.

DESVIGNES, Peter Hubert.—"Improvements in apparatuses "for exhibiting photographic, stereoscopic, and other pictures, "models, figures, and designs."

The object of this invention "is to show in one apparatus a "series of stereoscopic or other representations or models taken "in different views, as if the same or the things represented were "in motion."

The apparatus "are cylindrical or polygonal, and are caused to "rotate either vertically or horizontally."

To exhibit stereoscopic views, a certain number of them are placed within a cylinder whose periphery "has formed across it as "many slots as there are views." The glasses for viewing the pictures are "held stationary outside of the periphery of the "cylinder." "The views may represent, for instance, a steam "engine, and each view must be taken when the engine is at "different parts of its stroke. The views being placed in the "said cylinder, and the cylinder being caused to rotate, will show "to the eye the steam engine as if in actual movement in all its "parts."

To exhibit models and drawings, they are arranged "round a "cylinder placed and caused to rotate horizontally; vertical slots "are made in the periphery to see through." In some cases, a sort of fixed camera, in combination with a cylinder and disc or discs is used.

Endless bands and other combinations of moving slots and pictures are described and shown in detail; the theory of these apparatus is amply set forth, and methods of varying the optical effects are fully stated.

Various methods of shading extraneous objects from the observer, and of illuminating the views or figures under examination, are described. Electric sparks, elicited at the proper time, may be used to make every change of figure, "visible at its proper time "and place."

[Printed, 1*s*. Drawing.]

### A.D. 1860, February 28.—N° 543.

ASSER, Edouard Isaac.—"A process of photographic proofs "with printing or authographic ink, for the purpose of either "using them as such, or placing them back on lithographic stones "or on metal."

"I take paper prepared without size, and steep it in a solution "of bichromate of potash; I then dry it in the dark, and by a "negative on glass or other transparent negative, placed in an "ordinary frame and exposed to the light with the paper, pre-"pared as above described, I bring the object upon the paper. "The impression thus obtained, I dissolve in water the bichro-"mate of potash upon which the light has not struck. After "perfectly drying the paper bearing the impreseion, I wet it on "the back, and place it upon a paper not sized, but damped and "stretched upon a glass or other flat and hard object, the im-"pression being uppermost; I then pass over it a roller charged "with printing ink until the impression shews well in black. I

" then steep the impression in water containing a little nitric or " other acid, which will dissolve the bichromate of potash still " upon the positive, and remove it from the paper. After drying " the positive is ready." " By passing a roller charged with " authographic ink over the impression obtained by bichromate " of potash, and treated as above described, I obtain a stereotype " suitable for transfer by pressure on to a lithographic stone, " from which I draw, by the process known as lithography, auto- " graphic prints; this transfer may be made in like manner upon " any of the matters which may be used on the same principle " for every description of autographic printing, as, for example, " zinc or copper, either for reproduction after what is usually " termed zincography, or by the aid of galvanoplastic, hollow, " or in relief. In some cases the inking may be effected with " a rubber instead of a roller, as above described."

[Printed, 4*d*. No Drawings.]

A.D. 1860, February 29.—N° 559.

SWAN, HENRY.—" Improvements in stereoscopes, stereoscopic " pictures, and cameras for taking the same."

According to this invention stereoscopes are constructed with lenses of different magnifying powers, and are used " in conjunc- " tion with pictures in which the two views forming the stereo- " scopic combination are of different sizes, corresponding with " the different powers of the prisms or lenses opposite the two " eyes;" these pictures are connected together " by attaching " them both in suitable positions to the same mounting." In this manner a stereoscopic effect is obtained by the use of one large picture only, the said effect being " combined with minute- " ness of detail, such as a large picture alone can give;" the instrument is arranged in a more convenient form than that of the instrument heretofore employed for viewing large pictures.

" Instruments may be constructed having one prism or lens " only, the large picture being then seen without the aid of " a prism or lens. Magnifying mirrors may be employed in place " of prisms and lenses."

" My improvement in cameras has for its object the so arranging " the same that a large and a small picture forming a stereoscopic " combination, as above explained, may be taken at the same time " and on the same plate or surface. For this purpose the camera " is fitted with two lenses, the focal length of the lens for taking

" the larger picture being greater than that of the lens for taking " the smaller picture; these lenses are so placed as to focus on " the same plane or flat surface, and are furnished with separate " adjusting screws or means of adjustment."

[Printed, 4*d*. No Drawings.]

## A.D. 1860, March 2.—N° 581.

ALBITES, PIERRE MODESTE TITUS OCTAVE COEN.—(*Provisional protection only.*)—" Improvements in photographic ap- " paratus."

" The object of this apparatus is to simplify the work of the " photographic manipulator, for which purpose the ordinary dark " chamber and travelling tent are dispensed with. The whole of " the process being done in the dark chamber of the apparatus, " and as it is done mechanically, the work is performed more " regularly, and it is more accessible to every one."

The drawings show the sensitizing bath, a movable slide to hold the sensitive plate and put the ground glass in position, two grooves "fixed to the drawer of the apparatus" to move the said slide backwards and forwards, the positive developing bath " placed in the interior of the apparatus," a cord and pulleys to manipulate the slide, and the negative developing bath.

[Printed, 6*d*. Drawing.]

## A.D. 1860, March 21.—N° 735.

NEWTON, WILLIAM EDWARD.—(*A communication from Joseph Chase Rutherford and Benjamin Hinman Steele.*)—(*Provisional protection only.*)—" An improved preparation or solution for " toning photographic pictures."

" The improved preparation or solution for toning photographic " pictures which forms the subject of the present invention is " composed of corrosive sublimate, tartaric acid, sal soda, hydro- " chloric acid, and aqua distillata. This solution after well " mixing is left standing for forty-eight hours, and after filtering " is ready for use. The picture when taken from the printing " board is passed through the toning bath, then immediately " immersed in a clean water bath to give it a thorough rinsing. " As a rule this will be sufficient to tone a picture, but when it is " printed down very dark it may be necessary to go through the " same manipulation the second time. Care should be taken to " have the print well washed before it is put into the fixing bath

" The advantages of the above toning solution are 1st., that " it greatly cheapens the process of toning pictures; 2nd., that " the toning of a print with this solution is an instantaneous " process thereby saving a great amount of valuable time; 3rd., " that it gives a sharpness and distinctness to the picture; 4th., " that it can be worked in a clear light without injury to the " toning solution or the print; and 5th., that the colors are " permanent."

[Printed, 4*d*. No Drawings.]

A.D. 1860, March 28.—N° 801.

DAGRON, Aené Prudent Patrice. — " An improved micro- " scope to be used for exhibiting photographic views and " productions."

The eye glass is fixed in a socket placed in a slide holder, the socket being surrounded by a ring, by which the eye glass can be adjusted to the proper focus. A stud in the said socket, working into an inclined notch in the slide holder, regulates and adjusts the focus. The slide holder is movable, and consists of two tubes, which support the glass in which the microscopic image is to be placed.

A microscope in solid glass is " of a cylindrical form, rounded " at both ends, or it may be divided into two parts, one part " serving for the eye glass, and the other for the object glass; " the image to be viewed being placed between the two glasses, " may be fixed thereto by means of any suitable adhesive sub- " stance or material. In adapting the microscope to a ring to be " worn on the finger the stone or centre ornament of the ring is " made to open on a hinge, and the microscope is fixed inside the " lid, or cover of the hinged part." Two modifications of the miniature microscope are thus obtained, " in which the image may " be either permanent or variable, one being mounted in a me- " tallic setting the other formed of solid glass." Either of these instruments, furnished with a permanent object, may be set in jewellery, &c., " so as to obtain artistic or useful observations with " or without stereoscopic effect."

In a double-image microscope, the eye glasses are fitted at each end of a cylinder, and the glasses carrying the images are placed at the focal point of their respective lenses within the cylinder;

the image only occupies the centre of the glass to which it is fastened.

[Printed, 8*d*. Drawing.]

A.D. 1860, March 30.—N° 823.

BEAU, Paul Adolphe Auguste.—(*Provisional protection only.*)—" A neomonoscope or apparatus for viewing photographic " and other like pictures."

" My invention consists in constructing a neomonoscope or " apparatus with one glass or several glasses superposed for the " purpose of obtaining a similar effect to that derived from view- " ing pictures in or through a stereoscope. The monoscope is a " pyramidal or conical-shaped case, with a part of one of the " sides removed to admit light, fitted with a flap or not as desired. " The glass is fitted in the top of the apparatus, and in some " cases flaps for forming, when raised, a dark chamber between " the eye and the glass, are added. The bottom of the apparatus " is made to slide to admit of its being entirely removed, in order " to view transparent objects or others apart from the apparatus " itself. The sides of the apparatus are either made rigid or to " fold. I sometimes add a pocket which forms part of the sides " or bottom of the apparatus to contain photographic or other " representations."

[Printed, 4*d*. No Drawings.]

A.D. 1860, April 9.—N° 893.

EIDLITZ, Leopold. — (*Provisional protection only.*)—" Im- " provements in producing printing and other irregular surfaces " by the aid of photography."

A hot solution, containing gelatine and bichromate of potash, is poured upon a sheet of glass, so that when dry it will form a thin film. The edges of the dried plate are coated with varnish, and the plate put into the printing frame under a photographic negative. When the surface of the plate is of a brown colour, it is " immersed for about 5 minutes in a bath of cold water. " The effect of this will be that where the lights were strongest " the gelatine will have been rendered incapable of being ex- " panded by the action of the water, while where the shades were " deepest the amount of that expansion will be the greatest, and " this difference will be on all parts of the plate just as the plate "at each part has been exposed to more or less light." The plate is now left in an aqueous solution of sulphate of copper for some time, washed, dried on the surface, and metallized; this is effected by pouring upon it a solution of chloride of gold, and exposing the apparently dry surface to the fumes of phosphorus dissolved in sulphuric ether; the layer of gold thus obtained enables the plate to receive an electro-deposit of copper or other metal upon it. "When removed from the trough, the glass will readily "separate from the gelatine, and that may be washed off, using "warm water and a soft brush."

"If the plate is to be used for letter-press printing, a positive "photograph should be employed in the printing frame, or a cast "in wax may be taken from the gelatine plate, and the metal be "deposited upon that." An engraving or other design may be employed in the printing frame instead of the photograph.

[Printed, 4*d*. No Drawings.]

## A.D. 1860, May 22.—N° 1260.

SHAW, William Thomas.—"Improvements in thaumatropes "or phenakistoscopes."

This invention "consists in the application of the principle of "the stereoscope" to instruments "which depend for their "results on 'persistence of vision.'"

"The thaumatrope and the phenakistoscope, as is well known, "give to objects shown by them an appearance of motion; by "my improvements I cause these objects to appear also in solid "relief, as when seen in the ordinary stereoscope. I take from a "suitable object a number of photographic pictures, the pictures "being taken in pairs having a stereoscopic relation the one to "the other as is well understood. Between the taking of each "pair of pictures the object is caused to perform a portion of the "movement which it is desired to represent, the first pair of "pictures being taken at the commencing point of the movement, "and each successive pair showing the same somewhat further pro-"gressed than the previous pair, until the last pair shows the "object at a point just before it regains its first position. "Having obtained these pictures, I view them by means of a "stereoscope, and by mechanism I cause the pairs of pictures to "be changed rapidly, each pair being submitted to the eyes in "succession for a moment of time. When thus seen an appear-"ance is obtained as of a solid body in motion."

An apparatus "on the principle of the reflecting stereoscope" has (besides the synchronously revolving "picture" discs) revolving eye discs, which limit the time during which the pictures can be seen. In a refracting instrument, a revolving octagonal drum carries the pictures, and a revolving slotted drum limits the time of sight of the pictures as well as the sight of each eye to the picture which is intended for it.

[Printed, 1*s.* 4*d.* Drawings.]

### A.D. 1860, May 25.—N° 1299.

WALLIS, George.—"New or improved methods of preparing "drawings, writings, designs, prints, or impressions of engrav- "ings and photographs, for the purpose of impressing or "engraving the same in or upon metallic surfaces, and thereby "producing printing or embossing surfaces or ornamental "metallic surfaces for such purposes as the same are or may "be applicable to; also new or improved machinery to be "employed in the said impressing or engraving."

In treating photographs according to the first part of the invention, two methods are used; by one, all the parts it is wished to impress in metal are drawn on with a certain drawing material; by the other method, the photographic image itself is caused to stand out in relief. The drawing material by which the relief is obtained in the first method "is composed of finely "pulverized or levigated hard or granular substances in combi- "nation with soft mineral substances and glutinous or adhesive "materials, and a soluble chromic salt for fixing the same." In the second method, paper or other suitable material is coated with a compound containing bichromate of potash, gum arabic, or gelatine, and a granular powder; or a metal plate may be coated with a mixture of these with water. The coated surface is then dried, exposed to light, and washed; the parts to be impressed then stand out in relief.

The second part of the invention "consists essentially of a "pair of plain rolls mounted upon horizontal axes." Between the rolls a horizontal table with a hard metal bed is made to slide. To impress the photograph upon a soft metal plate, it is placed on the said hard metal bed with the prepared surface uppermost; upon this the soft metal plate is placed, and the table is passed between the rolls. The impression thus produced may either be used as a printing surface or as an ornamental metallic surface.

[Printed, 1*s*. Drawing.]

A.D. 1860, June 14.—N° 1454.

HENRY, Michael.—(*A communication from Louis Henry Obert, Jean Baptiste Vasseur, and Auguste Houbigant.*)—" Improvements " in treating vegetable substances so as to obtain paper pulp and " other useful products therefrom."

The products obtained by treatment of the said substances according to this invention, consist of " materials for textile and " fibrous manufacturing purposes, alcohol, essential, and other oils, " fecula and gums, colouring, saponaceous, and tanning matters, " &c." By means of the " successive and simultaneous " processes and mechanical operations comprised in this invention, the Specification sets forth that the parties communicating " have obtained from all, or nearly all plants the materials of photographic " paper met with in cotton, and thus collodion and photographic " paper can be manufactured from textile and fibrous plants " other than cotton."

The said substances are first crushed " by means of an arrangement of rollers one pair over another, in such manner " that the substances may pass down and between them, together " with a continuous stream of liquid, either clear or charged with " matters, whereby they are caused to travel along and away " more readily ;" the fibrous substances, materials for paper pulp, and the liquid products are then separately collected. The disintegration and decolorization are accomplished " more especially " by the action of steam," in a " closed receiver or boiler, in " which also " the vegetable substances " are distilled by steam, " when desired, and the liquors used in the operations performed " in the boiler are " " run into " a " soaking vessel to be used for " soaking or steeping the substances, instead of alkaline " salts, and for being again returned into the boiler or closed " receiver for fresh operations therein, until sufficiently concen- " trated."

The machinery for carrying out this invention is shown as it would be arranged in a factory, and the use of each apparatus in the successive stages of operations is distinctly described.

[Printed, 1*s*. Drawing.]

A.D. 1860, June 21.—N° 1503.

SMITH, John.—" Improvements in the manufacture of com-
" position jewellery and ornaments, and in cases for jewellery,
" photographs, and for other similar purposes."

This invention consists of certain improvements upon that set forth in No. 2891 (A.D. 1859).

The said composition consists of shellac dissolved by heat and mixed with ebony dust; the whole is then treated with the colouring matter. If a very intense black is wanted, and at the same time hardness, a mixture of black asphaltum and ivory black is used; if hardness be not desirable, charcoal or lamp-black takes the place of ivory black. A mixture of brown asphaltum and rouge gives chocolate. To produce light colours, the dust of boxwood is employed " as a colouring matter in itself;" if a still lighter tint be required, barytes may be added. Veined composition may be produced by twisting or rolling together masses of different coloured compositions.

The components of the composition having been well mixed whilst the lac is in a plastic state and under heat, the composition is stamped between suitable dies.

The quality of density may be imparted to the composition by barytes or oxide of manganese. The admixture of vegetable fibre will give great tenacity.

The dies and presses employed " are of the same kind pre-
" cisely as those now in use by the manufacturers of horn
" buttons."

[Printed, 4*d*. No Drawings.]

A.D. 1860, July 24.—N° 1788.

MACAIRE, Louis Cyrus.—(*Provisional protection only.*)—
" A substitute for nitrate of silver, particularly applicable to
" photographic purposes."

" In forming my substitute for nitrate of silver, I propose
" using, in combination with the pure silver and nitric acid now
" employed, certain quantities of azotate of potash and a small
" portion of a baser metal, such as copper, tin, zinc, or bismuth.
" These materials mixed in the proportions below mentioned I
" have found to answer well: pure silver 1·, nitrid acic 1·50,
" azotate of potash 1·50, zinc or other metal ·1, this combination
" producing 3·1 " [4·1 ?].

" The substitute thus produced is cheaper and more efficient " for photographic purposes than the ordinary nitrate of silver " now in use."

[Printed, 4*d*. No Drawings.]

A.D. 1860, July 31.—N° 1860.

WILLCOCK, Joseph.—(*A communication from Pierre M. T. O. Coen Albites.*)—" Improvements in photographic apparatus."

According to this invention " the ordinary dark chamber and " travelling tent are dispensed with." The whole of the process is gone through " in the dark chamber of the apparatus, and as " it is done mechanically, the work is performed more regularly, " and it is more accessible to every one."

A movable slide drops into the sensitizing bath; the said slide serves to hold the sensitive plate and to put the ground glass in position; it can be moved backwards and forwards by means of " two grooves fixed to the drawer of the apparatus." " The " positive bath" is made of coloured glass, " it is placed in the " interior of the apparatus, and receives the glass after the im- " pression by the light, and developes it instantaneously." A cord passing over two pulleys, is employed " to move the slide " carrying the prepared glass, and to dip it into the sensitive " bath." An india-rubber framework serves " to prevent the " light from penetrating into the apparatus, and to detach the " silver hook that holds the ground glass, and slide it into the " bath." The negative and positive developing bath " permits " you to judge the progress of the developement of the image." " When required, a hole may be made in top partition of the dark " chamber of the apparatus, and in the objective or in the last " only because a ray of light penetrating by this hole upon a " mirror that should be placed in the objective, and which reflects " this ray upon the ground glass without destroying the ray upon " the first glass upon which is the image, thus rendering the " production more instantaneous."

[Printed, 8*d*. Drawings.]

A.D. 1860, August 8.—N° 1922.

FLOUNDERS, Charles Fontayne.—" An improvement or " improvements in duplicating photographic impressions, and " also for certain machinery for the same."

This invention consists of a machine for photographic printing. The sensitive material may be exposed to light "until the impres-" sion is fully made," or it may be exposed for a shorter period and then developed by chemical means.

The sensitive paper or other material "is confined in a dark " space, and, by mechanical means, is made to traverse an aper-" ture through which ordinary or concentrated light, or the " image of the negative is received. The light is admitted and " shut off by a perforated slide, which passes over a similar or " other suitable perforation in a fixed part." The sensitive paper may be applied "to a cylinder or polygonal prism within the dark " space, and the same is then made both to revolve and to move " axially" until the paper is entirely covered with photographic impressions. "Springs, slides, levers, and various mechanical " devices are employed, either separately or mechanically com-" bined, to impart the proper motions to the several parts, so that " the sensitive material shall be caused to traverse during the " interval while the light is excluded, and to present a fresh sur-" face firmly held in its proper position during the period that " the light is admitted."

[Printed, 8*d*. Drawing.]

A.D. 1860, August 22.—N° 2022. (* *)

GRUMEL, François Remy.—"Improvements in the albums of " collection of photographic and lithographic proofs, engravings, " and other drawings."

Each leaf of the album is composed of three layers, by preference of thin cardboard; the outside ones "have a portion of " their central parts cut out in form a little less than the cards" to be inserted, and the middle one "has also a portion cut out in " the centre, but larger than the other two, and in addition it " has a portion cut away from the bottom." The edges of the layers are glued or cemented all round, "save where the portion " of the middle layer at its bottom edge is cut away;" care must be taken that the central spaces coincide. The card is pushed through the bottom opening to the central spaces, and if two cards are inserted, they must of course be placed back to back. "A small slip of cotton or skin attached to one side forms a " hinge or substance to bind the leaves so as to form a book." Each leaf may be made to "contain several spaces, each holding " cards."

[Printed, 6*d*. Drawing.]

A.D. 1860, October 13.—N° 2496.

BROOMAN, RICHARD ARCHIBALD.—(*A communication from Charles C. Harrison.*)—" Improvements in lenses."

This invention consists of a compound lens, which is " de-" signed primarily for use in cameras for taking photographs, " daguerreotypes, and other pictures, but is also adapted to " microscopes, telescopes, and similar instruments."

The lens is formed by " an achromatic combination of two " lenses which are meniscuses, the outer surfaces of which are " segments of similar spheres, and which lenses are so placed in " the combination that their axes shall coincide, and their outer " surfaces form portions of a sphere of like diameter."

The achromatic combination, of which each meniscus consists, is a meniscus of flint glass, to the outer surface of which is cemented a concavo-convex lens of crown glass; the achromatic combinations are so mounted for use that their axes coincide, " and their surfaces if produced would form a complete globe;" the convex surface of the crown glass lens forms a portion of the said " complete globe." The centre of the sphere is the focal point of each compound lens, and at this point a diaphragm is fixed.

Any object being properly placed, will be exactly reproduced, on the opposite side of the combination, of a size proportional to the respective distances of the object and of the image from the lens; for, " whatever be the angle at which the rays are refracted to the " centre " " of the sphere, they will so pass out through the " opposite lens as to restore the angle of incidence of the entering " rays, and will thus unite to produce the image upon a plane " parallel with that from which the incident rays proceeded."

A flat field of 90° range is obtained by this lens.

The mounted combination " forms the object glass of the " camera."

[Printed, 8*d*. Drawings.]

A.D. 1860, October 29.—N° 2639.

KNIGHT, JOHN ADAMS.—(*A communication from Charles Felix Korn.*)—(*Provisional protection only.*)—" A new system of photo-" graphic or daguerrean apparatus or objective to be called Korn's " polyograph."

" This invention relates to photographic or daguerre apparatus " or objective which may be used for both near views (such as

" portraits, &c.) and for distant views (such as landscapes and " reproductions) by means of an easy change in the relative " position, or removal of the different parts of the apparatus or " objective.

" The apparatus or objective used for taking near views (or " portrays) in the ordinary way having two lenses, the outer one " of which has a wide broad inlet, and for distant views (such as " landscapes and reproductions) having only one lens with a " smaller inlet of light limited by diaphragms of various sizes. " I do not change the system of objectives used in these two " different cases.

" But the invention I have made consists in so mounting the " said outer lens on suitable tubes fitted with diaphragms and " screw threads that {supposing said apparatus or objective to be " fixed on the camera obscura in the usual way for taking near " views (such as portrays)} the outer lens with its tubing and " diaphragms or place for the same, may be unscrewed from the " tubing appertaining to the inner lens, and the inner lens with " its tubing and appurtenances being unscrewed from the camera " obscura, the outer lens with its tubing, diaphragms, and appur- " tenances may be inverted or turned round bodily and screwed on " the camera obscura, thus forming or procuring at once a perfect " apparatus or objective for taking distant or extensive views, such " as buildings, landscapes, and reproductions of large objects or " paintings and vice versa."

[Printed, 4*d*. No Drawings.]

A.D. 1860, November 13.—N° 2772. (* *)

WILLIAMS, VALENTINE VAUGHAN.—(*Provisional protection only.*)—" An improved method of constructing stands for cameras, " telescopes, surveying and other instruments, parts of which are " applicable to other purposes."

This invention " consists in constructing a supporting tripod " stand, so designed that it is capable of being closed up as a " walking stick." " A supporting tripod is formed out of a " metal tube divided lengthways into three equal parts which are " united by suitable joints at their upper ends to a small tubular " socket through which socket passes a metal tube, this tube is " continued at its lower end by a small metal rod so as to form a " central or fourth leg." At the end of the central leg is a tapering screw, and the extremities of the other legs consist of points and shoulders. A hollow metal slide works in the central tube or leg, the said slide having a rack. The skeleton table top is fitted to the upper end of the slide.

The mode of action of the instrument is as follows:—

When the case, &c. is removed the tripod legs are opened, and the middle leg is shifted up and screwed into the ground by hand, being made fast by a thumb screw; the top is then opened out and secured by an elastic band. "The camera can then be raised "by the metal slide which gives the horizontal movement while "the vertical movement is obtained by means of a joint situated "between the table top and the socket."

A slight modification is made to adapt the top to telescopes of various diameters and descriptions.

[Printed, 4*d*. No Drawings.]

A.D. 1860, November 17.—N° 2832.

MACFARLANE, HUGH.—(*Provisional protection only.*)—"Improvements in cameras such as are used by photographers."

"This invention relates to improvements in cameras such as are "used by photographers, such improvements whilst of a simple "and inexpensive character permitting of a slight inclination "being given to the picture plate, which is in many cases of great "benefit as is well understood by photographers. In a camera "embodying the improvements the back portion of the box, "instead of being as ordinarily rigidly fixed to the inner shell "which is drawn out with the back, has such inner shell entered "loosely a short distance and attached by a single screw or pin at "the top, and another at the bottom, so that the back can be "inclined slightly in either direction about a vertical axis. The "back of the box can be fixed at any inclination by means of a "screw working in connection with the guide in the back supporting flap. A slight inclination about a horizontal axis is "obtained by means of a pair of levers inside the box, against "which levers the frame for holding the picture plate bears, these "levers being inclined by means of pins projecting through "slots to the outside, and being fixed by means of external screw "nuts."

[Printed, 4*d*. No Drawings.]

A.D. 1860, November 24.—N° 2881.

DALGLISH, ANDREW ADIE.—"Improvements in engraving or "for producing printed surfaces."

This invention relates to the application of photography, in conjunction with certain other processes, to the above-named purposes.

In producing printed surfaces upon the "cylindrical dies used "in preparing rollers for calico printers," a photograph of the design is transferred to the circumference of the die, which is coated with etching varnish for that purpose. The design is "then etched or engraved through the film," the coating of varnish entirely removed, and a fresh coating put upon the die up to the etched outline. Lastly, the die is immersed in acid "which eats away the metal to the required depth," and, when necessary, is finished by an engraver. When there are several colours in the design, an impression is taken from an original die (to which the photograph has been transferred, and which has been etched), "upon as many dies as there are colours to be "printed, the surfaces of these several dies having been pre-"viously varnished;" each die is then prepared, as set forth above.

In another method a "mill" is made from the above-mentioned original die, "which is caused slightly to impress the surfaces of "the several dies required for the separate colours," and those outlines are erased which are not required.

In a third method a photograph is taken for each colour.

In applying this invention to ordinary printing from metal plates, the metal is re-varnished after the outlines are engraved in order to engrave the shading.

An apparatus for transferring photographic films to cylindrical surfaces is described and shown, which consists of a plane table on which the photograph is fixed, and over which the die or roller is made to traverse uniformly by means of a rack and pinion movement.

[Printed, 8*d*. Drawing.]

### A.D. 1860, November 27.—N° 2913.

BEATTY, Francis Stewart, and ALEXANDER, Thomas.—"Improvements in the production of photographic proofs and "their application to printing purposes."

This process is called "heliographic printing."

Lithographic transfer paper is treated with a solution containing bichromate of potash and gelatine. The photographic image having been impressed, the surface is carefully covered with printer's ink and allowed to dry; it is then placed in hot water, which swells the composition, and dissolves the bichromate of potash and gelatine which has not been exposed to the light; the photographic image is thus fixed and retains the printing ink on its surface. The greasy ink on the unlighted portions of the proof is removed by rubbing with a sponge in warm water. The printer's ink left upon the proof, and forming the design, may be transferred to zinc or stone for printing.

According to a modification of the above process, a polished surface is coated with the transfer medium and sensitive coating, exposed to light, and washed. A plaster cast of the resulting proof is hardened by immersion in "liquid silex," and employed as a printing surface.

Printing surfaces may also be obtained by floating the sensitive coating alone upon the surface of a metal plate, exposing it to light, covering the whole plate with an acid-resisting composition (printer's varnish, asphalte and lampblack), washing the said composition from the illuminated portions of the proof, and acting upon those parts with acid.

When a sensitive solution is applied to a lithographic stone, the said stone is hardened by means of "liquid silex," or "silicate "zopissa."

The transfer and sensitive compositions may be applied to other surfaces besides those of paper with the admixture of pulverized mineral substances. The adaptation of grained plates to the invention, and printing in colours, are set forth in detail. Other details and modifications are described.

[Printed, 6*d*. Drawing.]

## A.D. 1860, November 28.—N° 2923.

GILLETT, Henry.—(*Provisional protection only.*)—" Improve-" ments in the ornamentation of the edges of the leaves of pho-" tographic albums especially intended for 'cartes de visite.' "

" The object of this invention is to ornament or illuminate the " edges of the leaves of albums generally, but more especially " those containing photographic pictures known as 'cartes de " visite.'

" I propose ornamenting such edges with various designs, " such as flowers, designs in the style of Arabesque, Italian, Gre-" cian, Alhambran, or any other style suitable for such purpose,

" either plain or in colours, or gold, or gold intermixed with " colours, such ornament being either printed, stencilled, painted " by hand, transferred, or placed upon the edges by any conve- " nient process."

[Printed, 4*d*. No Drawings.]

A.D. 1860, December 10.—N° 3024.

CLARK, William.—(*A communication from Charles Gustave Anthoni.*)—" Improvements in photographic apparatus."

This invention comprises a " mechanical laboratory " connected to the dark chamber of the camera used in photographic operations; by means of this invention damp collodion can be operated upon in the open air. The nitrate of silver or sensitizing bath " is at the opposite side to the object," and the sulphate of iron or developing bath " is placed at the under part of the dark " chamber." The collodionized glass, after having been fixed on certain hooks, is treated with the nitrate of silver solution by inclining the bath, the bath being hinged to the dark chamber for that purpose. The sensitized collodion film is then submitted to the action of light. In order to dip the impressed plate into the sulphate of iron bath, the apparatus is inclined, the hooks lowered and again lifted, and the plate allowed to descend so as to lie in the said bath, being guided in its descent by lateral guides; a small door in front of the apparatus is then opened, the trough containing the sulphate of iron solution removed, and the plate taken out therefrom by means of a hook. The sheet of glass is finally washed with water, or placed in a grooved box previously filled with water. The motion of the above-mentioned hooks, and a certain rocking motion imparted to the baths, is accomplished from the outside of the dark chamber, by means of spiral springs and ratchets connected with the axes of the hooks, and by means of slotted links.

An arrangement for taking stereoscopic views is described, in which certain partitions are added to the collapsible chamber.

[Printed, 8*d*. Drawing.]

A.D. 1860, December 13.—N° 3073.

MELLO, John Arnold.—(*Provisional protection only.*)—" Im- " provements in the manufacture of stereoscopic slides."

" This invention has for its object improvements in the manu- " facture of stereoscopic slides. In order to give a more natural

" appearance to the sky of opaque stereoscopic slides which " usually appear white, or nearly so, I cut away such portions of " the pictures as represent sky, going very carefully up to the out- " line of the objects represented in the pictures. Then in viewing " stereoscopic slides thus manufactured I employ a stereoscope " with an open or transparent stage, and I allow light to pass " through this stage, as is usual in viewing transparent stereo- " scopic slides; this light I color by causing it to pass through a " colored medium, and in this manner I obtain the appearance of " a transparent sky, either blue or of other tint or tints, which " may be desired. By the same method I give a more natural " appearance when water is represented, or when objects are repre- " sented which produce or reflect a bright light."

[Printed, 4*d*. No Drawings.]

A.D, 1860, December 24.—N° 3153.

GIBBONS, William John.—(*Complete Specification, but no Letters Patent.*)—" Improvements in stereoscopes and their cases."

1st. An instrument which combines the stereoscope with the case that contains the slides.—A box having a lid, which may be secured by a lock, contains the slides, and has such fittings as are necessary to view them stereoscopically. A "lens plate" is hinged to the inside of the front edge of the lid, and a " bracket " piece" or stay (serving " also as a separator for the two pictures " to be viewed") is hinged to the lid so as to unfold itself at right angles to the lens plate. A sliding carrier, furnished with spring holders, receives the picture; it "is adjustable, sliding in " grooves at each end of the case; it can be advanced or with- " drawn at will, and thus this form of stereoscope accommodates " itself to different visions, allowing a long or short focus, as may " be desired." In closing the instrument, the bracket is folded down into the lid of the case, the lens plate is folded upon the bracket, the carrier is advanced to the back of the box, and the pictures are laid in the bottom of the box.

2nd. An improved "folding stereoscope" without the case.— " It consists simply of a back plate" "upon which is hinged a " bracket" " and lens plate," " the carrier" " for the slides being " placed at the bottom edge of the back plate. On packing away " this instrument" it folds in manner precisely similar to that set forth in the 1st improvement, the bracket "folding first upon " the back plate, and then the lens plate upon the bracket."

The sliding carrier described in the 1st improvement may be used in the "folding stereoscope."

[Printed, 6*d*. Drawing.]

A.D. 1860, December 28.—N° 3181.

PALLU, CLAUDE.—(*Provisional protection only.*)—" Improvements in the apparatuses and process for producing photographic pictures without working in dark rooms."

" The camera obscura which I make use of is composed of two boxes exactly adjusted in one another, so that the smaller one can slide in the larger, the latter carrying the object glass, and the first one the ocular, so that when the prepared plate is placed vertically in the sliding box it can be easily brought to the very point or focus of the object glass,. The plate intended to receive the picture, whether glass, metal, or paper, is prepared as usual with the chimical agents used by photographers, but the last preparation for producing the action of the light is made in a closed gutta percha box or vessel impervious to light, of a size corresponding to the one of the camera obscura; diluted nitrate of silver is placed in that vessel, which is closed on the top by a narrow sliding board; the plate is placed in a small case closed at its lower part by a little sliding plate, carrying the plate prepared for dipping in the solution contained in the gutta percha vessel, over which the case is placed, instead of its sliding lid, which is removed; then the lower slide being drawn out, the prepared plate drops down in the solution, where it is left as long as required, and raised up back in the case by means of a thin hooked silver wire running through a small hole. When up, the lower slide is pushed back to hold it and hide the light, the upper case is then taken out and carried over the camera obscura, where a similar disposition of sliding lids permits also exclusion of light; now the lower slide of the plate case being drawn out, the photographic plate drops down into the camera obscura, falling into its right place, when, by uncovering the object glass for a few seconds, a positive picture is produced with great correctness, even to the most minute size."

[Printed, 4*d*. No Drawings.]

# 1861.

A.D. 1861, February 13.—N° 360.

BROWN, William.—" Improvements in the manufacture of " frames suitable for containing photographic and other portraits " and pictures."

" My said improvements consist in making the said frames " whole or entire from clay, and by preference I use that kind of " clay used in the making of tobacco pipes, and known as pipe " clay; but compounds of clay and other substances which are " capable of receiving a very sharp impression of the mould may " be used. The clay is taken in a moist state, and is forced into " a mould to give it the desired shape and pattern. The said " mould should be slightly oiled on the inside, and the back of " the clay should be covered with a tin plate, to which the clay " will adhere and facilitate the withdrawal of the frame from the " mould. The moulds when filled are subjected to pressure in a " suitable press, to give strength to the frames and sharpness to " the impression; after leaving the press the frames are removed " from the moulds and placed in a kiln and burnt to give them " the necessary hardness, after which they may be gilded, or other- " wise ornamented as desired."

[Printed, 4*d*. No Drawings.]

A.D. 1861, February 23.—N° 469.

POHL, Ludwig.—(*A communication from Paul Luz.*)—" Improve- " ments in albums or books for holding photographs, engravings, " and other representations, and in binding together sheets or " pieces of pasteboard or such like stiff materials, especially for " the purpose of forming such albums or books."

The pages have no raised parts on their surfaces, so that when the book is closed the pages lie flat over one another.

1st. To facilitate the sliding out of the photographs.—" At the " under edge of the slit or opening contrived for the purpose, and " on the inner side, I attach a strip or piece of paper or equivalent " material by another strip placed over their junction, (that is, the " part where they touch,) in such manner that this arrangement " destroys the obstructive effect of the edge, the strip first named " merging as it were into the same plane with the cardboard or "material of the sheet to which it is attached, and acting as if it "were a continuous surface with it, so that the picture or such like "article may slide smoothly over it when being drawn through."

When there are pictures or the like on both sides of the sheet, "this arrangement also serves to keep them separate."

2nd. To bind together sheets of pasteboard.—The following mode is adopted, which is especially applicable to photographic albums:—"I attach the thick sheets to be bound to folded strips "or pieces of leather, cloth, or similar flexible material, which I "call connecting strips, each strip connecting one side of one "sheet with the adjoining side of the next; I bind or connect "together a required number of folds of paper or equivalent "material; I attach the back of the leather or other flexible con-"necting strips to and between the sets of paper (or equivalent) "folds. The order of the operations may be varied."

[Printed, 8*d*. Drawing.]

## A.D. 1861, February 27.—Nº 508.

HENRY, Michael. — (*A communication from Gaspard Felix Tournachon called Nadar.*)—(*Provisional protection only.*)—"Im-"provements in photography."

"According to this invention I combine the employment of "electric light or of gaslight with Moitessier's mode of obtaining "positives." Effects dependent upon the control of the artificial light may be obtained, and the operator may dispense "with a "separate room for preparing and developing his plate as he can "direct the circumscribed rays of the reflector on the objectivity "(objectivité), for the use of this artificial light permits the per-"formance of various operations that have hitherto been per-"formed away from the room exposed to the full daylight. I "propose to modify, as follows, Moitessier's method to adapt it "for purposes of taking positives by artificial light. Hitherto in "known processes frames only have been employed, whereas I "propose to use two dark chambers, in one of which are placed "the negative plates, or stereotype plates, or clichés, and in the "other chamber are placed the glasses treated with collodion "intended for the reproduction of the positive stereotype or "cliché. The two dark chambers are placed opposite one another, "and the space between them is kept dark by a covering; the "chambers can be brought mutually nearer or further according "to the sizes to be produced; the light is placed in front of the

" first dark chamber and concentrated by a reflector on the " negative plate, stereotype plate, or cliché."

[Printed, 4*d.* No Drawings.]

A.D. 1861, April 6.—N° 852.

KNIGHT, James.—(*Provisional protection only.*)—" Improve- " ments in the manufacture of baths and trays, and other vessels " for photographic purposes, which improvements are also applic- " able in the manufacture of galvanic battery and other gal- " vanic chambers or cells, and other vessels to contain chemical " solutions."

" Baths, trays, and other vessels for photographic uses, as also " for galvanic battery and other galvanic chambers or cells and " vessels for other chemical solutions, are now commonly made " of gutta percha. But in addition to the cost of the material, " vessels made entirely of such material are very liable to leak at " the joints and elsewhere, and my improvements consist in form- " ing such vessels of a comparatively thin lining of gutta percha " or india-rubber, or compounds thereof, and then coating the " same with layers of paper or other fabric or material united " together by means of solutions of india-rubber, gutta percha, " shellac, or other suitable adhesive material.

" By these means economy of manufacture and increased " strength and durability are obtained."

[Printed, 4*d.* No Drawings.]

A.D. 1861, April 18.—N° 955.

BROOMAN, Richard Archibald.—(*A communication from Adolphe Fargier and Nicolas Charavet.*) — " Improvements in " producing photographic pictures."

" A coat of gelatine, holding in suspension a carbonaceous or " other powder," is spread upon a surface of glass, and sensitized by means of bichromate of potash. The coating is then dried, the picture impressed by exposure to light, and a coating of " non-iodized collodion " poured upon the gelatine; the collo- dion film is removed from the glass by immersion in hot water, " and the sheet of collodion floats in the water, showing the " picture which has now become visible." The picture, so pro- duced is washed, received upon a sheet of gelatinized paper, and, finally, the coat of collodion is " removed from the picture by " alcoholized ether or other solvent."

A modification of the above-described process consists in applying a coat of the said gelatinous solution to a sheet of thin paper made from homogeneous pulp. The coating is dried, exposed to light on the ungelatinized side, and the resulting reversed picture detached by washings in warm water.

" A further modification is as follows :—Spread upon a glass a " first coating of the gelatinous solution, but without powdered " coal, dry and expose for a few minutes to the light, and then in " the dark; apply " [and then, in the dark, apply?] " over the " first coating a second coating with the powdered coal, allow to " dry, place the glass in the frame upon the plate, not on the " gelatinized but on the opposite side, expose to solar rays, and " the picture is obtained upon glass, from which it is removed by " washings in warm water."

" The action of light coagulates the gelatine and retains it on " the surface of the collodion, of the paper, or of the glass with " the powdered coal."

[Printed, 4*d*. No Drawings.]

## A.D. 1861, April 29.—N° 1074.

DIXON, HENRY.—(*Provisional protection only.*)—" Improve- " ments in photography."

" This invention consists in mixing nitrate of silver with collo- " dion or other coating medium prepared in the ordinary manner, " in the proportion of about five grains of silver to each ounce " of collodion or other coating medium. The mode ordinarily " practised for coating plates of glass with collodion or other " coating medium has been first to pour the prepared collodion " on the plate, and then to immerse it in a nitrate of silver bath " to coat it with silver. Now, the intention of this invention is " to dispense with the use of said bath by mixing silver with the " collodion or other coating medium, so that after pouring the " collodion or other coating compound on the plate intended to " receive the picture, it is simply necessary to wash the plate in " water to remove the grease therefrom. The plate is then ready " for exposure in the camera. The usual developing solutions " may be employed. By this invention the numerous objections " attending the use of a nitrate of silver bath are entirely dis- " pensed with, and the process rendered more certain, convenient, " and considerably less expensive than heretofore."

[Printed, 4*d*. No Drawings.]

A.D. 1861, May 1.—N° 1089.

HOOMAN, THOMAS, and MALISZEWSKI, JOHN.—"Improvements in photographic printing upon the interior of any glass or other transparent vessel."

A waxed paper photographic negative is fixed "upon the exterior" of that part of the vessel wherein the picture is intended to be made; for this purpose a gutta percha mould completely covers the vessel, "with the exception of an aperture of the size of the negative wherein it is placed, the light being thus wholly excluded from all other parts of such vessel." Collodion is then poured into the vessel, sensitized, and exposed "to the even light in such manner that the power of light shall act equally around and through the transparency of the vessel upon the sensitized substance fixed on the interior of such glass or other transparent vessel;" the picture is developed in the usual manner. "When dry, we back the interior of such glass or other transparent vessel with oil colors in imitation of marble, alabaster, wood, or any other fancy decoration, by which means the portrait, landscape, view, picture, arms, device, or ornament is rendered permanent, and thus protected on the inside thereof by such body of oil paint, and on the outside it is protected by the glass or other transparent vessel."

[Printed, 4*d*. No Drawings.]

A.D. 1861, June 8.—N° 1457.

DU MONT, HENRY.—"A photographic apparatus, having for object to reproduce the successive phases and shiftings of a motion."

A succession of sensitive surfaces "succeeding each other at regulated intervals" are exposed to light "at regulated intervals, and in due time, that is to say, when the plane of the sensitive layer is perpendicular to the axis of the ray."

A prismatic drum, whose periphery is mounted with sensitive layers, is caused to revolve so that the sensitive layers "succeed each other in the focus of the obscure chamber."

"Another means of obtaining the quick succession of the sensitive layers is to place them together side by side on an ordinary sash, but much longer, and moving intermittently in its groove, vertically or horizontally."

A third method of accomplishing the said object "is to set "them behind each other in a long box with vertical grooves, "and moving in a contrary direction to a similar box situated "under it, and intended to receive the layers impressed by the "light." A ratchet wheel and ratchet movement enable the sensitive layers to fall into the lower box as they are impressed.

The exposure of the sensitive layers to light at the proper instant is accomplished by means of "a black moveable screen, "the regulated motions of which are connected with the motions "of the system bearing the sensitive layers."

"The series of images thus produced, viz., the series of motions "of a dancer, of one or more soldiers, of a machine, &ca., &ca.," may be utilized, "both for the pleasure of the eyes and for other "purposes."

[Printed, 6*d*. Drawing.]

### A.D. 1861, July 12.—N° 1756.

SMITH, Thomas John.—"Improvements in photographic "albums."

This invention "relates to an improved manufacture of photo-"graphic albums, the improvements consisting in the manufac-"ture of the leaves of such albums of parchment, vellum, textile "fabric, or paper mounted upon textile fabric, such materials "being more durable and less liable to be torn by the insertion "and removal of the photographic pictures than the paper or "cardboard hitherto employed in the manufacture of the leaves "of books for this purpose.

"In addition to the advantage of greater strength and dur-"ability, a further advantage arises from the use of the materials "above mentioned, inasmuch as the sewing or stitching of the "leaves when binding the album is entirely dispensed with, there "being sufficient strength in the leaves to admit of their being "simply glued to the back piece."

The Specification describes, and the drawings show a sheet of parchment pasted "over the two contiguous faces of any two of "the leaves, the body or inside part of such leaves being com-"posed of cardboard" "cut out in the centre to allow for the "thickness of the photograph in the usual manner." The series of leaves being united by the cementing of the above-mentioned sheets of parchment thereon "are glued to the back piece," "and "thus the necessity for sewing is avoided."

[Printed, 6*d*. Drawing.]

A.D. 1861, July 24.—N° 1855.

NEVILLE, Henry.—(*Provisional protection only.*)—" Improved " apparatus for taking photographs."

" My invention consists of apparatus made of wood or metal, " or both, for taking micro-photographic and other minute " photographs; also for taking magnified microscopic photo- " graphs on metal, glass, or paper, by means of which apparatus " the reduction of the photographs as at present performed is " rendered unnecessary, the photograph being obtainable from " the object itself. The instrument is in shape similar to a " common camera, about eighteen inches in length; at one end " is attached a brass telescopic tube, forming what is termed 'the " 'coarse adjustment.' Outside of the last-mentioned tube is " placed a larger tube having a fine threaded screw or rack " and pinion forming 'the fine adjustment.' Attached to these " tubes is a piece of metal or wood, dovetailed into which slides a " box containing the prepared plate, which rests upon silver pins, " and is secured in its proper position by means of springs " attached to a lid covering it, made of metal or wood. Inserted " in the centre of this lid is a microscope for the purpose of " focusing the object to be copied, which, when once adjusted, " enables the operator to take any number of photographs with- " out further adjustment or alteration."

[Printed, 4*d*. No Drawings.]

A.D. 1861, August 3.—N° 1936.

LEWIS, Joseph.—" Improvements in producing and treating " printing surfaces, in producing and preparing transferring sur- " faces, in transferring, in producing impressions on an altered " scale, in preparing or treating surfaces of lithographic stones, " and in obtaining devices or designs; also in agents and " apparatus used in some of such improvements, parts of " the invention being also applicable to photography, and to " ornamenting pottery, porcelain, and glass."

Of the 26 heads under which this invention is set forth, the following have the most immediate reference to photography:—

5th head.—An "automaton register," consisting of certain fastening frames by which etchings, exposures to light, &c., can be repeated until perfect results are obtained.

6th head.—" Obtaining photographs, photo-lithographs," and photo-prints by means of the " automaton register."

7th head.—To transfer photographs to a printing or transferring surface.—" Oil of lavender and asphaltum " is the sensitizing mixture used. When the image is impressed, it is fixed by means of a mixture of turpentine and bisulphide of carbon. Successive sensitizings and fixings take place until a perfectly developed copy is produced, such of the fine tints as are sufficiently developed from time to time being protected from further action by means of an opaque powder.

10th head.—Using " oil of lavender and asphaltum as a sensitive " medium in photographic operations."

11th head.—" Producing printing surfaces by means of re- " peated photographic or other impressions on extended india- " rubber or gutta percha, which is afterwards allowed to collapse " or not."

13th head.—Bichromate of potash is employed "as a photo- " actinic medium," "in combination with gelatine or other sub- " stance, which will" "retard the crystallization of such bichromate " using the compound either in solution or in powder upon stone, " cardboard, paper, or metal, previously covered with transfer ink, " also as a sensitive mixture instead of asphaltum and oil of " lavender."

23rd head.—Photographic negatives are obtained by means of " a modification of the automaton register" in which the plate is placed for the negative. A first picture having been produced thereon, it is again coated with a sensitive medium and the picture reproduced. This is repeated as often as desired.

24th head.—The use of successive sensitive coatings and reproductions as a means of raising the design to a sufficient height to be stereotyped or electrotyped.

25th head.—The application of the " automaton register" "to " the production of printing surfaces by the photoglyphic and " other processes."

[Printed, 4*s*. 2*d*. Drawings.]

## A.D. 1861, August 20.—N° 2073.

SUTTON, Thomas.—" An improved camera for taking photo- " graphic portraits and instantaneous pictures."

"The body of the camera is an oblong box of the usual shape, " having the end open for the reception of the dark slide, and

" having an orifice in the front for the lens. At the upper edge " of the open end of the camera is hinged a moveable plane re- " flector." " When this reflector is placed within the camera at " an angle of 45° with the top or bottom, and the sensitive plate " is put in its place behind it, with the front shutter of the dark " slide open, the reflector must be large enough to shield the " sensitive plate from the light which passes through the lens. " Thus, when the reflector is turned up on its hinge till its " lower edge touches the top of the camera, the image formed by " the lens will be thrown upon the sensitive plate; but when the " reflector is turned down till its lower edge touches the bottom " of the camera, and makes an angle of 45° with it, none of the " light from the lens can fall upon the sensitive plate. The re- " flector may be turned up and down on its hinge instantaneously " by any suitable contrivance with an external handle.

" Instead of a focussing screen, which in an ordinary camera " fits into the open end of the camera, the focussing screen is " fitted into an open panel in the top of the camera, so that when " the face of the reflector is turned towards the lens at an angle of " 45° with its axis, an erect image is thrown upon the screen." " The plates of the focussing screen, the sensitive plate, and the " reflector must have a common line of intersection, and the latter " plane must bisect the angle between the two former planes."

" The focussing screen may be made either of plain ground glass " or of ground glass of an orange tint, so as to intercept the " entrance of actinic rays into the camera."

[Printed, 6*d*. Drawing.]

A.D. 1861, September 19.—Nº 2347.

DAGRON, Réné Prudent Patrice.—" An improved micro- " scope to be used for exhibiting photographic views and produc- " tions."

This invention is based upon certain parts of Nº 801 (A.D. 1860); it consists of a double-image tubular microscope. Each lens is fixed to its own tube which slides on a smaller tube, the smaller tube serving to connect the lens tubes together and to enable them to be placed properly for focussing; the said smaller tube " can be furnished with a diaphgram " [diaphragm ?]. The photograph to be viewed, by means of the lens at one extremity of the apparatus, is placed on the lens at the other extremity.

According to another arrangement the said microscope may be made entirely of glass. When three pieces of glass are used, the external pieces are plano-convex, and are glued to a cylindrical piece at opposite ends, with the photographs placed within the joint. The cylindrical portion of the microscope may consist of three separate portions, so as to render the instrument achromatic. The intermediate glasses may be dispensed with, a cylindrical piece convex at each end being employed; in this case a photograph placed on each end and is protected from injury by varnish. The photographs may be fixed on a piece of glass placed between the lenses; the distance between the lenses should be such that each lens renders the image in front of it distinctly enlarged.

[Printed, 8*d*. Drawing.]

A.D. 1861, September 30.—N° 2434.

GEORGE, Benjamin George.—" Improvements in the mounting of tablets, show bills, prints, photographs, and drawings, " and of producing embossed ornamentation applicable to various " purposes."

This invention consists in improvements in the processes set forth in No. 3004 (A.D. 1860), "and relates chiefly to the " mounting of tradesmen's show bills and other similar articles."

The said article is mounted upon a slab of plaster of Paris or composition " by means of any suitable cement." To put an ornamental border to the article, the inventor uses " moulds " formed from a model of the pattern or design required, and " from which casts may be taken in plaster of Paris or other " composition; and these ornamental portions, which may re- " present the frame, or any other part of the show bill or other " design, may be gilt or painted or otherwise ornamented, as " desired, and the show bill or design may be mounted upon the " smooth or other portion of the cast, and the whole or any part " thereof may then be covered with varnish, gelatine glass, or " other suitable preserving material."

[Printed, 4*d*. No Drawings.]

A.D. 1861, October 8.—N° 2513.

GRISDALE, John Edwin. — (*Provisional protection only.*) — " Improvements in certain tickets or passes for railway and other " purposes."

"This invention consists in the production of a new description "of tickets or passes, particularly where such tickets or passes "are granted by railway companies and others to persons travel-"ling or having rights of admissions for stated periods of time, "and the object of the improvements is a safeguard to the com-"panies or others issuing them against the transfer of such "tickets or passes to others than the persons entitled to use the "same; this object is effected by manufacturing or producing in "any simple and convenient manner, and by the aid of photo-"graphy, tickets or passes, each of which bears upon a con-"spicuous part of it a portrait of the person to whom the ticket "or pass is issued, so that upon the customary production of "such ticket or pass the person appointed to check the same "would at once see by comparing the portrait on the ticket or "pass with the person presenting the same, that such ticket or "pass was presented by its true owner."

[Printed 4*d*. No Drawings.]

A.D. 1861, November 1.—N° 2739.

CLARK, William.—(*A communication from Claude Mamès Augustin Marion.*)—(*Provisional protection only.*)—"Improve-"ments in photograph albums."

"This invention relates to improvements in albums for holding "photographic portraits, the object of which is to facilitate the "introduction and removal of said portrait cards from the frames "or cases forming the leaves or divisions of said albums. At "the present time the cards are introduced at an opening, of the "same width as the card, made at the lower part of the leaf, "which being made in the thickness of the card leaves a project-"ing edge, and renders the picture difficult of removal. I "obviate this disadvantage by covering the projecting edge with "a sheet of paper, which guides the picture card over the pro-"jection, and prevents its catching."

The drawings show the ordinary arrangement at present in use as well as the improved arrangement.

[Printed, 6*d*. Drawing.]

A.D. 1861, November 5.—N° 2781.

BOURQUIN, John Peter.—(*Provisional protection only.*)—"Improvements in ornamenting the covers of photographic "albums, books, writing cases, and other like articles."

"The object of this invention is to ornament the covers of " albums, books, writing or blotting cases, and other similar " articles, by applying thereto decorations resembling inlaid " woods, known as marqueterie and wood mosaics. This I pro- " pose to effect by printing upon veneer mosaic and other " designs, and thereby imitating differently colored woods " arranged in patterns. Having selected, say, a veneer of white " wood, I attach the same to cloth by cement before printing, to " protect it from breaking or splitting under the pressure to " which it will be subjected. The wood being damped is placed " in the lithographic press to receive the prepared design, which " is then transferred thereto in the same manner as colored " designs are applied to paper. When printed with the requisite " number of colors to complete the pattern the wood is laid " aside for the colors to harden, and when this is accomplished " the wood is polished and varnished, and may then be applied " to the book or other cover in the form of a border, corner pieces, " lines, panels, or even as an entire covering for the lids or for the " back, as may be thought desirable."

[Printed, 4*d*. No Drawings.]

A.D. 1861, November 13.—N° 2860.

BROOMAN, Richard Archibald.—(*A communication from Auguste Marion.*) — " Improvements in albums for containing " photographic and other pictures."

The two parts which, when united, form one sheet of the album, are attached to a zig-zag backing of parchment, in the recessed parts of the said backing, through certain hinge pieces; " the " intermediate folds" " of the backing render the sheets inde- " pendent of one another, and enable them to be fully opened " out without one sheet dragging on another." "The two parts" " are formed of two leaves" " hollowed out at the centre to " allow of the pictures being seen." " An incision" " rounded " at its two ends, allows of the insertion of the picture, and to " avoid any stop or impediment to its withdrawal, the lower end " of a sheet," " which does not extend beyond the edge of the " frame, is attached to the tongue formed by the incision." " Intermediate pieces of thin cardboard" " are pasted or glued " behind the leaves" " and the sheet;" " thus the finished sheet " shews on each side a central opening for viewing the pictures, " with a slot for their introduction and withdrawal." The edges of the hinge pieces are inserted between the intermediate pieces and the leaves, "and the backs of the hinge pieces are sewn to "the backing." "The threads which secure each sheet of the "album pass through the middle of the hinge pieces, and when "the two parts" "have been fixed together the sewing is entirely "hidden."

[Printed, 10*d.* Drawings.]

### A.D. 1861, November 21.—N° 2924.

POLYBLANK, George Henry.—(*Letters Patent void for want of Final Specification.*)—"A new or improved method of "protecting and preserving photographic and other prints, water-"color drawings, and other works of art from injury and "decay."

"If the paper, material, or fabric upon which the print or work "of art has been produced be of such a nature as to absorb the "preserving or any liquid put upon it, I first saturate the material "or fabric with some clear size which will not injure the work of "art printed or otherwise put upon it, and when the sized print "or work of art is dry I press it, when necessary, in order to "make it flat and level and it is then ready for the next process. "I take the print or other work of art to be preserved, either "previously sized or not, as may be necessary, and entirely coat it "with paraffin, made fluid by means of heat, so as to be capable "of being spread over the print or work of art without injuring "it. The coating is then to be allowed to set or become dry, and "the covering thus put upon the print or other work of art will "protect and preserve it from the injurious effects of atmospheric "and other influences."

[Printed, 4*d.* No Drawings.]

### A.D. 1861, November 27.—N° 2996.

AMPHLET, Samuel.—"An improvement or improvements in "ornamenting surfaces."

This invention "consists in ornamenting surfaces by the com-"bination herein-after described of tartan or Scotch plaid patterns "or imitation tortoise-shell, with photographs of objects without "any other background than the said tartan or Scotch plaid "patterns or imitation tortoiseshell.

"In carrying my invention into effect I employ photographs " taken on paper or other flexible material on which positive " photographs are or may be taken, and I prefer photographic " portraits or groups or photographs of sculpture or statuary. I " cut away all the background of the photograph, leaving only " that portion of the said photograph which consists of the por- " trait or group, or picture of the sculpture or statuary or other " detached object or objects. I rule the tartan or Scotch plaid " patterns on the surface to be ornamented, or I paint, print, " stain or otherwise colour the said surface in imitation of tortoise- " shell, and I attach the photograph thereto by paste or size or " other adhesive material. I then size, varnish and polish the " surface in the manner commonly practised in finishing surfaces " ornamented with tartan or Scotch plaid patterns. Or instead " of producing the tartan plaid patterns or imitation tortoise- " shell upon the surface of the article to be ornamented, I rule " paper with tartan or Scotch plaid patterns or paint, print, stain " or otherwise colour paper in imitation of tortoiseshell, and I " affix the photograph to the said ruled paper or paper colored in " imitation of tortoiseshell. I afterwards apply the said tartan " plaid or imitation tortoiseshell paper with the photograph upon " it to the surface to be ornamented."

[Printed, 4*d*. No Drawings.]

### A.D. 1861, December 4.—N° 3044.

BROOMAN, Richard Archibald.—(*A communication from Henry Strauss.*)—" Improvements in albums or books for con- " taining and showing photographic and other pictures, and in " slides for the same."

" I take two of such sheets of paper with apertures in the " centre as are generally used in albums for showing photo- " graphic pictures, and unite them at three sides, after having " interposed a strip or strips of thick paper, cardboard, or the " like, equal or nearly so to the thickness of two of the pictures " to be contained and shewn. Thus the sheets will be united at " their edges on three sides to interposed filling pieces, and one " side will be left open for the insertion of the slide hereafter " described."

" The slides consist of a frame of cardboard or thick paper " equal in thickness to the interposed filling pieces. To one " edge of the frame on both sides a sheet of gelatine, talc, or

" other transparent material is attached, the space left in the " frame is calculated to receive two pictures back to back. After " being inserted in the space in the frame, each picture is " entirely covered by the gelatine or talc. The slide is formed " with angle pieces, and when inserted between the sheets of " paper, the outer edge of the slide fills up the space between the " unclosed edges of the two sheets."

The drawings show a card placed in the frame which "may be " used as a support to the pictures which are placed back to back, " one on each side of the card."

[Printed, 6*d*. Drawing.]

### A.D. 1861, December 9.—N° 3080.

MENNONS, Marc Antoine François.—(*A communication from Louis Henri Bouillette and Jean Amable Hyvelin.*)—" A new " or improved combination of microscopic photographs and lenses " with certain precious stones or imitations thereof."

" This invention consists in the application of microscopic " photographs and lenses or lenticular surfaces to such precious " stones and imitations thereof as possess a sufficient degree of " transparency to allow the passage of the visual rays.

" To this end the microscopic photograph is applied either " directly or by transfer to the under surface " of a section of the gem, cut in the form of a thin zone or frustrum of a cone. "To " this surface is then adapted, by Canada balsam, Venice turpen-" tine, or other suitable adhesive matter, the pyramidal counter-" part," "to the apex of which is cemented a small magnifying " lens." "The complete gem" "is then mounted in a ring, " watch key, breast pin, or other setting, the point of which is " pierced to form eye piece in the axial line of the lens correspond-" ing with the microscopic object in the centre of the front plate.

" Instead of adapting a separate lens as above, the apex of the " gem may, in certain cases, be ground to a lenticular surface, " but this process is generally more difficult and less economical " than that just described.

" Or, again, the photograph, instead of being applied directly " to the plane surface of the front plate, may be formed on a " separate disc of transparent matter, which is interposed between " the sections of the gem, and secured by cement as above."

[Printed, 4*d*. No Drawings.]

A.D. 1861, December 11.—No 3105.

SCHLOSS, Joseph.—(*A communication from Simeon Schloss.*)—(*Provisional protection only.*)—" An improvement in forming the " leaves of albums, and books for containing photographic portraits " and views."

" In the leaves as now made, the pictures are introduced into, " and when required removed from each leaf through an aperture " or space made or left at the side, top, or bottom, or through a " slit made above the top or under the bottom of the window, or " space, made for showing the picture. Now in this invention " neither such space, aperture, nor slit is made, but the leaves are " so put together that the pictures are introduced and removed " through the windows themselves.

" The invention consists in interposing between two sheets of " paper with windows cut therein, and which are intended to form " the front and back of a leaf of the album, or book, one or two " thicknesses of stout paper, equal in size to the window sheets, " but without any parts removed, then in fixing along the edge " of these interposed pieces of paper, and on both sides thereof, a " frame or beading of card, or millboard, or other suitable material " extending inwards from the edge, only about a quarter of an " inch more or less. The two first named sheets of paper are " next attached to the cardboard frames, and the leaf for the " album or book is complete. The photographic picture is intro- " duced cornerwise through the window, and is worked into " position by pressure of the finger."

[Printed, 4*d*. No Drawings.]

A.D. 1861, December 14.—No 3147.

DEBENHAM, William Elliott. — (*Provisional protection only.*)—" An improved plate holder for photographic purposes."

" The invention consists in so forming the plate holder for the " purpose of holding photographic plates while undergoing " chemical action, that there shall be no risk of the hands of the " operator being stained or soiled by the spilling of the chemicals " used, nor danger of the plates falling off the holder."

The Specification describes and the drawing shows a holder having a handle " in which is inserted a stout metal wire or other " material," which is bent into such a form as to afford three points of support to the plate; the wire also branches into two parts from the handle; one of the parts being straight may be used as an additional support to the plate, the other part carries the points of support or "guards" above mentioned. The "guards" "may be made of gutta percha, wood, glass, metal, "caoutchouc, or other suitable material," and "serve to prevent "the plate from falling off the holder. The guards are so formed "as to present a surface slightly slanting outwards, in order that "the lower edge only of the plate touches and rests on the holder "and guards (the upper edge never coming in contact therewith), "so that the danger of staining the plates which might arise "if any part of the holder were to touch their upper or coated "edge is entirely obviated. The handle is raised above the level "of the surface of the plate, the wire" "being bent upwards "previously to its insertion into the handle. The shape of the "holder may be varied according to circumstances, and the "guards may be arranged in different positions to those shewn in "the drawing, the principle of the invention being nevertheless "adhered to."

[Printed, 6*d*. Drawing.

### A.D. 1861, December 26.—N° 3232.

SCHLOSS, Joseph.—(*A communication from Simeon Schloss.*) —(*Provisional protection only.*)—"Improvements in envelopes for "containing photographic portraits and pictures."

"This invention consists in forming envelopes, intended chiefly "for containing 'carte de visite' portraits, and for their being "forwarded conveniently by post as hereafter described. I take "a sheet of paper or paper cloth, or other material suitable for "envelopes, and of a breadth equal to the length of the envelope "when finished. I form the usual flap at one end of the paper, "and form a frame in or on the opposite end, either by folding "the paper over on itself, or by the employment of a separate "frame of paper. In the first case, that is, where the paper is "folded over, I first cut the 'window' or aperture therein, for "showing the picture, and cement the top and bottom edges "to the other part of the sheet; one side wlll thus remain "unattached for the insertion of the picture. Where the frame "is in a detached piece, I attach it at top and bottom and one "side. After insertion of the picture, that part of the sheet on "which the frame is applied, together with the picture and frame, "is folded over, and the flap is folded over the back of that part "on which the frame is applied. When several pictures are to "be enclosed, I use a greater length of paper, and apply additional frames, and where only two pictures are to be enclosed, "I apply an additional frame on that part of the paper which "when the envelope is folded covers the face of the first picture."

[Printed, 4*d*. No Drawings.]

---

## 1862.

### A.D. 1862, January 13.—N° 95.

SCHOTTLANDER, Henry.—(*Provisional protection only.*)— "Improvements in albums for containing photographic and other "pictures."

"My invention consists in so forming the leaves of albums "that the picture or pictures, together with part of the leaf "holding the same, may be placed and held at an inclination so "as to obtain the most suitable light for viewing the pictures."

"I apply on each side of every leaf, or on one side only, a frame "for holding one or two pictures, and unite it at one side or end "only, and I make a tongue or frame in or on the holding frame. "When the album is closed, or when the leaves and pictures are "in their ordinary position, the holding frames, pictures, and "tongues do not protrude beyond the thickness of the leaves. "When the album is opened, and any picture is to be viewed, "the holding frame is raised and the tongue or back frame is "protruded outwards and the frame resting partly upon it is "maintained inclined. If desired, pictures may be held in the "leaves which are covered by the holding frames, and which can "only be seen when such frames are raised.

"Again, instead of the tongue or supporting frame before "mentioned, I sometimes form the support by hinging a piece of "cardboard, or other suitable material to the top of the holding "frame which is pushed outwards for supporting the holding "frame, and which is folded inwards and under the holding "frame when the pictures are not required to be raised to be "viewed."

[Printed, 4*d*. No Drawings.]

A.D. 1862, February 7.—N° 322.

BROOMAN, Richard Archibald.—(*A communication from Louis François Saugrin.*)—" Improvements in stereoscopic albums, " books, and cases."

" This invention consists in combining a stereoscope with an " album case or book in such manner that both may be folded " and occupy a small space."

The stereoscopic case contains the photographic pictures, and its hinged cover has a lock for securing the case when closed. By means of flaps united to a hinge, "the rod of which is sur- " rounded by an adjusting screw" or helical spring, on opening the case the instrument is placed ready for observation. The base of the stereoscope "forms an ordinary slide;" the sides are united by flexible joints, so that they may fold easily. The top plate carries the glasses, which sink down with the sides when the case is closed. This stereoscope may be fitted in the lid of any book or album, or "in a slot in each intermediate leaf." The pictures are all placed one on another in the case on a frame, or movable bottom plate, which rests on helical springs, the pictures are thus raised within the focus of the stereoscope; as they are done with, an axis with a disc at each end and carrying two rollers, causes the pictures to fall into a drawer. So soon as the upper picture has thus slidden away, the spring frame raises the succeeding picture.

When the stereoscope is applied to an album, the slight projection which forms the stereoscope when folded, lies in a recess formed in the cover; a " compass joint " " maintains the opening " of the album at the desired inclination."

In "a stereoscopic pocket book," the bottom is made in two parts, and a vertical partition " acts as a support to the apparatus " when open." When the instrument has been partially opened, spring hinges wholly place the parts ready for use.

[Printed, 8*d*. Drawing.]

A.D. 1862, February 18.—N° 429.

SÉGOFFIN, Charles Denis.—(*Provisional protection only.*)— " An improved apparatus for the purpose of viewing photographs " on cards."

" My invention consists in the construction of an apparatus of " a rectangular form, three sides of which are closed, the fourth

" being partly open for the admission of light on to the card " placed at the back, the smaller side opposite, having a mag- " nifying glass which allows of a good view of all the details of " the photographic card.

" For viewing proofs placed in albums I make the apparatus " without a back so that all that is required to be done is to place " the apparatus over the proofs to see them distinctly. The sides " of these instruments are formed with stiff or flexible materials, " and of a bellows form, that is to say, with parts folding on " themselves. To these small apparatus I have given the name " of microphore."

[Printed, 4*d*. No Drawings.]

A.D. 1862, February 25.—N° 504.

BLISS, EDWIN, and LAMPLOUGH, HENRY.—" Improved " means for viewing microscopic photographs and other minute " objects."

" Our invention consists in having a circular disc of metal or " other suitable material, with small sight holes perforated therein " at a convenient distance from the edge, and at a proper space " apart, and placing opposite to each hole the photograph or " object to be inspected, and having such disc and objects en- " closed in a case with a microscopic lens fixed therein in the line " of the holes in the disc, and a hole through the other side of " the case directly opposite the lens, so that by causing the disc " to revolve the objects placed on or connected with the disc may " be brought seriatim in front of the lens in the case, and viewed " through it. Of course if the disc be of clear glass no sight " holes will be required in it. Tbe disc may be caused to revolve " by means of toothed wheels or other known means."

The drawings represent a disc turned by means of an ornamental boss placed on the side of the case opposite to the eye piece. " At the outside edge of the disc, and in perfect truth " with each object, holes are made that will in conjunction with " a bolt and spring," " stop each object exactly opposite the eye " piece."

[Printed, 8*d*. Drawing.]

A.D. 1862, March 8.—N° 636. (* *)

GEBHARDT, JOHN JAMES HENRY.—(*A communication from Herr Kugler.*)—(*Provisional protection only.*)—" An improved

" fastening for albums and other books, bags, reticules, and " other articles."

On the inner side of a plate attached to one side or edge of the article to be fastened, is secured a sliding piece, wherein are fitted spiral or other springs in grooves or recesses, " and " on the inner surface of the outer plate are studs, stops, or " abutments so placed that the springs may abut against them " at one end." The sliding piece carries a knob or finger piece " presenting itself on the outer side of the first-named plate " through a slot or opening " in which it or its shank pin works. The edges of the outer plate are by preference bent inward to form guides for the sliding piece, which carries or is formed with a hook or catch intended to engage with an eye, staple, or its equivalent on the opposite side or edge of the article. " The " fastening may be somewhat modified by using a single spring " (and stop) instead of two or more, and also by having the " travelling plate outside instead of inside the other plate, and " the stop on the travelling, the spring on the other plate."

[Printed, 4*d*. No Drawings.]

### A.D. 1862, March 10.—N° 640.

BROOMAN, RICHARD ARCHIBALD.—(*A communication from Arthur Guy Morvan.*)—" Improvements in producing by the aid " of photography copies of maps, charts, plans, and drawings."

Obtaining a positive image on cloth, paper, or glass, " without " the aid of a negative."—A tracing of the design, on transparent drawing cloth, is placed in contact with the sensitized paper, or other surface, the face of the design being next to the sensitive surface; the frame containing this arrangement is exposed to the action of light, and the image is fixed and finished in the usual manner. The sensitizing solution described in the Provisional Specification contains iodide of silver, that set forth in the Final Specification contains " azotate " of silver.

Transferring the design upon the transparent cloth to stone.—The stone is prepared with acid, washed in water, dried, coated with a sensitive varnish, allowed to dry, and exposed to the action of light; it is then washed with white wine, with water, allowed to dry, washed with soapy water, again washed, " and then dried " in a stove or otherwise." Ink is then applied, the stone is allowed to rest 24 hours, and washed with essence of turpentine.

Again ink is applied, and the subsequent inking operations are facilitated with phosphoric acid and gum, so as to bring out the design; impressions may then be taken from the stone. By this process the stone gives off positive impressions. The varnish contains white of egg and bichromate of ammonium, and those parts exposed to light become more or less insoluble. The ink first used contains transfer ink and lithographic crayon, that employed the second time consists of transfer ink and phosphoric acid.

[Printed, 4*d*. No Drawings.]

A.D. 1862, March 12.—N° 677.

GRISDALE, JOHN EDWIN.—(*Provisional protection only.*)—"Improvements in photographic cameras, and in the mode of "fixing the lens therein."

The two ends of the camera "are connected together on the "four sides" by "lazy-tongs levers," which "are braced together" by "metal braces which pass round and enclose them. "The whole may be covered by the ordinary bellows covering or "by a bag."

The lens tube is secured in front of the camera by three or more pairs of hinged radial flaps or arms. By this means different sizes of lenses may be readily fitted into the same camera, provision being made "for excluding the light which would otherwise "enter the camera when fitted with this arrangement for securing "the lens.

"Another part of my invention consists in the substitution of "a peculiar hinged cover for the dark slide in lieu of the ordinary "sliding cover, the improved cover being made in four parts, and "so arranged that each part will fold back inside the camera, the "divisions in the cover being so disposed as to admit of each part "folding back into a tapered camera when required to do so."

[Printed, 4*d*. No Drawings.]

A.D. 1862, March 13.—N° 681.

FONTAINE, FORTUNE HIPPOLYTE.—(*Provisional protection only.*)—"A process for reproducing photographies, drawings, "paintings, and engravings engraved on metal."

To reproduce engraved plates by "negative stereotypes."—A gelatinous solution of bichromate of potash is poured upon a zinc plate; when dry, the gelatinous surface is placed in contact with the printed side of the negative, and exposed to light. The zinc is then removed and placed in lukewarm water to dissolve the bichromate; the picture appears engraved, and is covered with a solution of gallic or pyrogallic acid and washed. A gutta percha cast is then taken of the plate and a mould in relief is made from the cast; the latter is suitably prepared, and an electrotype is obtained from it.

To reproduce engraved plates by "positive stereotypes."—A prepared copper plate is covered with a gelatinous solution of bichromate of potash and gum arabic; when dry, the gelatinous surface is placed in contact with the printed side of the positive, and exposed to light. The copper plate is then removed and covered with a solution of perchloride of iron; this solution expands "the unimpressed or soluble parts of the gelatine dissolves "the bichromate and prepares the way for the acid, by which I "hollow out the picture or image in presenting it in relief to its "action;" when the picture is developed and engraved by the said method, the acid is poured off, "and the plate is ready for "impression and the reproduction of copies identical with the "original."

[Printed, 4*d*. No Drawings.]

## A.D. 1862, April 3.—N° 948.

MANN, Alexander.—"Improvements in photographic apparatus."

1st. A camera shutter for covering and uncovering the lenses instantaneously.—The covering consists of two pieces, which, by means of a spring and elastic bands, are separated in front of the centre of the lens, and their motion takes place "in opposite "directions towards the edges of the lens."

2nd. "Changing the direction of the lens and picture without "moving the apparatus."—A "turntable movement" is arranged under the front part of the instrument. The Final Specification does not mention this improvement.

3rd. "Affording a longer period of time for the action of light "on the foreground of the picture than on the sky or upper part." —A curtain of dark paper, with the horizon line cut away upon its upper part, is situated in the interior of the camera near the picture, and is capable of being removed downwards by means of an elastic cord. Another apparatus for effecting the same object consists of a small rectangular box, placed upon the front of the lens tube, which contains a frame carrying a number of uniform pieces of metal capable of sliding up and down with friction; a light-proof covering is thus formed to the upper part of the front of the box.

4th. Apparatus for finishing paper photographs.—The sliding table on which the picture rests is heated by means of gas, so as to afford "the advantages of hot-pressing with perfect safety from "charring the picture." This arrangement is not alluded to in the Final Specification.

[Printed, 6*d*. Drawing.]

### A.D. 1862, April 8.—N° 996.

CARTER, Charles Pemberton.—(*Provisional protection only.*) —"An instrument for inserting photographic or other pictures "into or removing them from between the 'mounts' of photo- "graphic albums or other flat spaces, into which the fingers "cannot be inserted."

The instrument resembles a pair of pliers, "but with the jaws "gradually thinned down to an edge at their extremity, and more "or less widened out." The inner surface of the lower jaw is roughened, "whilst the corresponding surface of the upper jaw is "made quite smooth. When quite closed, these surfaces "meet "only at their extremity." "The ends of the jaws being made "quite thin they can be inserted with facility into the narrow "space of the mount, together with the picture, so as to adjust it "accurately in its position, or they can be easily inserted between "the surfaces of the picture and the 'mount' for the purpose of "removing the former from between the latter. The handles of "this instrument may be formed straight, and in a line with the "jaws, or, by preference, I bend them, so as to be at an angle "with the same for more convenient manipulation, or they may "be first bent up at an angle, and the extremities bent down again "parallel with the jaws. I sometimes also introduce a spring "between the handles to press them open when let go by the "hand; and in some cases I form one of the jaws with a sharp "knife-edge for cutting the pictures; or I form the handles close "to the hinge as scissor blades for the same purpose. This "instrument may be formed of any suitable material, and it may

" be employed for inserting pictures into any other flat spaces " besides those in the 'mounts' of photographic albums."

[Printed, 4*d*. No Drawings.]

A.D. 1862, April 11.—N° 1042.

GARNETT, John.—(*Provisional protection only.*)—" Improve- " ments in apparatus for washing photographic pictures."

" This invention relates to a peculiar construction of apparatus " for washing photographic pictures, and consists in the employ- " ment for that purpose of a shallow box perforated on its upper " and under side, into which box the prints or pictures requiring " to be rinsed or washed are deposited, suitable divisions being " placed in the box to separate the different sizes of prints, which " divisions or partitions are made moveable to suit different sizes. " The box with the photographs therein is placed inside a con- " venient reservoir of water, and a vertical motion is imparted to " the box whilst submerged by means of a handle or other " suitable device so as to cause the water to flush alternately " through the top and bottom perforations in the box and thereby " effectually rinse the pictures on both sides."

[Printed, 4*d*. No Drawings.]

A.D. 1862, April 25.—N° 1222.

McLACHLAN, Lachlan.—" Improvements in governing or " regulating light used for taking photographic portraits and other " photographic pictures, part of which improvements is also ap- " plicable to lighting picture galleries."

1st. " A method of preventing the direct rays of the sun from " falling upon the sitter or other object to be copied, and of mo- " difying or directing the light as occasion may require."—Strips of wood are mounted upon centres so as to hang downwards from the roof or to be inclined at any angle; cross pieces may be placed within the spaces formed by the strips. To regulate the side light, projecting slips, which extend sideways, are used " in combination with partitions situated within the spaces " between them."

2nd. " A method of preventing light from being reflected from " the surface used as a background in taking photographic " portraits."—Over the framework of the background is stretched a surface " of gauze or other material, which is partially trans-

" parent." Behind this is placed black velvet " or other " material which wlll absorb light," the effect of which is, that, instead of the light being reflected from the surface of the background, it passes through it and becomes absorbed by the black velvet screen ; or a second transparent surface placed at an angle may be used, one part of the background can thus be obtained of a darker shade than the other. The angle at which the last-mentioned surface is placed may be determined by means of bolts which pass " through rails at each side."

3rd. "The application of the above-mentioned method of re-" gulating light to the lighting of picture galleries."

[Printed, 8*d.* Drawing.]

A.D. 1862, April 30.—N° 1267.

HARRINGTON, John, and PERKINS, Thomas.—(*Provisional protection only.*)—"An improvement or improvements in mounting " photographic portraits for visiting cards, and in mounting pho-" tographs in general."

" Our invention consists in making a depression in that part of " the card at which the photograph is to be mounted; the said " depression being of the size and shape of the photograph, and " constituting a sunk recess, in which the photograph is placed, " and secured by gelatine or other adhesive material. The de-" pression or recess may either extend through only part of the " substance of the said card mount, or through the whole sub-" stance of the said card mount; that is, the depression or recess " may be made in the front of the card mount without any cor-" responding projection at the back or opposite. The depression " in front may be a corresponding projection at back. When we " make the depression in front without a projection at back, the " card is compressed at the sunk part into a less thickness than " that of the other parts of the card. When we make a pro-" jection at back, a thinner card may be employed than when there " is no projection at the back. The depression may be sur-" rounded with an ornamental border. The sinking of the " photograph in the card mount preserves the photograph, and " gives it an elegant appearance.

" Our invention is particularly applicable to small photographs " for visiting cards, but is also applicable to photographs of " larger size. We make the depression in the card mount by " means of dies and pressure."

[Printed, 4*d.* No Drawings.]

A.D. 1862, May 14.—N° 1453.

BROOMAN, RICHARD ARCHIBALD.—(*A communication from Leon Farrenc.*)—"An improved method and apparatus for the "production of photographic and stereoscopic portraits and "pictures."

"This invention consists, first, in the construction and employ-"ment of instruments containing lenses capable of producing "portraits or pictures, distorted in length, breadth, and obli-"quity;" "second, in the production of a new style of photo-"graphic portrait, distorted in length, breadth, and obliquity, "possessing all the clearness of an ordinary portrait;" "third, in "the application of these portraits to the stereoscope;" "fourth "in uniting and combining a pair of distorted portraits, accord-"ing to the plan adopted for stereoscopic pictures reciprocally "taken at the desired angles.

"The instrument is chiefly composed of two cylindrical lenses, "of different focus, both convergent, the generating lines of one "being perpendicular to the generating lines of the other, and "combined in such manner that the images they produce are "superposed on the same plane." This combination concentrates the rays of light so as to give an elliptical image. If the object is set on the small axis of the ellipse, the portrait will be short in comparison to its width; "if it is set on the chief axis it will "be lengthened, enlarged, and thin." If the two cylindrical lenses are not quite perpendicular, or if the object is not set on either axis of the ellipse, the figure will be distorted throughout, Ordinary lenses may be used in combination with the cylindrical lenses. "If two stereoscopic prints are distorted, they will remain "stereoscopic after the distortion."

[Printed, 8*d*. Drawing.]

A.D. 1862, May 19.—N° 1516.

MORRIS, TIMOTHY, WEARE, ROBERT, and MONCKTON, EDWARD HENRY CRADOCK.—The title of this invention is "Improvements in obtaining and applying light and heat by "electricity;" "light suitable for various kinds of photography" is mentioned in the Provisional Specification.

"Our invention consists in obtaining light by passing elec-"tricity through vacuum tubes or other suitable vessels, the "electricity employed being obtained from frictional, electro-

" magnetic, magneto-electric, or hydro-electric machines, and also " from other sources, but we prefer to obtain it from our induction " coils, as patented on October 24th 1861, No. 2661, in connection " with our patented batteries, and to obtain heat also with or " without such induction coil by means of such batteries."

After mentioning several applications of their methods of obtaining and applying light, the Patentees state:—" We also use our " light in conjunction with lenses and glass of various colors and " variously compounded, in order to produce light suitable for " various kinds of photography; and we also use ground glass " and other suitable media over our vacuum vessels, in order to " diminish the light; and we further use a metallic or other cover " of a suitable shape and contrivance to place or draw over the " light, in order to shut it off as required."

[Printed, 4*d*. No Drawings.]

### A.D. 1862, May 29.—N° 1611.

HIRST, John, junior, and WOOD, Joseph.—" Improvements " in stereoscopic apparatus."

" The object of the improvements is to neutralize the granular, " fibrous, and general coarseness of objects seen through stereo- " scopic glasses, especially when high magnifying powers are " employed, and also to produce various changes of effect to such " objects."

Tinted, transparent, or semi-transparent media are so applied that the light falling on the front of the picture may pass through them without intercepting the vision, When the picture is transparent, the said tinted media may be appled in combination with other tinted media at a short distance behind the picture.

When the picture is transparent, tinted transparent, or semi-transparent media are applied "at a short distance behind the " picture, so as to produce varied effects to the picture;" the media are connected " by cords, strings, or chains, or otherwise, " so that there may be intervals between the respective media, " when the natural color of the transparency may be seen."

In connection with stereoscopic apparatus, a lamp is applied, " with a shade or shades, lenses, and transparent or semi-trans- " parent media of various tints and shapes." Separate slides may also be added to the transparent picture.

The drawings show apparatus in which flexible transparent media are mounted on rollers, which are turned by means of suitably placed milled heads. Parts of the media are shown removed, to allow the light to fall direct on the picture.

[Printed, 1*s.* 8*d.* Drawings.]

## A.D. 1862, June 7.—Nº 1712.

HASELTINE, GEORGE.—(*A communication from Allen Benjamin Wilson.*)—" A new and improved photographic camera."

The camera box is made to serve as a dark chamber, besides being employed to expose the sensitive surface to the action of the light that has passed through the lenses of the instrument. Opposite to the camera tube is a door shutting tightly. Bearings, rotated by a key, enable the focus and negative glasses to be turned up and down. The negative glass holder and the bath are combined in one instrument. The journal is hollow and has a crooked stem funnel inserted from the outside of the box.

The focus glass holder being placed in the bearings, is turned into a vertical position, and the focus is obtained. The negative collodionized glass, secured to the plate holder, is then placed in the bearings with the collodion towards the lens. When the " tunnel," and the key are inserted, the nitrate of silver solution is poured through the tunnel on to the horizontal plate. The plate, being sensitized, is turned into a vertical position and the nitrate solution runs off, through the depressed tunnel, into its phial. The plate is then exposed, brought to the horizontal position, and the " developing bath is introduced and removed " as was the first." The tunnel and key may then be removed " and the negative cleaned in the light."

By means of tubes, these operations can be performed when the combined bath and plate holder is vertical.

[Printed, 8*d.* Drawing.]

## A.D. 1862, June 10.—Nº 1724.

SMITH, WILLIAM.—(*A communication from Henri Lissagaray.*) — (*Provisional protection only.*) — " Improvements in photo- " graphy."

" These improvements relate to the ornamentation of photo- " graphic works, and chiefly to the production of backgrounds, " borderings, and such like aids to the art of photographing

" objects. For these purposes I produce the design required by " employing a sheet of glass or other equivalent transparent sub- " stance, upon which by galvanic agency I deposit first a coat of " gold or other metal suited for the purpose, and next a coat of " platina or other equally suitable metallic surface. I then engrave " by means of a stile or point the required ornamental design, so " that the light may pass through the lines so marked, engraved " or cut into or through the metallic surface. An open space " may be left in the centre or other part or parts of the plate " prepared as described, so that a portrait or portraits, or other " subject or subjects intended to be reproduced may be introduced " behind or before such space or otherwise in combination with " such engraved plate.

" This process when used for ornamenting photographic " proofs, is employed by means of double stereotype plates of " glass or other equivalent transparent substance prepared as " described and used together by superposition or in succes- " sion."

[Printed, 4*d*. No Drawings.]

### A.D. 1862, June 25.—N° 1871.

**CLARK, William.**—(*A communication from Gustave Jeanne Julien.*)—" An improved frame for holding photographic pictures."

The "portable sliding frame" consists "of grooved bars in " which the pictures slide, and are held in position," thus the inventor is "enabled to combine in one frame of variable dimen- " sions, which may be of any desired form or ornamentation any " number of photographic pictures."

The drawings show an arrangement for holding three pictures, a second frame for holding five pictures, and a third frame for holding a still greater number of pictures.

" The facility for producing these combinations consists in the " arrangement of parallel bars forming the frame or holder, which " may be made of greater or less length, and placed at various " distances apart in any number of rows in such manner that the " whole shall present an artistic appearance."

The horizontal bars of the arrangement are grooved, so that the pictures are held securely, and in the grooves the said pictures may slide easily. The horizontal bars are connected together by vertical bars placed behind, which have perforations or screw rings at their upper ends to facilitate their suspension in any desired position. In cases where all the pictures may not be of the same size, the horizontal bars are furnished with a thumbscrew, and nut sliding in a slot in the vertical bars, "whereby pictures "may be fixed in any desired position."

[Printed, 8*d*. Drawing.]

A.D. 1862, June 27.—Nº 1888.

BROOMAN, Richard Archibald.—(*A communication from Jean Théodore Dupuy.*)—"A method or methods of preparing "paper for the reception of photographic pictures or impressions, "in order that the said pictures or impressions may be trans-"ferred to and fixed on wood, porcelain, and other surfaces."

The sized paper is covered with a solution of ammoniacal citrate of iron and bichromate of ammonium and dried in a dark room it is then submitted to light and damped; the dampness condenses only on those parts which have been shaded from the light. The impression is developed in a solution of iron and fixed by hyposulphite of soda and cyanide of potassium, it is then washed and placed on gummed paper, coloured with a vegetable or mineral colour, and varnished.

To transfer a paper photograph to any desired object, it is coated with a thin coating of transfer varnish, placed with its prepared side next to the object, covered with damp cloth, and rubbed evenly with a paper knife. The cloth is then raised, the paper is wetted with water, and when it (the paper) is peeled off, the photographic print will be found to be transferred to the surface on which it was placed. The print is then washed, dried, and varnished.

Other sizing materials, sensitizing agents, and varnishes, besides those mentioned above, are alluded to.

[Printed, 4*d*. No Drawings.]

A.D. 1862, July 4.—Nº 1944.

RUSSELL, Samuel.—(*Provisional protection only.*)—"Im-"provements in stereoscopes."

The stereoscopes made according to this invention are in the form of a box, "so that the interior of the instrument is conve-"niently used as a case to contain pictures." "Against the front "of the box and within it a mirror is fixed and the lenses or eye "pieces are mounted in apertures in the lower part of the back

" thereof, and they are inclined so as to be directed nearly towards " the horizontal centre line of the mirror. The picture to be " viewed is placed on a ledge at the back of the interior of the " case, and immediately over and close to the eye pieces, and it is " inclined somewhat so as to bring the reflected picture perpen- " dicular to or at right angles with the line of sight, and to " accomplish which, by my arrangement, the picture need not " form an angle of more than 16 or 17 degrees with the mirror. " The light falls directly on the picture entering at the open top " of the case, and to admit more light to the picture I make the " portion of the front of the case over the mirror, together with a " small portion of each of the sides, to turn back on hinges, and " this piece, when turned back, also serves to support the picture, " its upper edge resting against it." To give a tinted appearance to the pictures, the mirror or lenses may be made of coloured glass, or pieces of coloured glass may be placed in the eye pieces; " it is " convenient to place a disc carrying it may be five or other " number of pieces in front of each of the eye pieces, so that by " rotating the discs either of the colors may be brought opposite " the eye pieces, either the same colors may be brought opposite " each eye piece, or one eye piece may have one color brought " opposite to it, and the other another color."

[Printed, 4*d*. No Drawings.]

### A.D. 1862, July 8.—N° 1962.

GRUNER, Carl Bernhard.—" Improvements in photographic " apparatus."

By means of this apparatus the " dark chamber can be entirely " dispensed with, and the operations of preparing the plates and " of developing, strengthening, and fixing the image after expo- " sure in the camera, as well as that of washing the plate, can be " carried on in broad daylight."

" If the apparatus is to be detached from the camera it may " consist of as many double boxes as there are solutions or fluids " to be employed;" each box in its lower part contains the dipping bath, the upper part receiving the plate in its slide. To prevent the action of light from affecting the various operations, a slide divides the upper from the lower box, and the boxes may be made of yellow glass.

The drawings show an apparatus made so as to form a part of the camera. The body of the camera contains the ordinary adjustable sliding box, which has, however, a projecting part reaching above the body of the camera and fitted with a door. The uprights of the slide box acts as guides to a counterbalanced cross bar that supports the plate frame. The dipping baths are, one after another, placed immediately under the plate frame, and a slide is removed when the plate has to be lowered into a bath. Whilst the plate is being sensitized the focussing screen may be introduced for setting the lens.

The process is applicable, with suitable modifications, to dry plates as well as to wet.

[Printed, 10*d*. Drawing.]

## A.D. 1862, July 10.—N° 1988.

PONTI, Joseph.—(*A communication from Carlo Ponti.*)—(*Provisional protection only.*)—" An improved apparatus for viewing " photographic pictures, and the preparation of photographic " pictures to be used in such apparatus."

The apparatus resembles, in general form, that of a stereoscope, but it is of considerably larger dimensions, and is " provided with " only one large magnifying lens, and only one representation of " the picture " of large dimensions, the object of the invention being to present, by means of one lens, a magnified representation of the picture readily to both eyes of the observer, thus making the object appear of the natural size and with a certain amount of stereoscopic effect; a shade prevents the eyes from being affected by extraneous light. The instrument is placed in a horizontal position, and the end carrying the pictures is made to revolve, so as to bring its longest dimension either into a horizontal or into a vertical position at will. For pictures to be viewed by light reflected from their surfaces, two reflectors are provided; these are closed when transparent pictures are to be viewed, and the opaque end of the apparatus is let down, so that the light passes through the picture. The picture, when inserted, is caused to be slightly concave to suit the form of the lens. During the changing of the pictures, an opaque flap prevents the sudden accession of light from blinding the eyes.

The paper photograph that are to present day and night and other effects, are made so as to reflect the light and yet to be artially transparent. Tinted transparent surfaces are also used in connection with the photograph, when it is viewed by transmitted light.

[Printed, 4*d*. No Drawings.]

A.D. 1862, September. 5—N° 2459.

JOHNSON, John Robert, and HARRISON, John Ashworth.—" Improvements in apparatus for taking photographic " panoramic pictures."

In the curved-plate camera, the sensitive surface forms a part of a cylinder, and the panoramic picture is produced by rotating the instrument upon a vertical axis that passes through the centre of the lens, the sensitive surface being stationary. If the sensitive surface be flat, the panoramic picture is produced by making the plate or surface traverse in the opposite direction to the direction of rotation of the lens.

1st. The lens is in the centre of the camera when a flat plate is used, and rollers that support it are equally disposed round its centre of motion. The plate holder slides within the camera. It is preferred to mount a narrow box at the back of the lens holder to contain the plate holder and to allow of its motion.

2nd. The movement of the plate from the vertical axis is accomplished either by a toothed wheel and rack, or by means of cords and pulleys.

3rd. Adding a fly wheel to the gearing for working all forms of panoramic cameras.

4th. The adaptation of a spring or falling weight to give motion to such cameras.—A fly vane, the arms of which may be set at a varying angle, or a pendulum, may control the rate of motion; a piston working in a tube of water is also used.

5th. The exposure, in cameras moving at an invariable rate, is regulated by an expanding diaphragm which is placed near the lens.

6th. To regulate the form and dimensions of the aperture that admits light to the plate, two slips of sheet zinc, curved so as to admit of a suitable relative proportion of light to the sky and to the foreground respectively, are employed.

[Printed, 2*s*. Drawings.]

A.D. 1862, September 20.—N° 2579.

FORESTIER, Pierre Louis.—" Improvements in photographic " albums."

"Cartes or portraits are surrounded by a bevil edged border "having the appearance of a frame, and ornamented in any suit- "able manner."

"To this end, I commence by cutting in thin cardboard or "strong paper of the required dimensions an opening, correspond- "ing with the outline of the proof to be framed, and ornamented "or left plain at will. Over this first sheet I paste a second "formed of cardboard, corresponding in thickness with the depth "of bevil required, and having a centre opening somewhat larger "than the first. Over this I apply a third sheet of thin card. "with a smaller opening than the preceding provided with an "ornamental border, from which the bevil is formed by pressure "as in ordinary passe-partouts. Finally, I apply, when necessary. "a fourth sheet formed of Bristol board or other paper, the open- "ing of which precisely corresponds with the outline of the bevil "edge, and which may be gaufered or otherwise ornamented at "will. To form a double faced leaf, from each pair of single "frames completed as above, I glue them back to back, after "interposing at the centre two thicknesses of cardboard, which "represent the space to be occupied by the photographic proofs. "The slideway for the introduction of the proofs may be reserved "at any suitable point, top, bottom, or side. The leaves, pre- "pared as above, are bound into albums in the ordinary way, the "form, dimensions, and decorative details being varied according "to circumstances."

[Printed, 4*d*. No Drawings.]

## A.D. 1862, September 20.—N° 2582.

DIXEY, Lewis, and SMITH, George.—"A new or improved "method of tinting, by lithographic printing, photographic por- "traits and backgrounds, and embossing the same."

A photographic print is obtained from a negative in which the background is blocked out with opaque paper, thus producing a white background in the print. Wherever it is desired to preserve the photograph from the tint about to be applied, in the usual manner from a lithographic stone, a piece of tracing paper is gummed upon the photograph with its outline suitably placed. When the tint has been printed on to the photograph and tracing paper, the tracing paper is removed with a damp sponge, leaving a photograph with an evenly tinted background. The tone of the photograph may, in some instances, be improved by printing

" again with the lithographic press all over the picture unprotected " by the tracing paper."

Embossing the photograph gives it "the appearance and ad- " vantages of drawing paper." The damped photograph is laid upon a silk " or other fabric that has the resemblance of wire " mark, such as the drawing papers usually show impressions of," the whole is then placed under a press " with several thickness of " blotting paper next the photograph between it and the press," and submitted to severe pressure. "The photograph will have " the impression of the fabric or block used, and should then " have very much the appearance of a photograph taken upon " drawing paper with a fine even background tinted," and may be mounted on cardboard.

[Printed, 4*d*. No Drawings.]

A.D. 1862, September 23.—N° 2598.

BROOMAN, Richard Archibald.—(*A communication from Jean Jacques Leopold Rousseau de Lafarge.*)—(*Provisional protection only.*)—" Improvements in photographic apparatus."

" This apparatus enables the manipulator to operate in full " light upon damp collodion, to sensitize the glass, and to finish " the proof in the light. The apparatus does not alter the ne- " gative now used, but it serves as an auxiliary to it. In the " apparatus the following characteristic elements are combined :— " The employment of two vertical and independent bath vessels " arranged to allow of the glasses being easily plunged therein ; " one vessel, containing the bath of silver for sensitizing the " damp collodion, may be of gutta percha, hardened rubber, or " other material having no action on the bath ; the other, con- " taining the iron bath for developing the picture, should be of " yellow or orange colored glass, if made of any other material it " should be lined with yellow colored glass. There is a frame " for holding the glasses to be exposed, capable of being opened " and closed at bottom by withdrawing or inserting a sliding bar. " There is a second frame for holding the collodionized glass in " the former frame, which second frame is styled the plunging " frame because it is used for introducing the glass into the bath " vessels entering also witn it ; it is insulated from the first " frame and may oe passed through it ; a piece of cat-gut is

" attached to the plunging frame to raise and lower it into and
" out of the baths."

[Printed, 4*d*. No Drawings.]

A.D. 1862, October 4.—N° 2682.

AMPHLET, SAMUEL.—(*Provisional protection only.*)—" An improvement or improvements in ornamenting surfaces of wood."

" My invention consists in ornamenting surfaces of wood by " mounting directly thereon photographs of objects from which " the background has been cut away, the surface of the wood on " which the said photographs are mounted constituting the background to the said photographs. Or the said photographs " may be mounted upon plain, colored, or ornamented paper or " fabric secured to the middle or other part of the surface of " wood to be ornamented, the size of the said paper or fabric " being larger than that of the photograph to be mounted thereon, " and having any desired figure, but leaving part of the surface of " the wood exposed. The photographs I prefer to use are photographic portraits, or groups, or photographs of sculpture or " statuary, or other detached object or objects, the background " of which has been cut away. I use positive photographs taken " on paper or other flexible material, and I attach the photographs to the surface of the wood or the background on the " surface of the wood by paste or size, or other adhesive material. " The photographs are then sized and varnished, and may afterwards be polished in the manner commonly employed in finishing surfaces ornamented with tartan or Scotch plaid patterns. " The surfaces of wood which, among others, I ornament according to my invention are surfaces of plane tree or sycamore, " satin wood, maple, walnut, and mahogany."

[Printed, 4*d*. No Drawings.]

A.D. 1862, October 20.—N° 2820.

BROOMAN, RICHARD ARCHIBALD.—(*A communication from François Louis Marquier.*)—(*Provisional protection only.*)—" Improvements in transferring designs and prints produced by photography to stone or zinc."

A grained lithographic stone is treated with a composition containing " water bi-chromatized to saturation " and a " thick " solution of gum arabic." The face of a glass positive is then applied, in contact with the stone, and the arrangement is exposed to the action of light. The inventor then raises the positive, and throws "on the stone a solution of potass, for the purpose of "destroying the coat of bi-chromatized gum in those parts pro-"tected from the action of light, and to form a slight engraving; "in about a minute or more I incline the stone and allow it to "drain, then place it flat, and pass over it a sponge containing "some such fatty matter as soapsuds, which I cause to enter into "the engraving produced by the solution of potass, I next wipe "the surface with a soft linen or other rag, and when dry I gum "the stone, as in the ordinary lithographic process, after it has "been acidulated, I leave the stone to rest for about a quarter of "an hour, and then, inclining it, wash it with plenty of clear "water and allow it to drain, I again gum it and place it on a "lithographic press, allow about a quarter of an hour or more to "elapse, and then ink the stone and proceed to print as in the "ordinary process."

[Printed, 4*d*. No Drawings.]

## A.D. 1862, October 25.—N° 2882.

BOURQUIN, John Peter.—(*Provisional protection only.*)—"An improved manufacture of mount for photographic and other "albums, miniatures, and other pictures."

"The object of this invention is to manufacture mounts that "will not soil with the touch, and which when dirty may be "readily cleaned without injury thereto; this I propose to effect by "substituting veneers of choice woods for the cardboard facings "of mounts, which veneer facings will admit of being French "polished, and of receiving various kinds of ornamentation. In "making album mounts I first back leaves of veneer with paper "or cloth, and then cut out the openings for the pictures; I next "place two leaves back to back, so that the openings in each shall "correspond, and apply cardboard filling at certain parts so as to "leave space between the leaves of veneer for the insertion of the "photographic or other pictures, as is well understood; then by "glue or other cement applied to the filling pieces and leaves "where in contact therewith I connect the leaves of veneer to-"gether. When preparing mounts for passe-partouts and minia-"tures as well as albums mounts, I propose to ornament the "mounts either by guilding, painting, or printing, preferring to

" apply mosaic printing which I have already adapted to the or-
" namentation of book and album covers."

[Printed, 4*d*. No Drawings.]

A.D. 1862, October 28.—N° 2906.

SUTTON, Thomas.—" Improvements in preparing albumenized " paper for photographic purposes."

" In the ordinary method of preparing albumenized paper for " photographic printing purposes, the sheet of paper is floated " upon the surface of liquid albumen, but without having under- " gone any previous preparation, in order to render it impervious " to liquids; the albumen therefore soaks into it more or less ac- " cording to the nature of the common sizing which it contains. " Moreover, in the subsequent operations of printing, the various " solutions employed penetrate the paper in the same way. The " prints are therefore less vigorous and brilliant, and also less " permanent, than if the albumen had been applied to the sur- " face only or to a sheet of paper which had been previously " rendered impervious to liquids by the application of some water- " proof sizing. In order to remedy this defect it is proposed, " according to this invention, to prepare the paper before albu- " menizing it, by soaking it in a solution of india-rubber or gutta- " percha, in benzole, or any of the other suitable solvents of these " substances, for example, chloroform, kerosolene, or bisulphide " of carbon. A very good solution for the purpose consists of " five grains of india-rubber dissolved in one ounce of benzole. " After the paper has been immersed in this solution and dried, " it is to be albumenized in the usual manner."

[Printed, 4*d*. No Drawings.]

A.D. 1862, November 5.—N° 2997.

NEWTON, Alfred Vincent.—(*A communication from Paul Schulze and Frederick William Billing.*)—" A new process of " obtaining printing surfaces, dies, and substitutes for photo- " graphic negatives."

The objects of this invention are :—

1st. " To procure a cheap substitute for wood engraving," from which to obtain electrotype printing surfaces.

2nd. To make easy the process of etching metal surfaces.

3rd. To obtain an easy mode of making dies for seals.

4th. To obtain by drawing, "substitutes for photographic negatives, which may be printed from in the same manner as "those negatives by the action of light acting through them upon "sensitive paper."

The principal feature of the invention consists in, first, making a drawing in ink, which is soluble in water, upon a hard surface which has been previously coated with an alcoholic solution of shellac, then covering the whole surface with thin varnish; the plate is then acted upon by water, so as to wash off the latter coating only from the lines of the drawing, the ink of the drawing being itself washed away. The plate, in this condition, can be used for the above purposes.

To carry out the 4th part of the invention, a surface of finely-ground glass is used, the thin varnish consists of beeswax, asphalte, rosin, and lampblack, and when the ink is removed, the dry surface is dusted over with lampblack and varnished. The plate thus prepared "is used for printing upon prepared paper in "the same manner as a photographic negative obtained by the "camera."

[Printed, 4*d*. No Drawings.]

## A.D. 1862, November 7.—N° 3009.

MENNONS, Marc Antoine François.—(*A communication from Eugène Marie Mathieu Plessy.*)—(*Provisional protection only.*) —"Improvements in the manufacture of paper."

1st. "The addition to paper pulps of albumen, gluten, serum, "or other analogous substances coagulable by heat."

2nd. "The application to the surface of the manufactured paper "of one of these substances, either alone or in combination with "coloured or colourless powders."

"The object of these processes is to produce a glazed paper or "cardboard for general printing and photographic purposes, applicable in certain cases as paper hangings, which being thus "coated with a transparent and insoluble matter, may be readily "washed or otherwise cleaned at will. The albuminous matter "is incorporated with the pulp by any convenient means, the "remaining processes of manufacture being carried out in the "usual way. The coating of the manufactured paper may be "effected either by immersion in a bath of the albuminous matter, "or by simple brushing of the latter over the surface. The paper "thus treated is then dried and glazed in the ordinary manner.

"In the manufacture of paper hangings the albuminous matter " may be applied either by immersion, as above, or by means of " rollers operated by suitable machinery, the coagulation being " afterwards effected by steam, alcohol, or other appropriate re- " agents."

[Printed, 4*d*. No Drawings.]

A.D. 1862, November 14.—N° 3071.

CASSAIGNES, VICTOR JULIEN.—(*Provisional protection only.*) —" Improvements in stereoscopes."

"This invention has for its object the giving to the foreground " and background of a stereoscopic picture, when viewed through " the instrument, a natural and pleasing gradation of tints. For " this purpose I color the prisms or lenses of the instrument at " the lower parts, or those parts through which the foreground " of the picture is viewed with a tint suitable for the foreground, " I prefer a yellow or golden color; and the upper parts of the " prisms or lenses, or those parts through which the background " and sky of the picture is viewed, I color with a tint suitable for " the background and sky; I prefer blue for these parts. These " colors in passing from the bottom to the top of the prisms or " lenses I shade off the one into the other, in a regular gradation " of tints." A very good effect may be obtained by tinting either the tops or bottoms of the lenses only; one part of the stereoscopic image will thus "preserve its own proper tint, " whilst the other part will take the color given to it. In place " of applying the tints to the lenses, a tinted glass may be inter- " posed between the picture and the eye, either over or under " the prisms or lenses, the glass being, as already described, so " colored that the fore and the background of a picture viewed " through the instrument shall appear with appropriate and dif- " ferent tints. Or for glass tinted as above described strips of " stained glass may be substituted."

" This improvement can be applied to all sorts of stereoscopes; " and I name chromo-stereoscopes all stereoscopes provided " with it."

[Printed, 4*d*. No Drawings.]

A.D. 1862, November 20.—N° 3119.

BROOMAN, RICHARD ARCHIBALD.—(*A communication from Joseph Corradi.*)—(*Provisional protection only.*)—" A method of

" and apparatus for indicating and recording the course of ships " and vessels."

" This invention consists in recording upon paper the actual " course run by a ship by the combined agency of photography " with natural or artificial light, the compass, and clockwork. " The apparatus is called by the inventor a 'Loxodrograph.' In " the binnacle, under the compass, clockwork is fitted for the " purpose of unwinding paper sensitized in any manner followed " in sensitizing paper for the photographic process from a roll, and " of winding it on another roll. The rolls are placed on opposite " sides of the binnacle, and the paper is drawn horizontally across " and under the compass. A small object is placed upon the " disc of the compass, say, at the north pole, throwing down or " projecting a very intense luminous point upon the sensitized " paper. The effect produced is this, that if the ship proceeds, " for example, in a straight line from north to south, the luminous " point will also trace on the paper a straight line from north to " south, and so on for all changes in the ship's course. Should " a ship be compelled to turn round, the luminous point will " trace an elliptical figure. The action is infallible, as the paper " turns with the ship, while the object remains stationary under " the unfluence of the north pole. The speed at which the paper " is made to travel being known, it becomes easy to ascertain " exactly how long a ship has run on such and such a tack."

[Printed, 4*d*. No Drawings.]

A.D. 1862, December 1.—N° 3222.

JOHNSON, Edward Daniel.—" Improvements in pocket " watches."

1st. Constructing " a watch which will indicate the time in the " dark by the touch."

2nd. " An improved mode of making black dials." This improvement is not mentioned in the Final Specification.

3rd. Obtaining " photographic seconds." " This is a tem- " porarily operating seconds hand for the use of the photo- " grapher, who requires a seconds hand fixed at zero but ready " to move forward various numbers of seconds but to be found " always at zero unless in use," A wheel or " circular rack " is loosely mounted on the ordinary seconds pivot, and carries a hair spring, the end of which is fixed to the plate of the watch, " thus " the wheel on being turned round bends up the spring which

" returns the wheel to its original position when released." A pinion engages both fourth or seconds wheels, its pivot "next " the secondary fourth wheel being mounted in a bar moving on " a stud to and from the said secondary fourth wheel. This bar " is acted upon by a spring that keeps it out, thus when not in " action the spring detaches the pinion from the secondary while " it remains in gear with the primary seconds wheel. A small " stud is passed through the watch case and when pressed upon " moves the bar forward and makes connection with the second " wheel, which then starts and moves forward carrying the " seconds hand."

This improvement is called the 2nd improvement in the Final Specification.

[Printed, 10*d*. Drawing.]

A.D. 1862, December 3.—Nº 3247.

EDEN, ALFRED FREDERICK.—" Improvements in apparatus for " taking minute photographic pictures and magnified pictures of " microscopic objects."

1st. " Adapting to the object end of the camera a small re- " moveable box or dark chamber in which is placed the glass " plate with the sensitive surface when it is desired to obtain a " minute photographic picture. In order to obtain an enlarged " or magnified picture of a microscopic object the latter is placed " in the small removeable dark chamber, and light is admitted " through a proper aperture and allowed to pass through the " object and also through an arrangement of lenses, which will " throw the magnified image on to a sensitive surface placed at " the proper focal length behind."

2nd. In taking micro-photographs, a stop with an aperture of less than one-fiftieth of an inch is placed behind the lens and in front of the negative.

3rd. " Adapting to the object end of the camera a focussing " lens, which is screwed on to the small removeable box or dark " chamber, and is adjusted by means of a sliding tube. This " lens is only used for focussing when it is desired to produce " micro-photographs, but it will also serve as a condensing " lens for throwing light on the object when the apparatus is " employed for taking photographic pictures of microscopic " objects."

This apparatus will also serve for taking full-length microphotographs from life by removing the negative from the large end of the camera and placing the sitter opposite to the said end.

[Printed, 8*d*. Drawing.]

A.D. 1862, December 4.—N° 3249.

SWAN, HENRY.—" Improvements in stereoscopic apparatus."

Stereoscopic pictures are combined with glass prisms in such manner that the pictures, which are in fact depicted on the exterior of the glass, shall produce an image apparently solid and imbedded in the glass. Two prisms are placed " with a side or " surface of the one in contact with a side or surface of the other. " The prisms each have one of their angles a right angle, or " nearly so, and they are so placed together that they produce a " solid having parallel sides. The stereoscopic pictures are either " produced on, applied to, or placed near the adjacent sides of " this solid, one picture on one prism, and the other on the " other. On looking at one of the sides of the solid unoccupied " by a picture the image is seen, the light from one picture " coming from it through the two prisms and direct to one of " the eyes of the observer, whilst the light from the other picture " is reflected at the surfaces of the prisms which are in contact, " and is thus directed to the other eye of the observer." "The " side of the prism through which the pictures are viewed is con- " veniently made convex, the image is then seen magnified more " or less."

The drawings show various methods of mounting the prisms.

If the apparatus is of large size " prisms of water contained in " glass vessels " may be employed. When a certain size is exceeded, the illumination of the pictures is regulated " so that the " light coming from a direction directly or nearly directly opposite " to each picture is confined to such a space as to reach that " particular eye only for which it is intended."

[Printed, 6*d*. Drawing.]

A.D. 1862, December 9.—N° 3298.

CLARK, WILLIAM. — (*A communication from Jean Baptiste Sabatier Blot.*)—(*Provisional protection only.*)—" Improvements " in photographic apparatus."

" This invention relates to an improved arrangement of dark " chamber for photographic purposes, having a laboratory com-

" bined therewith, which permits of its being used in the open " air. This laboratory, which is completely closed, is placed " under the chamber, and is formed of several compartments con- " taining the different baths, in which the plates descend succes- " sively by varying the position of said compartments or baths."

A "lower rectangular box" is placed under the dark chamber of the camera, and has a smaller one "in its interior, open at top, " and containing a series of vertical divisions" "ranged parallel " to each other, their number being varied according to the number " of operations to which the plates are to be subjected." The smaller box is moved into a position suitable for the manipulation to be performed, by means of a rod which passes through the end of the rectangular box, the plate holder is then made to descend into the requisite bath by means of a button at the back of the dark chamber. A spring catch holds the plate holder at its highest elevation. The groove which the above-mentioned button moves in, is closed by a strip of india-rubber, so that no light is admitted into the interior of the apparatus. A space in the rectangular box is set apart "for depositing the plates, bottles, " and other requisites."

[Printed, 8*d*. Drawing.]

## A.D. 1862, December 20.—N° 3410.

PERKINS, William. — (*Provisional protection only.*) — "Im- " provements in the manufacture of a substitute for turpentine, " which is also applicable to the manufacture of varnishes and to " purposes to which turpentine is now ordinarily applied."

"I propose to effect the purification of amyle or fusel oil, or " of the oxides of amyle by means of hypo-chloride of calcium, " chromic acid, sulphuric acid, or any mixture of them, in about " equal proportions, and afterwards to effect the rectification or " distillation of the same at a temperature of not less than 270 " degrees Fahrenheit upon such bodies." An admixture of the said amyles, with their equivalent proportions of liquid hydro-carbons, constitutes the said substitute for turpentine. The amyle, treated as described above, is united "with liquid hydro- " carbon in such proportions as to make it analogous to the " ultimate analysis of spirits of turpentine, say one part amyle to " ten parts of liquid hydrocarbon. In some cases I propose to " use other deoxidizing agents for reducing the amyle as above, " such as potassium or sodium. This substitute for turpentine

" will dissolve gum copal, damar, mastic, lac, or other gums, and " will form an improved varnish, the natural texture of the gums " not being destroyed by fusion, which is not necessary. The " varnishes will be elastic and capable of receiving any amount of " polish, they will also be especially applicable to photographs."

[Printed, 4*d*. No Drawings.]

A.D. 1862, December 23.—N° 3429.

RUSSELL, SAMUEL.—" Improvements in stereoscopes."

The instrument is in the form of a box, and, when out of use, contains the stereoscopic pictures. A cover or covers closes or close the apparatus, and against one side mirrors are fixed. "The " stereoscopic pictures or slides are, when about to be observed, " placed in suitable grooves or holders opposite the mirrors, and " on the side where the lenses or eye-pieces are applied they are " held in a position inclined to the plane of the lenses and also " to the plane of the mirrors;" the light thus falls directly upon the pictures. In some cases the box is made "in two similar " parts, and each part, when separated from the other, becomes " a stereoscope." Sometimes a stereoscope, fitted as described above, is arranged with both top and bottom open, and a handle with spring arms is added, so that the instrument "can be turned " as on an axis between the ends of the two arms," concave cups being applied " to the outer ends of the arms and convex projec- " tions to the ends of the instrument."

To determine the respective angles of the prisms, the picture, and the reflectors :—The distance of the image which is to be produced is marked opposite and parallel to the eye piece. The lower edge of the picture is then placed immediately over the prism, this point and the bottom of the image already drawn is joined by a line, and a perpendicular is raised to the said line, at its bisection. The reflector is placed in the perpendicular, and " the line of the image " being produced upwards so as to cut the perpendicular, will give a point which, if joined by a straight line to the bottom of the picture, determines the position of the picture.

Modifications of the above are set forth.

[Printed, 8*d*. Drawing.]

## 1863.

### A.D. 1863, January 21.—N° 185.

CLARK, WILLIAM.—(*A communication from Adolphe Teissonnière.*) —" Improvements in preparing and obtaining photogenic pictures " or representations."

A positive image is obtained " direct from a positive by the use " of ammoniacal salts combined with the organic matter," and by throwing down black precipitates from certain metallic solutions. When tannate of iron is precipitated from a solution of sulphate of iron, by means of tannic acid, a positive impression is obtained in common black ink.

" A sheet of paper, albumenized with chloride of sodium or " hydrochlorate of ammonia," is sensitized by means of " bichro-" mate of potash (or other ammoniacal salt)," exposed to the action of light, immersed in water, and its whole surface treated with the precipitant. The excess of the precipitant is then removed by means of water, and the " proof " is plunged into a bath of the metallic solution. When the image has acquired a certain distinctness, the " proof " is washed, dried, and coated " with white " varnish dissolved in alcohol."

The principle upon which the process is based is as follows :— The " bichromate of potash or bichromate of ammonia," when combined with organic matter, becomes insoluble under actinic influence, the parts of the paper which are unaffected by actinic force remaining soluble. The soluble parts, on being washed, " give up the substance and leave the paper exposed in its pri-" mitive condition." When the precipitant is applied, the " chlo-" ride of sodium or hydrochlorate of ammonia " acts " as a mor-" dant in the formation of peroxide of iron," and only those parts which have been unaffected by actinism absorb the precipitant, and become black by immersion in the metallic solution.

The sensitizing solution is applied, by means of a sponge, on to the sheet of albumenized paper.

[Printed, 4*d*. No Drawings.]

### A.D. 1863, January 28.—N° 256.

CLARK, WILLIAM.—(*A communication from François Willème.*) —" Improvements in the means and apparatus for copying and " reproducing sculpture and other objects of art."

This invention relates to "photo-sculpture," in which photography is employed in connection with the photograph. This process enables sculpture to be produced of any size, from either a living or inert subject.

The model is placed "on the centre of a ring furnished with "object glasses" or photographic cameras, "placed at equal "distances apart, at the same height." Of the series of photographs thus obtained, two are taken that represent views of the model at right angles one to the other, and on these views two pantographs are employed to act upon the material to be sculptured, one pantograph acting in a plane at right angles to the plane in which the other pantograph delineates. When the said two photographs have yielded their contour to the solid figure, by means of the pantographs, the two next photographs, at right angles to one another, in the series are taken, and so on, until the whole of the photographs have been similarly employed. The figure is revolved, in the requisite degree, by means of a suitably divided platform, between each application of the pantograph.

To obtain bas-reliefs the cameras form a semicircle.

The photographs are contracted or extended, by means of india-rubber, in order to produce sculpture in caricature.

The photographs may be enlarged by means of a solar microscope.

Thin strips of wood, &c., may be used in conjunction with a plastic material, to give the requisite sections of the model and thus to produce mechanical sculpture.

When one pantograph only is used, the platform has a backward and forward as well as a rotating motion.

[Printed, 1*s*. 2*d*. Drawings.]

A.D. 1863, January 29.—N° 267.

POUNCY, JOHN.—"Improvements in obtaining, transferring, "and printing from photographic pictures or images, also in "preparing materials for the same."

This invention consists in the employment of a sensitive ink, which may be transferred or printed from. The surfaces used for the reception of the pictures "may be paper, silk, linen, cotton, "or mixed fabrics, leather, wood, ivory, glass, porcelain, or stone, "or surfaces of metal or metallic alloys," or any other suitable surface. The ink or composition employed to coat the surface selected consists of colouring matter, "fat, tallow, or oil, bichro-

" mate of potash, or bitumen of Judæa, or both of such last-" mentioned substances, and benzole, turpentine, or other hydro-" carbon or analogous solvent;" these ingredients are used in suitable proportions.

When the photograph is to be transferred on to a printing surface, a larger quantity of fatty matter is used in the preparation of the ink than when the picture is not to be transferred from the surface on which it is taken.

To produce a positive photograph by means of a negative picture, the negative is placed on the uncoated surface of a transparent material prepared as described above; the parts not acted on by light, remaining soluble, are dissolved off with benzole or other solvent. The picture is left in printing ink.

The pictures obtained as described may be transferred to a porcelain surface and " burnt in."

When a lithographic stone is the surface operated upon, it must be " grained " before the application of the ink, " and the surface " of the coating should be also 'grained' after it has been laid " on the surface of the stone."

[Printed, 4*d*. No Drawings.]

## A.D. 1863, January 31.—N° 286.

BENNETT, THOMAS.—" Improved arrangements for obtaining " pictorial backgrounds, foregrounds, and perspectives when " taking photographic portraits or sun pictures."

" In taking photographic portraits or pictures in which it is " desired to adopt suitable backgrounds, foregrounds, or perspec-" tives, it has been hitherto the practice to use a painted canvas " descending to, and terminating at the floor or standing place " of the apartment or place. Now, according to my invention, " I use a canvas (or equivalent substance) " suspended upon a roller, " with pulleys and cords," by which the canvas can be raised or lowered, the said canvas " being always of sufficient " length to allow it to be brought down to the floor, then stretched " along the same, and kept in position (if desired) by catches or " grippers, so that the person or persons, or principal object or " objects to be pourtrayed may stand upon a part of the canvas." The background, " which is to remain perpendicular," is to be " painted in the ordinary manner, and that part of the canvas " which is to lay upon the floor is to be painted so as to have the "same concurrent effect as the upright part, and so as imperceptibly to 'run into' the same, for which purpose I have "hitherto adopted the plan of first painting the upright part, "and fitting the same to a corresponding position to that in "which the same is intended to be used, and then stretching the "horizontal part upon the floor, drawing lines or objects by the "aid of camera."

The arrangements have to be varied for landscapes, in this case the perpendicular part being "continued in the horizontal part "with a curve," and not with an angle as is the case with "interiors."

[Printed, 8*d*. Drawing.]

### A.D. 1863, February 11.—N° 376.

BROOMAN, RICHARD ARCHIBALD.—(*A communication from Jean Jacques Leopold Rousseau de Lefarge.*)—"Improvements in "photographic apparatus."

"This apparatus enables the manipulator to operate in full "light upon damp collodion to sensitize the glass, and to finish "the proof in the light. The apparatus does not alter the "negative now used, but it serves as an auxiliary to it. In the "apparatus the following characteristic elements are combined:— "The employment of two vertical and independent bath vessels "arranged to allow of the glasses being easily plunged therein; "one vessel, containing the bath of silver for sensitizing the "damp collodion, may be of gutta percha, hardened rubber, "or other material having no action on the bath; the other "containing the iron bath for developing the picture, should "be of yellow or orange colored glass; if made of any other "material it should be lined with yellow colored glass. There "is a frame for holding the glasses to be exposed capable of "being opened and closed at bottom by withdrawing or inserting a sliding bar. There is a second frame for holding the "collodionized glass in the former frame, which second frame is "styled the plunging frame, because it is used for introducing "the glass into the bath vessels entering also with it, it is insulated from the first frame, and may be passed through it. A "piece of catgut is attached to the plunging frame to raise and "lower it into and out of the baths."

[Printed, 8*d*. Drawing.]

A.D. 1863, February 21.—N° 478.

CEILEUR, Albert.—(*Provisional protection only.*)—" Improve-" ments in apparatus for taking photographic impressions or " likenesses by means of the camera."

These improvements " consist of an arrangement of parts for " holding the negative plate in camera obscuras while taking the " impression therein by means of the lens in such manner that " the plate can be moved laterally and also vertically for the pur-" pose of changing its position and presenting a new surface to " receive an impression."

For the above-mentioned purpose, a plate holder is arranged " with a series of divisions having suitable holes to form catches " at regular intervals according to the size of the impression to be " taken." Slides and stops are employed to traverse and fix the slide holder in a horizontal or a vertical direction, so as to suitably adjust the position of the impression.

" The opening of the plate carrier to the dark chamber is ar-" ranged according to the size of the impressions to be produced, " while the movements of the plate holder must be regulated to " shift accordingly. The plate holder is made of a size and " arranged for a number commensurate with the size of the " impressions to be taken and convenience of operation, as for " instance, for small impressions the plate holder may be arranged " for a hundred or more impressions on the same plate without " removing it from the plate holder, while for large impressions " two dozen may be a convenient limit.

" Instead of or in addition to the plate holder sliding up and " down," the lens or lenses may be mounted on a vertical slide, and the camera may contain two or more dark chambers, " so " that the lenses may be moved to a position opposite a part of " the negative plate and unoccupied by impressions previously " taken, and the plate holder then traversed in its transverse " slide so as to produce another horizontal row of impressions on " the plate."

[Printed, 4*d*. No Drawings.]

A.D. 1863, March 2.—N° 586.

CLARK, William. — (*A communication from Alphonse Louis Poitevin.*) — (*Provisional protection only.*)—" Improvements in " preparing and obtaining photographic impressions and in the " application of such impressions."

In this invention a positive image may be produced from a positive.

1st. Obtaining permanent images on paper.—The surface is sensitized by means of a mixture of perchloride of iron and tartaric acid, the impression is produced through a positive, and a developing bath of caseine or albumen is used; when the photograph is plunged into the coagulating mixture of carbon or other inert colour, the image appears; the sheet is then washed, dried, treated with weak hydrochloric acid, again washed, and, finally, dried. If a negative be used to produce the impression, a developing bath of gelatine is employed. Thus plates either for relief or for copper-plate printing may be obtained.

2nd. Obtaining permanent positive photographs on paper, glass, or other surface.—A uniform layer of gelatine containing the colour is sensitized by means of a solution of perchloride of iron and tartaric acid. When the sensitive surface is dried, the organic matter is insoluble in water, and, in order to take proofs, the coloured coating is impressed by means of "a positive image "(reversed)." The image is developed by plunging the sheet "into warm water, when all the parts which have received the "action of light will disolve." The sheet is then dried, washed in acidulated water, rinsed, and, finally, dried. The organic matter may also be fixed "either with alum, perchloride of mer-"cury, tannin, or other body in solution." Ceramic manufactures may be ornamented, and printing surfaces, and surfaces in relief produced by means of this invention.

[Printed, 4*d*. No Drawings.]

## A.D. 1863, March 17.—N° 711.

BRIERLEY, John Henry, and GREENWOOD, Barzillai.—(*Provisional protection not allowed.*)—"An album belt."

"Having cut a piece of leather or any other material to the "requisite dimensions or size for a lady's or a boy's belt, we cut "out pieces of the material and insert in the place from which "the material has been cut any selection of photographs, say "'the royal family,' or any persons of distinction, and afterwards "complete the manufacture of the belt in the usual manner; or "having by means of a die and an embossing press embossed "the belt at certain distances, we insert or attach the photo-

" graphs to the embossed portion of the belt by means of gum, " paste, or glue, or any solution of the like nature."

[Printed, 4*d*. No Drawings.]

### A.D. 1863, May 13.—N° 1204.

CASSAIGNES, Victor Julien.—" Improvements in stereo- " scopes."

This invention " consists in colouring the prisms or lenses of " stereoscopes, so that each prism or lens presents several different " tints or colors, and so that consequently the pictures or images " seen through these glasses will appear under different aspects, " giving, for example, the effects of morning, of sunrise, or of " moonlight. Thus, for example, to have the foreground of a " golden color, the distance neutral, and the sky blue, the upper " parts of the prisms or lenses should be colored blue and " the lower parts yellow, and between the two there should be " a gradual gradation of the two colors, the one passing into the " other."

Another arrangement consists in placing a flat glass, coloured as above described, over or under the ordinary lenses.

The coloured glasses or lenses may be either fixed or moveable.

The way preferred for colouring the said prisms, lenses or glasses is that set forth in No. 2618 (A.D. 1863), " that is to say " I employ plano-convex prisms or lenses, and I paint on to their " flat sides a cement (I prefer a resinous cement) colored with " transparent colors to the tints required, and by means of the " cement so applied I attach to the prism or lens a piece of " flat glass which serves to cover and protect the painted " surface."

The above improved stereoscopes are called " chromo-stereo- " scopes."

" In place of glass, other transparent substances capable of " being colored may be employed, such as gelatine and others."

[Printed, 4*d*. No Drawings.]

### A.D. 1863, June 9.—N° 1434.

MURRAY, John. — (*Provisional protection only.*)—" Improve- " ments in clips or holders for inserting and fixing photographic " pictures in albums."

"The clip holder or instrument according to my invention
" consists of two thin plates of metal or other material superposed
" and fixed one upon the other with a sufficient interval between
" to receive and hold loosely the photographic picture. These
" plates I form of considerable breadth, say of about half the
" whole breadth of the picture, and form them somewhat of
" a pointed spade-like form, the principal and lower one being
" longer than the other; I also thin them at the point so as to
" effect easy access in the slits of the album. The lower and main
" plate is extended to a convenient length to form a handle, and
" of a suitable shape. Instead of superposed plates this clip or
" holder may be made of one piece, and of ivory, or other suitable
" material; in either case suitably ornamented and finished. The
" great breadth of the clip gives considerable command over the
" photographic picture in inserting and placing or fixing it
" in position, while at same time the slight hold taken of it
" between the spade-like plates enables it to be freely withdrawn
" after adjusting the picture."

[Printed, 4*d*. No Drawings.]

### A.D. 1863, June 23.—N° 1588.

TOOVEY, William.—"Improvements in photolithography, " photozincography, and photographic engraving on copper or " steel plates, or on any other suitable substances."

"To produce on a lithographic stone an impression suitable to " be printed from."—Sized paper, coated with a solution containing gum arabic and bichromate of potash, is exposed "to " light behind a negative." The sheet of prepared paper is placed face downwards on a lithographic stone previously arranged in a percussion press. Several sheets of damped paper are then placed on the stone over the photograph, and a heavy pressure is applied; such parts of the gum as are soluble are thus dissolved and attached to the surface of the stone, producing thereon a negative image. The stone is then dried, coated with greasy ink (the said ink being brought into contact with all the parts of the stone untouched by gum), the coating "removed by passing " through the lithographic press, by spirits of turpentine, or other- " wise," and all the gum washed off. "The stone is then rolled " in with ordinary printing ink, and the positive image appears " in black; it is then printed as every lithographic drawing."

In photo-zincography a zinc plate is substituted for a lithographic stone.

In photographic engraving on copper, &c., the operations are the same as those set forth above, except that a positive is used to impress the image. When the plate is dried, after the photograph has been detached, it (the plate) is varnished. When the varnish is dry, the gum is removed by means of water, and the plate is etched in the usual way; "wherever the plate is protected "from the varnish by the gum it will be attacked by the acid."

[Printed, 4*d*. No Drawings.]

### A.D. 1863, July 10.—N° 1729.

BOURQUIN, John Peter. — (*Provisional protection only.*)— "An improved construction of rolling press."

"The chief object of this invention is to improve the rolling "presses used by photographers for glazing photographic paper "and pictures. Hitherto the travelling table of such presses has "been made with a surface of polished iron or steel, which, "besides being expensive to manufacture, is liable to injury from "rust and scratches. In my improved press I retain the steel "pressing roll, but the travelling table I fit with a slab of polished "glass, which will retain its smoothness for an indefinite time. "In order however to prevent the fracture of the glass under "the severe pressure of the pressing roller, I support the table "upon rollers in such a manner that the glass will not be subject "to a direct bite from the upper and under rollers. To this end "I employ two rollers to carry the table, and mount them in such "a position that a vertical line, drawn from the axis of the "pressing roller, will cut about the middle of the space between "the supporting rollers. This allows of a slight yielding of the "glass to any excessive pressure of the upper roller, sufficient to "prevent the fracture of the glass slab, without interfering with "the glazing action of the press. The surface to be glazed will, "of course, be laid in contact with the glass slab. This arrangement of glazing press is susceptible of extended use in the "arts."

[Printed, 4*d*. No Drawings.]

### A.D. 1863, August 15.—N° 2028.

LUDEKE, Johann Ernst Friedrich.—(*Provisional protection only.*)—"Improvements in the means of keeping cameras or other "apparatus steady when suspended to balloons."

My invention consists in keeping cameras or other apparatus
" steady when suspended to balloons at certain distances from the
" earth, which is effected by means of a peculiar arrangement of
" two or more ropes attached to the car of a balloon, so that
" when the balloon is waving about the camera will be kept
" comparatively steady."

One of the ropes has a weight attached, another rope is
" fastened to the ground at one end by any suitable contrivance,"
and then passes over a pulley to a roller or drum " for the purpose
" of raising and lowering the balloon, although other means for
" obtaining the same result may be employed in place of the
" roller. By this arrangement the camera will be kept steady and
" in its place." The camera " may be raised or lowered by means
" of a rope, and is opened and shut by suitable mechanism; or I
" can make use of a wire in connection with a battery for the
" same purpose. I do not restrict myself to the use of any pre-
" cise number of ropes, as they may be varied if considered
" necessary."

[Printed, 8*d*. Drawing.]

A.D. 1863, August 19.—N° 2063.

BONELLI, Gaetano, and COOK, Henry.—(*Provisional protection only.*)—" An improved mode of and apparatus for pro-
" ducing by the aid of photography optical illusions of moving
" animals and bodies."

1st. " Taking a series of representations by means of photo-
" graphy of men or animals with limbs in motion, and then
" arranging them round a disc, to which rapid rotary motion is
" communicated by suitable mechanism, so that by the different
" figures in the series being brought in rapid succession before
" the eye of the observer, the effect of figures in motion will be
" produced. In carrying out this part of the invention the
" figure or body in motion to be represented is photographed in a
" variety of positions, that is to say, supposing the effect of rais-
" ing an arm or leg is to be produced the limb must first be
" represented in its lowest position, then slightly raised, then
" raised still higher, and so on, until the last figure represents it
" as raised to its fullest extent." " We prefer to produce these
" representations on a microscopic scale arranged in a circular
" series round a small glass or other disc, which may be mounted
" on a central spindle, and adapted to the optical apparatus
" which forms the next feature of the invention."

2nd. The said apparatus "consists of an ordinary compound "microscope mounted in a brass or other tube, to which is "adapted a spindle for carrying the figure disc at one end, and a "perforated rotating eye-piece at the other. Rapid rotary motion "is communicated to this spindle by means of a small winch "provided with a screw wheel, which gears into and drives a "snail or screw on the spindle. The figure disc is placed at the "proper focal distance from the object glass, and the lens at the "eye end is capable of adjustment in the ordinary manner."

[Printed, 4*d*. No Drawings.]

### A.D. 1863, August 22.—N° 2083.

PEGRAM, THOMAS.—(*Provisional protection only.*)—"A plate "holder for the photographic camera adapted to carry different "sized plates, each plate being on the same plane, and in the "same focus."

"This invention consists in the use of two moveable slips of "wood or other material or materials to carry different sized "plates instead of the usual carriers. Two pieces of grooved "wood, or other material or materials, are fitted into the camera "back or dark slide, in the top and bottom, or in the sides, in "which grooves two strips of wood or other material or materials "are fitted, having the sides towards the centre curved or straight, "or notched and rabbited, if notched they can either be rabbited "or have wire attached. The strips move in the grooves, and "can be drawn together or extended according to the size of the "plate so as to receive the plate in the rabbits or on the wires as "the case may be. The slips may be made to slide either by "hand or by means of a left and right screw. If made to slide "by hand a piece of elastic or a spring fixed on the top or at "each end of each slip will hold the slips in position. A spring "fixed upon the door of the dark slide, so as when the door is "closed to press lightly upon the plates will keep them in "focus."

[Printed, 4*d*. No Drawings.]

### A.D. 1863, August 22.—N° 2085.

WATSON, ALFRED.—"An improved method of and apparatus "for inserting pictures in and withdrawing them from photo-"graphic albums."

This invention "consists in forming and employing a card-
"board or other suitable frame, in which the picture is first made "to slide, or is placed in such manner that the edges of the "picture or card carrying the picture are protected. This frame "with its picture is then slidden in through an aperture in one of "the sides or top and bottom of the compound leaf of the album. "The thickening border usually employed between every two "leaves is removed to allow of the frame and picture entering "without making the compound leaf thicker than is allowed for "by the thickening border."

The drawings show one of the frames in which "the picture is "inserted at the top of the frame, the inner edges of the sides of "which are slit or made double" "to receive the side edges of "the picture. The frame with the picture is then slidden in "through an aperture in the bottom of the compound leaf of the "album." "I prefer to fill the blank frames in the leaves of the "album with pieces of cardboard of the same size as the pictures, "and which I remove before inserting the pictures. Each frame "is capable of holding two pictures placed back to back."

[Printed, 8*d*. Drawing.]

### A.D. 1863, September 12.—N° 2252.

WHIPPLE, JOHN ADAMS.—"Improvements in apparatus for "supporting photographic cameras."

This invention consists in applying spring mechanism to sustain the photographic camera; "in the employment of three "supporting pillars, to each of which said spring mechanism "may be applied; and in the use of india rubber rollers." The pillars are hollow, and rods work in them, the camera being attached to the upper ends of the rods. The bar, connecting the lower ends of the rods, works in grooves in the inner sides of the pillars. From each end of a bar connecting the pillars at their upper ends, bearing plates depend, for the fusee pulleys. Suitable cords, connecting the sliding rods with the fusees, and springs, fastened at one end to their respective pulley shafts, and at the other to the bearing plates, are used. Depressing the upper bar unwinds the cords, coils the springs, and lowers the camera. A spring, connected to toggle arms, is placed on the upper bar, and arrests the upward movement of the camera. "These arms "are situated below the said cross bar. The opposite ends of "the toggle arms work in the grooves cut on the inner side of

" the supporting pillars, and are made to impinge the sliding " camera rods by the action of the spring. When it is required " to adjust the camera the toggle arms are withdrawn from the " sliding rods by means of a hand lever operating against this " spring."

The object of this invention is to obtain " an easy and quick " adaptability of adjustment, and noiseless movement, so as to " make the least possible jarring, and so as to readily catch the " proper time of expression or pose of the subject."

[Printed, 8*d*. Drawing.]

## A.D. 1863, September 24.—N° 2354.

HELSBY, William George.—" Improvements in mounting " or setting transparent photographic pictures."

1st. " Mounting or setting at the back and front, but preferably " at the back only, of photographic pictures, taken on a so-called " transparent substance or material ' vignette ' coloured or shaded, " or other coloured, shaded, or ornamented transparent or trans- " lucent sheets, or one sheet only, of glass or other material, to " give to the pictures a mellow and artistic tone."

2nd. " Mounting or setting transparent photographic pictures " arranged as above described, or transparent photographic pic- " tures of any kind, in hinged or folding cases, somewhat similar " in appearance to those at present employed, but having oval or " other shaped openings in or through one or both of the sides " or leaves thereof, so that such pictures when so mounted or set " are in a conveniently portable and ornamental state."

Two transparent sheets may be used, according to a modification of this invention, " the vignette one being in front of that on " which the picture is taken."

Another modification consists in only using one transparent sheet, " the picture being taken on a vignette coloured or other " illuminated or shaded or ornamented sheet, and the picture " side protected by varnish or other means."

A thin heet of paper or other suitable material, that may represent colour paint, may be placed between the front and middle plates of glass, and either before or behind the photograph. " The edges of such sheet may be plain, pectinated, or of any " other ornamental form, and may have any desired figure or " design thereon."

[Printed, 6*d*. Drawing.]

A.D. 1863, October 23.—N° 2618.

CASSAIGNES, VICTOR JULIEN.—(*Provisional protection only.*) —" Improvements in the manufacture of the prisms, lenses, and " glasses of stereoscopes, and in ornamenting glass."

The tinted prisms described in No. 1204 (A.D. 1863) are, according to this invention, produced by applying to the flat surface of the prism transparent cement of different colours, laid on so as to produce the required gradation of tints. "The cement colors " being thus laid on I place a flat glass over them, and this by " the cement becomes attached to the prism or lens and serves " to protect the painted surface from injury. Or the prisms or " lenses may be made each in two or more parts, as in forming " an achromatic combination, and in place of cementing the " parts together with colorless cement, as now practiced, one " part is tinted with graduated colors, as above described, and " the other part is then attached to the first by the cement so " applied. Flat plates of tinted glass and glasses of other forms " I make for use in stereoscopes in a similar manner.

" In some cases I obtain the gradation of tint by varying the " thickness of the layer of cement, thus, if two surfaces are put " together with a colored cement between them, and the surfaces " be pressed together more nearly into contact on one side than " the other, a regular gradation of color will be obtained, the " color increasing in intensity with the thickness of the layer of " cement."

In ornamenting glass, the ornament is painted on the glass in transparent cement colors, and, by the cement so applied, a second sheet of plain glass is fixed to the first. Resinous cements are preferred.

[Printed, 4*d*. No Drawings.]

A.D. 1863, November 24.—N° 2954.

DAVIES, GEORGE.—(*A communication from Mathieu Risler, the younger.*)—(*Provisional protection only.*) — " Improvements in " photography."

The squares of glass used to obtain collodion negative proofs being " fragile, heavy, and expensive," are replaced by talc or mica in sheets. "The improvement then forming the subject " of the present invention is characterized by the employment of " talc or mica in sheets as a substitute for the glass, such sheets "being prepared in the following manner:—A sheet of talc or "mica is to be cut of a corresponding size to the dimension of "the negative which it is desired to produce, and fixed on a piece "of ordinary glass by smearing the edges with a thick gum, the "talc is then to be cleaned simply by a tuft of carded cotton-"wool, the collodion is then poured on the surface, over which "it flows freely. The other operations of sensitizing, developing, "washing, and fixing are effected in the usual manner, and "by means of the ordinary accessories. When the negative is "finished it is only requisite to remove the talc from the glass, "and to place it in contact with the prepared paper to obtain the "positive proof. The advantages resulting from this substitution "of talc or mica for ordinary glass will be readily perceived from "the previous explanation, in fact these sheets are better suited "to all the manipulations, not only from their lightness, but also "from the small space which they occupy, and also from their "non-liability to fracture."

[Printed, 4*d*. No Drawings.]

### A.D. 1863, December 8.—N° 3089.

DESVIGNES, Peter Hubert.—"Improvements in apparatus "for exhibiting dissolving views."

"My invention consists in constructing an apparatus which I "may term a double stereoscope with screens for admitting and "excluding light when and as it may be necessary, and with two "holders for containing the stereoscopic or other pictures or "objects to be viewed. In addition to the ordinary glasses I "employ a diaphanous reflector placed at an angle to those "glasses in such manner that when one screen is raised to admit "light, the picture in a line with the ordinary glasses shall be "viewed, then, upon gradually closing that screen and opening "the other, the picture first seen gradually fades away or dis-"solves, while another picture gradually appears in its place, and "so on alternating for any given number."

The apparatus consists of "two boxes or cases similar to that "of an ordinary stereoscopic," placed at right angles to one another, one carrying the ordinary lenses and partition, and having the picture placed in the line of direct vision, the other without lenses, but meeting the first box in such a manner that the diaphanous reflector reflects the picture to the eye as if it were placed in the position of the picture in the first box, when the shutter of the first box is shut, and that of the second box open. Buttons, in connection with rods, levers, and springs, are used "to drive out or remove the views or slides." A hood is used to exclude false light from the eyes of the observer. In this arrangement the reflected view is placed downwards; in another arrangement the reflected view is placed upwards.

[Printed, 10*d*. Drawing.]

A.D. 1863, December 17.—N° 3187.

JEFFREYS, Charles.—(*Provisional protection only.*)—"Improvements in jewel, photograph, instrument, and other cases."

"My improvements in jewel, photograph, instrument, and "other like cases for holding fancy and other goods, consist in "the application of a spring thereto to throw up the lid or cover "when it is unlocked or the snap released. For light leather "goods such as jewellery cases, I find a small piece of watch "spring answers the purpose, which I simply place within the "lining and the body of the case, carrying the double or middle "part of it round outside the hinge, the one end terminating in "the lower part of the case, and the other in the cover wherein "they are properly secured. The spring is enclosed and hidden "from view by the lining and covering of the case. Springs "may be otherwise applied to answer a like purpose, for instance, "it may be a spiral or other spring, and act on the hinge with "like effect, or it may be a spring compressed between the cover "and the lower part of the case, in closing which again throws "it up when released. In all cases the spring is put in tension "by the closing of the case."

[Printed, 4*d*. No Drawings.]

---

## 1864.

A.D. 1864, February 1.—N° 270.

ROWSELL, Charles John.—(*Provisional protection only.*)—"Improvements in apparatus for viewing photographic and other "pictures, coins, and medals, which is also applicable in the pro- "duction of drawings and paintings."

This apparatus consists of an easel for viewing the above-mentioned objects "by the aid of a magnifying glass." Boards are "hinged together at one end, the lower one forming the base, " while the other may be raised at one end, and inclined at any " angle." "Towards the lower end" of the inclined board a rest is mounted and hinged, against which the objects to be viewed are placed; at the higher end is mounted a hinged transverse bar, on which a magnifying glass is placed. "The object of hinging " these pieces is to fold them down flat." To view single pictures a large magnifying glass is used, "through which a person can " look with both eyes." "With this apparatus a person desiring " to view the picture places it on its rest, raises or lowers the " inclined board to suit his convenience, and in looking through " the glass regulates the position of the picture to the focus best " adapted to his sight by moving it toward or further from his " magnifying glass. The same apparatus I adapt for viewing " stereoscopic pictures." "To the inclined board I hinge a thin " black partition piece, disposed at right angles to the plane of " the pictures or glasses," and "hinged, so as to be laid down " flat when viewing single pictures," but the partition "is raised " in a vertical position for stereoscopic pictures, and I substitute " double eye glasses in lieu of the single one before mentioned, " when the apparatus becomes in fact an open stereoscope."

[Printed, 4*d*. No Drawings.]

## A.D. 1864, February 5.—N° 305.

LEE, Joseph, and THOMSON, James.—(*Partly a communication from Malcolm Mouat.*)—"Improvements in mounting " photographic and other pictures."

"The object of this invention is so to mount or combine pic-" tures" that softness of tone and agreeable depth" shall be produced.

Two pictures of the same subject are used "to produce or make " one picture. We mount or place one immediately over or in " front of the other so that the characteristics of both pictures " are blended together, the front picture being in all cases of a " semi-transparent character." "The back picture may in some " instances have only some of the outlines of the front picture, " and such back picture to produce artistic effect might be " variously shaded, coloured, or tinted."

To adapt these improvements to photographs, an ordinary photograph is suitably tinted for the back picture; the front picture is taken on transparent paper and placed "over or in front of the "back picture either in immediate contact therewith, or, which "we prefer, a short distance therefrom. When the front picture "is to be kept a short pistance from the back one we have found "it convenient to attach the front picture to the back of a sheet "of glass by a transparent varnish and to place a separating piece "between the front and back pictures round the edges thereof, or "we may place a thin sheet of glass between the pictures."

[Printed, 4*d.* No Drawings.]

## A.D. 1864, February 11.—N° 363.

FONTAINEMOREAU, Peter Armand le Comte de.—(*A communication from Alphonse Liébert and Jean Lafon-Saint-Cyr.*)—"Improvements in photographic apparatus."

This invention "consists in the construction of cameras for "magnifying, in which the solar rays are received direct on the "condenser without the use of a reflector." A sensitive sheet placed at one extremity of a wooden cone, receives a magnified image from a negative plate placed at the other extremity, the solar rays passing through the condensing lens, through a convexo-concave lens, then through the negative plate, and through "a "double ¼th usual size objective glass" to the frame carrying the sensitive sheet. Wheel and rack movements, that are worked by hand, are attached to the stand of the instrument to enable "the solar chamber" to receive "the rays of the sun perpen-"dicularly to its axis during the magnifying operation;" a heliostat may replace the hand instrument.

"I make use also of a triplet actinic lens with diaphragmed "centers, and replacing the one-half, one-fourth, one-sixth, and "one-ninth objectives."

The above-mentioned convexo-concave lens renders the converging solar rays parallel and achromatic.

This chamber can be used for the production of positives from a small negative, of a large negative from a small positive, of "reproductions" either reduced or enlarged, and "of direct "images of large dimensions."

[Printed, 8*d.* Drawing.]

A.D. 1864, February 22.—N° 441.

GEDGE, WILLIAM EDWARD. — (*A communication from Jules Abelous.*)—(*Provisional protection not allowed.*)—" A kind of pho-" tographic album forming a new method of publicity, or photo-" graphic and commercial propagator."

" This album will contain, firstly, photographic pictures either " of illustrious personages, artists, reproductions of ancient or " modern paintings, drawings, and engravings, carriages, horses, " objects of art, merchandise, shops, workshops, private houses, " public edifices, and establishments, factories, and new inven-" tions; secondly, inscriptions placed on the drawings or around " them on the margins behind and facing the photographs, which " will be pasted or otherwise secured in the album which will be " designated the photographic and commercial propagator. One " or more photographs may be placed on each leaf of the album; " these photographs may be square, round, oval, octagonal, " hexagonal, or triangular, at pleasure. The album to be of any " form which may be found suitable in folio, quarto, octavo, " 12mo, or other of any desired thickness, and the binding to " vary in design and colour at pleasure."

[Printed, 4*d*. No Drawings.]

A.D. 1864, February 22.—N° 446.

NEWTON, ALFRED VINCENT.—(*A communication from George Gardnier White and Charles Alden.*)—(*Provisional protection only.*) —" An improvement in photography."

In order " to prevent the reproduction of photographic pictures " it is proposed to apply to paper such colors as picric acid, " which by proper mordants can be combined with the fibres of " the paper in a manner that will prevent their removal without " destroying the texture of the paper, and any picture which may " have been produced on the same either before or after it had " been tinted. The yellow color of the picric acid absorbs the " chemical rays of light or fails to transmit or reflect the same, " and a picture produced on paper tinted or dyed with such color " either before or after the picture is taken cannot be reproduced. " By the application of such a picture to a bank bill or other " paper of value, the reproduction by photographic process or the " counterfeiting of such paper is effectually prevented.

"It should be remarked that the invention is not confined to "the use of the picric acid, as other tints or dyes possessing the "like chemical qualities may serve for the same purpose."

[Printed, 4*d*. No Drawings.]

### A.D. 1864, February 29.—N° 503.

SWAN, JOSEPH WILSON.—"Improvements in photography."

This invention relates to "carbon or pigment printing," a "gelatinous photographic tissue" being used instead of paper, or only in conjunction with paper as a temporary packing.

The said tissue consists of a mixture containing gelatine, water, sugar, colouring matter, and bichromate of ammonium. When the tissue is not required for immediate use, the bichromate of ammonium—the sensitizing medium—may be applied subsequently to the tissue. Either a surface of glass or of paper may be used to form the tissue upon, by pouring the above-mentioned mixture (suitably heated) on to the said surface, and allowing the film thus produced to solidify. The tissue is then exposed to light in the usual way and mounted temporarily upon paper, with the photographically impressed surface next to the paper. The mounted tissue is submitted to the action of warm water, so as to remove those portions of the coloured gelatinous matter of the tissue which have not been rendered insoluble by the action of light. The tissue is then removed from the water, dried, and transferred to the surface to be permanently ornamented, the temporary paper mounting having been removed by the soaking in water.

The above-mentioned temporary mounting of the undeveloped prints may be done by means of a solution of india-rubber in benzole; albumen is the cementing medium, when the undeveloped print is mounted upon the surface to which it is to be permanently attached. Gelatine is the cement for remounting the print after development, if such remounting be necessary.

[Printed, 6*d*. No Drawings.]

### A.D. 1864, April 5.—N° 843.

SARONY, NAPOLEON.—(*Provisional protection only.*)—"Improvements in photography."

"My invention has for its object the production of 'cartes de "'visite,' or other photographs similar in appearance to crayon

" drawings, ordinarily known as Richmond drawings, and crayon " heads. I effect this by a double process of vignetting direct " within the camera; firstly, I produce the bust of a person " sitting, or other object required, vignetting that portion only " that is seen in the finished picture by the use of a properly " graduated glass suspended within the non-focal rays of the " posterior of the lens, thus protecting certain portions of the " sensitive plate from the action of the light; secondly, I place " before such portion of the said plate as has already been acted " upon by the light an inverted or reverse vignetting glass in " such a manner as to secure the said part from any further " action of the light, whilst that portion of the plate previously " unacted upon is now brought into the focus of a drawing " which is placed before the lens, and which represents the " afore-named Richmond or scroll drawing, the result being a " scroll vignette or photograph. The same effect may also be " obtained by the process known as 'double printing.' "

[Printed, 4*d*. No Drawings.]

A.D. 1864, April 21.—N° 1000.

BONNEVILLE, Henri Adrien.—(*A communication from Desiré Charles Emanuel Van Monckhoven.*)—" Improvements in pho- " tographic apparatus."

" This apparatus is a solar microscope, which serves to project " upon a sensitized surface the enlarged image of a photographic " negative on glass." This apparatus is composed of two distinct parts—the condenser, and " the enlarging objective."

The condenser, although not achromatic, has no spherical aberration. A crown glass outer double-convex lens having " a very " large ratio of diameter to its focal length " is employed; exactly midway between the said crown glass lens and its focal point, is placed a concavo-convex lens, " which completely corrects the " spherical aberration of the condenser." At the back of the latter lens the photographic negative is posited. This arrangement ensures that any one point of the negative will only be traversed by one luminous ray, of a given refrangibility, from the condenser.

" The enlarging objective " is free from spherical and chromatic aberration—so far as the yellow and indigo rays are concerned; it is an ordinary photographic lens with a negative achromatic lens behind the objective to render the field flat.

The complete apparatus is mounted so that the photographic negative "is placed in the cone of solar rays so as to to be "completely surrounded and equally heated."

[Printed, 8*d*. Drawing.]

A.D. 1864, April 27.—N° 1060.

BROOMAN, Richard Archibald.—(*A communication from Charles Raphael Maréchal, junior, and Cyprien Marie Tessié du Motay.*)—"Improvements in producing photographic pictures "photogenically indelible."

The following process is employed to obtain photographically indelible pictures on siliceous substances generally:—A coating of sensitive collodion is laid upon collodionized caoutchouc; the picture being produced and developed upon the said sensitive collodion, is fixed "by the consecutive action of soluble iodides "or iodo-cyanides, cyanides, or hyposulphites;" the salts or "subsalts of silver are then reduced upon the picture by means "of a solution of sulphate of protoxide of iron." The picture is then "strengthened" by the action of pyrogallic acid or other reducing agent on an acidulated solution of silver. "The strength-"ening may be continued successively in a number of baths." The "strengthened" picture is then washed "in the baths of "cyanides, iodo-cyanides, or hyposulphites, or in ammoniacal "baths."

By plunging the cleansed picture into baths of salts of platinum and gold, the metal of the bath "becomes substituted galvanically "for the coating of silver." The substituted picture is again washed in the above-mentioned cleansing solutions, coated with caoutchouc, submitted to heat in a muffle, covered with a siliceous, boracic, or lead flux, and finally "submitted to the action of the "fire, which fixes and solidifies it."

[Printed, 4*d*. No Drawings.]

A.D. 1864, April 30.—N° 1093.

PRANG, Louis.—(*A communication from Lewis A. Roberts.*)—(*Provisional protection not allowed.*)—"Improvements in preparing "sheets of paper or cardboard for holding photographs and other "articles."

This invention consists in stamping or otherwise making a "recess or recesses of about the depth and size of the photo-

"graph or other article to be held in sheets of paper or card-"board, and in cutting or stamping a slit in each corner of each "recess; the slits are made either at the same time, or before or "after the recesses, and the corner of the photograph or other "article to be held are inserted into the said slits. The sheets "of paper or cardboard so prepared may be used in the con-"struction of books for the preservation of photographs, cards, "pictures, or other articles, the recesses not only serving to "protect the photographs or other articles from injury, but "also to aid with the slits in keeping them in their places."

[Printed, 4*d*. No Drawings.]

### A.D. 1864, May 3.—N° 1111.

GITTENS, Alfred Watson.—(*Provisional protection only.*)— "An improved method of finishing photographic pictures and "marbled papers, and improvements in pressing rolls employed "therein, and applicable to other pressing rolls."

"My invention consists in subjecting photographic pictures, "after being developed, and marbled papers, to pressure between "one or more heated rolls or plates."

My improvements in pressing rolls consist in heating "the "upper or lower roll by means of a spirit or other lamp extending "the length of the roll; the lamp is maintained stationary "while the roll is free to revolve; this part of my invention "also consists in raising or lowering one or other or both rolls by "means of screws which work in nuts in boxes in which the axles "of the rolls are carried."

[Printed, 4*d*. No Drawings.]

### A.D. 1864, May 6.—N° 1149.

RIEDER, Aimé.—"Improvements in shades, spectacles, and eye "glasses, which can also be applied to telescopes and stereo-"scopes."

"My improvements relate to the manufacture of shades, with "or without glasses, in such manner that they can be folded into "a convenient size for holding in the pocket, and for this purpose "I construct the framework of the shade of four different parts, "viz., the top, the two sides, and the front, in which the glasses "can be inserted, and to which I also attach a handle and unite "the edges of the three latter parts to those of the former either

" by a metallic joint or by a joint of flexible material, so that they " can be easily folded flat, one on the other; and in order that " such article be further reduced in size, I divide the top and " front parts into two halves down the centre, which parts are " connected as above-mentioned so as to be easily folded one on " the other."

" Another improvement in folding opera glasses consists in " constructing them so that they pivot on a central axis passing " through the line of their diameter."

The first-mentioned improvements are applied " to stereoscopic " boxes, the bottom and back pieces of which being united to the " four other parts mentioned on one or both of their edges by a " fixed flexible joint, or a fixed metallic joint, or by a metallic " attachment, such as a hook capable of being disconnected."

[Printed, 10*d*. Drawing.]

### A.D. 1864, June 9.—Nº 1438.

SARONY, NAPOLEON.—" Improvements in the production and " treatment of photographic portraits or pictures."

An ordinary negative vignette is combined with a " second or " etching negative photograph " in such a manner as to produce (by the action of light) a photograph having the appearance of a drawing or engraving, the said second negative consiting " merely " of the hatching or sketchy lines photographed upon the glass " from a chalk, pencil, or other drawing of such lines."

Another process for producing the effect of etched vignettes consists of " etching the lines by acid upon the usual vignetting " glass, employing if necessary thin varnish or any other suitable " medium as a coating for the etching lines, and to produce half " tints, if the crude lines should print too darkly; or I can coat " the vignetting glass with japan varnish, then scratch the etch- " ing lines through the varnish, and produce the half tint, as " above described, by coating the lines with thin varnish or any " other suitable material. To produce the same effect by other " modes of printing, such as lithography, copper-plate, or wood " engraving, I merely print the lines from the stone, plate, or " block upon the paper upon which the photographic picture is " to be taken, or after it has been taken."

[Printed, 4*d*. No Drawings.]

A.D. 1864, July 6.—N° 1676.

GEDGE, William Edward. — (*A communication from Jules Abelous.*)—" An improved album."

"The improved album, the subject of this invention," contains : " In the first place photographic pictures, either inserted in the " usual way of inserting cartes de visite, or permanently photo- " printed on the leaves of the album."

" In the second place the above-mentioned pictures are sur- " rounded with ornamental, printed, engraved or embossed frame- " work, leaving when engraved or embossed blank spaces in the " grouping, in which spaces are printed in typographical cha- " racters, inscriptions of every character, whether analogous with " the object they surround or distinct therefrom, and intended " for other purposes."

" It is to be understood that one or more photographs may be " placed on each leaf of the album, and that the photographs " themselves may be of any form, square, round, oval, hexagonal, " or triangular at pleasure. The album itself to be of the form " most suitable to the pictures or objects illustrated on or in its " leaves, and its size may be folio, quarto, octavo, or duodecimo, " or any intermediate or other size at will, the thickness depend- " ant partly on the size of the pictures, but not confined to any " proportion, the binding to be varied as taste may direct in " design, material, or colour."

[Printed, 4*d*. No Drawings.]

A.D. 1864, August 23.—N° 2079.

GRISDALE, John Edwin.—" Improvements in apparatus for " washing photographic prints."

Centrifugal machinery is employed for washing photographic prints.

In one arrangement, a revolving drum is used in connection with a trough in which the drum is partially immersed. The prints (fresh from the water in which they have been placed on their removal from the fixing or other bath) are placed in piles round the circumference of the drum, each pile being composed of alternate prints and sheets of wire gauze. The piles are held in their places by means of gratings that form a part of the drum itself.

"According to another arrangement, the piles above described " may be laid flat upon a disc, which is made to revolve either " vertically or horizontally in a trough or cistern, provision being " made in the horizontal arrangement for allowing the piles to be " brought in or out of contact with the water as required; or in " lieu of the photographic prints being disposed in the form of " piles or packs round a drum or revolving disc, they may be laid " separately and individually round the surface of a drum, a " webbing of open or reticulated fabric being wound on such " drum simultaneously with the placing of the prints thereon, so " as to interpose a thickness of the fabric between each succeeding " layer of prints. The process of washing consists in alter- " nately driving out the moisture from the prints by the centri- " fugal action of the revolving drum or disc, and saturating the " prints again," the trough being supplied with water during the latter operation.

[Printed, 10*d*. Drawing.]

A.D. 1864, August 29.—N° 2122.

THOMAS, RICHARD WHEELER.—" Improvements in tents and " apparatus employed in taking photographic pictures."

This invention consists of a certain " combined arrangement of " parts," " whereby a box or case combined with flexible material " is arranged to open into a tent, in which the apparatus and " materials employed in photography are fixed in position suitable " for use as soon as the tent is opened out."

The apparatus and materials " retain their positions when the " box or case is contracted for transport, thus dispensing with " the packing and unpacking of the apparatus and materials."

The tent, when in use, is fixed at the top of a tripod stand, and " is constructed of a box, " having the two ends and back fixed " to the bottom, whilst the front is hinged at its bottom edge in " such manner that it may fold down." The inclined bath is secured in the position in which it is used, either during transport, or when the tent is in use. The top or cover of the box is in two parts, respectively hinged to the two ends of the box; " when " closed the cover comes just above the top of the inclined bath." The flexible material is fixed to the edges of the cover in such manner that, when its two parts are open, the flexible material forms a covering to the tent as well as a part of the back of the tent; two sliding bolts hold the upper part of the front above the head of the manipulator. "The gutta percha sink is placed in "the centre of the bottom of the box or case, and the camera is "arranged to be strapped or fixed to the bottom of the tent when "placed on the cover of this sink."

[Printed, 4*d*, No Drawings.]

A.D. 1864, August 31.—N° 2143.

ROLLASON, Alexander. — (*Provisional protection only.*) — "Improvements in glazing and varnishing paintings, prints, and "photographs, and which improvements are also applicable to "plain paper, woven fabrics, leather, and other substances."

"I take collodion, either alone or in combination with any "other suitable gum, such as gum animi, common resin, balsam "of aniseed, or oils, such as linseed, nut, or castor, that will "dissolve therein, and in about the proportion of one pound "of collodion to half an ounce of gum, resin, or oil, and having "poured or otherwise placed the same into a polished surface of "glass or metal, I allow it to become quite dry. Upon this "collodion or its compounds when so dry I brush or pour a "cement of gum arabic, dextrine, albumen, gelatine, sugar, "honey, isinglass, or of a nature similar to these, either separately or combined, and if combined in about equal proportions, upon which again, when sufficiently dry, I place my "painting or photographic or other material, having previously "wetted or moistened the same. When quite dry I cut rounds "its outside, upon which it may be removed from the glass or "metal together with the cement and collodion or its compounds "all firmly attached together, producing a beautifully glazed and "varnished surface."

[Printed, 4*d*. No Drawings.]

A.D. 1864, September 8.—N° 2197.

FRUWIRTH, Daniel.—"An improved apparatus for cutting "photographic impressions, cartes de visite, stereographs, and "other such like purposes."

By this apparatus "all four sides of the card or paper are cut "away by one stroke of the cutter."

"The apparatus consists of a table plate of metal or other "material on which a block or rest of iron or steel is fixed, its "dimensions and form corresponding to the size and pattern to

" which the card or paper is to be reduced; over this block is a " four-sided or other shaped cutter or shears, the internal edges " of which fit the outer edges of the block; this cutter is held by " an arm whose extremity is attached to a hinge or fulcrum on " the end of the table furthest from the block; underneath the " table a flat spring is set by a screw, and the moveable end of " the spring is connected by a vertical pin (passing through the " table) to the arm of the cutter, so as always to exercise a down- " ward powerful pressure or strain upon the same. Near the " centre of the table a horizontal cam is set in bearings, and is " provided with a handle by which it may be turned by the " operator. The surface of this cam is made to assume an " oblique or incline during its entire circumference, and at its " highest point a vertical ledge descends to its lowest surface; in " its revolution it acts upon a vertical pin underneath the arm of " the cutter, and raises the same against the action of the spring; " the card or paper is then arranged on the block and under the " cutter, and the cam being turned slightly further, the pin " escapes over the ledge and the cutter is forced down, thus " shearing all sides of the paper or card with clean and even " edges."

[Printed, 6*d*. Drawing.]

A.D. 1864, September 20.—N° 2300.

SCHOTT, JOHN BERNARD.—(*Provisional protection only.*)— " An improved envelope frame for the use of photographers and " others."

Upon opening the envelope, the uppermost photograph, if several are enclosed, " will be at once visible encased in a frame."

" I purpose effecting this by taking, say, an oblong sheet of " paper, millboard, or other suitable material (paper stretched on " linen for example), one end of which I cut to the usual shape " of the overlap of an envelope, and near the extremity of which " I cut out an oval, oblong, or other shaped piece, leaving an " an opening of the shape of the photograph I purpose enclosing " (this opening or frame may be surrounded by embossing, gild- " ing, or other ornamentation). I fold the part carrying this " opening over on to the body of the sheet of paper or other " material, to which I fasten this overlap at the top and bottom " with gum, paste, or other adhesive substance, thus forming at " one end of my oblong sheet a case open only on one side, by "which the photographs are inserted, the upper one appearing "framed in the opening cut out of the overlap. The part con-'taining the photographs is then folded over on to the sheet, the "end cut and gummed like an ordinary envelope, overlap brought "over, and the envelope is closed, leaving the usual back for "address. The paper on the right of the photograph and upon "which the case or frame folds, that is to say, the inner face of "that part which carries the address, gives space for writing a "letter or note. I have stated that I use an oblong sheet of "paper or other material, but it is evident that this shape is not "obligatory."

[Printed, 4*d*. No Drawings.]

## A.D. 1864, September 23.—Nº 2338.

WOODBURY, WALTER BENTLEY.—"An improved method of "producing or obtaining by the aid of photography, surfaces in "'relievo' and 'intaglio' upon aluminous, vitreous, metallic, or "other suitable materials."

When the said surfaces are viewed as a transparency, the varying thicknesses in the material will show the design.

This invention may also be used to produce any design in relievo or intaglio, or copies of such articles.

A mould is first obtained, according to the chief feature in the invention, in which the thicknesses of material "are a tran-"script of the light and shadows in a photographic negative;" castings may be obtained from an electrotype of this mould.

A film, one-eighth of an inch thick, is spread upon a sheet of glass and allowed to solidify, the said film containing sugar, gelatine, albumen, and "bichromate of ammonia." The film, so produced, is cut into pieces, stripped from the glass, and has its polished side laid on the negative; the said film and negative are then placed in the focus of a solar camera, so that the design is produced clearly at the back of the film. Another mode (not mentioned in the Provisional Specification) consists in the interposing talc between the glass and the film, and placing the talc side next the negative in printing. According to another method (also not mentioned in the Provisional Specification), the film is thoroughly dried in a dessicating box and then used. When sufficiently exposed, the non-actinized portions are washed away by means of hot water, and an image is produced in which the lights are depressed and the shadows raised.

By printing from a deeply-coloured negative, a gelatine mould is obtained which will bear casting from.

[Printed, 4*d*. No Drawings.]

A.D. 1864, September 24.—N° 2347.

WORTLEY, Archibald Henry Plantagenet Stuart, and VERNON, William Warren.—(*A communication from Jacob Wothlij.*)—" A new chemical process for producing photographic " pictures," also " an invention " " in the preparation and manner " of using the materials in such process."

1st. Preparing photographic paper.—The surface of ordinary photographic paper is enveloped with a solution of starch, albumen, or other simple medium, in combination with any other material suitable for any special purpose, and is either placed under pressure between two polished surfaces, or subjected to a suitable rubbing process, or the prepared paper may be passed between specially constructed rollers; in either case the fibre of the paper is brought into actual contact with the starch or other medium, and, after being fully charged, the fibre is laid down again.

2nd. Sensitizing photographic paper.—A mixture containing collodion, nitrate of uranium, and nitrate of silver, is employed for this purpose. The sensitive solution is poured on to the surface of the paper, in a properly lighted room; the paper is then floated, and the superfluous quantity of solution poured off; it is then dried in a dark room. When this process is used, the positive is printed to the actual intensity of colour required, and not over-printed.

3rd. Producing on photographic paper a "matt" or "dead" surface.—The sensitive salts are dissolved in alcohol and water, instead of in collodion; the paper is prepared by covering it with the said solutions. Saccharine substances may be added to the solutions if necessary. The paper, thus prepared, is hung up in a dark room to dry, when it is ready for the process of printing.

[Printed, 4*d*. No Drawings.]

A.D. 1864, September 27.—N° 2373.

LANE, Kedgwin Hoskins.—" Improvements in cases or re " ceptacles for photographs."

"I form a square, polygonal, or cylindrical box, inside which I "place a case or chamber mounted upon a spindle, which has its "bearings in the top and bottom of the outer box; upon this "inner case the photographs are placed, which case can thus, by "means of the spindle before named, be caused to revolve so as "to bring each photograph in rotation opposite to a door or "opening which is formed in one portion of the outer box. "Lockets may be thus made to contain many miniature photo-"graphs in small compass, whilst in larger forms, such as may be "required for drawing-room tables, the case may assume very "ornamental shapes, such as a pagoda, which could be made "by mounting several cases in tiers one over the other; each tier "containing its revolving chamber caused to rotate by the same "spindle.

"Another modification of my invention consists in forming "lockets or cases for the same purpose upon the plan of a ribbon "measure, hinging together a number of frames for photographs, "which frames may be made, by preference, in thin metal, in such "manner that the ribbon or chain thus formed could be wound "round a spindle within an outer box, so that upon the ribbon "being drawn through an opening in the outer box, the photo-"graphs may be successively exposed to view, and by turning the "spindle they may be again enclosed."

The drawings show this invention applied to a locket, a cabinet, and a pagoda; a locket constructed upon the principle of a ribbon measure is also shown.

[Printed, 10*d*. Drawing.]

### A.D. 1864, October 6.—N° 2465.

FONTAINEMOREAU, PETER ARMAND LE COMTE DE.—(*A communication from Jean Nicolas Truchelut.*)—"Certain improve-"ments in photography for obtaining images direct on cloth and "other materials."

Instead of employing dessicated oil for preparing the cloth, virgin wax or spermaceti, together with resin and elemi, is dissolved in essence of lavender, and mixed with carbonate of lead, and used for the said purpose. When the cloth is dry, it is heated, and its surface is spread over with a thin layer of white wax, to which has been added the above-mentioned resinous substances. The collodion is spread on the resulting waterproof surface, and "is employed to make a negative or a positive." "To "render the proof indestructible," after the development and fixation of the picture, it is "allowed to dry for some days; then "a flat iron, with a long handle is moderately heated and passed "behind the proof over all its surface;" the proof thereby becomes "an entire homogeneous substance with the layers of paint, oil, "and collodion."

The method of operating "at the moment of producing the "image" is:—The glass—"the negative and objective"—being in their places, the cloth is fixed, the focal distance ascertained, and "the objective" is covered. The cloth is taken off, and the collodion spread over it; when the latter is dry, the sensitive plate is put into the silver bath, the cloth replaced, and "the objective" uncovered. "When the proof is slightly indicated," it is developed, fixed, and washed. "For the proofs on glass I use collodion "preferably to the iodites" [iodides?] "of ammonia and potassium mixed in equal quantities"—this is not mentioned in the Provisional Specification. "The collodion is spread with rectified "alcohol." The silver bath contains nitrate of silver and acetic acid. The developing solution contains gallic acid and acetic acid. The fixing is done by means of hyposulphite of soda. A blue shade may be given to the proof by means of a solution of chloride of gold.

[Printed, 4*d*. No Drawings.]

### A.D. 1864, October 13.—N° 2526.

BROOMAN, Richard Archibald. — (*A communication from Arthur Baudesson and Paul Houzeau.*)—"Improvements in the "manufacture of prussiates of ammonia, and the application of "prussiates of ammonia to dyeing, printing, and to photography."

"This invention consists in depositing under the influence of "light upon paper, textile fabrics, wood, glass, porcelain, and "other suitable substances, a matter colored or ready to become "colored, by the aid of chemical agents, by immersion, dyeing, "printing, or otherwise."

The invention also consists in the manufacture of "prussiates "of ammonia." For this purpose sulphate of ammonium acts upon prussiates of potash; the proportions, together with the salts of potash used, determine the salt obtained.

In applying the said prussiates to photography, the following sensitizing solutions are used:—1. A solution of ferricyanide, or ferrocyanide of ammonium, or a mixture of these salts. 2. A solution "of prussiate of potash or other prussiates." 3. A solution "of prussiate with a base of iron at the maximum, and of organic "acid salts of iron;" the image appears in blue. 4. A solution "of double salt of iron and of ammonia, an oxalate, a tartrate, a "citrate, or otherwise;" acid prussiate of potash developes blue tints, tannin yields black tints. Sulphocyanide of ammonium or of potassium yields blood red tints.

In the "application upon textile fabrics," for blue, the prussiates of ammonia or of potash, with or without citrate of iron, are used; for black, the blue image is dyed with logwood and glue; for violet, the blue is decolorized by carbonate of soda, then treated with madder, chloride of lime, and soap; for bronze, the decolorized image is dyed with madder and quercitron, and for olive and brown the decolorized image is dyed with quercitron, and is brought up in a soap bath.

[Printed, 4*d*. No Drawings.]

### A.D. 1864, October 14.—N° 2539.

DALLMEYER, John Henry.—"Improvements in the con-"struction of lenses, which improvements are especially applicable "to lenses for photographic purposes."

This lens "is chiefly intended for landscape photography."

"I form two lenses of crown or plate glass, by preference of "two different kinds of crown or plate glass, and divide for a "given focal length the focal power between the two lenses, by "preference in the ratio of 1 to 3, both acting therefore as posi-"tive or collecting lenses. The form of the first or anterior crown "or plate lens is by preference a concave meniscus. The second "or posterior crown or plate lens is also by preference a concave "meniscus. The central lens situated between the two crown or "plate glass lenses is made of flint glass and concavo-convex, "and of such focal power as when placed between the two crown "or plate glass lenses the adjacent surfaces having the same radii "of curvature and being cemented produce an achromatic whole "or nearly so."

When architectural objects are to be photographed, the straightness of the lines will be better maintained if the anterior lens b placed at a distance from the other parts of the compound lens.

The diaphragm is placed in front of the anteror lens.

" By my improvements a more perfect correction than hereto- " fore of both the spherical and chromatic aberrations of the " oblique or excentric pencils," [is ?] " obtained, whilst at the " same time the flint glass lens is protected from injury by the " crown or plate glass lenses on either side of it.

[Printed, 6*d*. Drawing.]

A.D. 1864, November 3.—N° 2717.

FOX, THOMAS.—(*Provisional protection only.*)—" An improved " photographic process."

" This invention relates to an improved photographic process " in which bichromate of potash is brought into contact with " logwood; and in practising the invention paper is sensitised by " floating it upon or steeping it in a solution of bichromate of " potash and sulphate of copper. The paper having been dried " in the dark is exposed to the action of light in a printing frame " or in the camera, and is afterwards floated upon a decoction of " logwood, which is by preference warm, and in half a minute to " a minute thereafter the print is carefully removed and dipped in " hot water, which carries off the superfluous logwood. The result " is a very distinct print with the original lights of the picture " or transparency printed from, or of the object before the camera " reproduced as lights, and with the shades of a deep black tone. " The print is completed by drying and varnishing. A good " sensitising solution is made by taking one part by weight of " bichromate of potash to two parts of sulphate of copper; " whilst the tone of the print may be greatly varied by varying " the strengths of the sensitising and logwood solutions. The " print may be transferred to paper, cloth, or similar substances, " by passing it through a press with the printed side in contact " therewith; and it may be transferred in a similar way to wood, " stone, and the like, if a coating of albumen is first applied " thereto."

[Printed, 4*d*. No Drawings.]

A.D. 1864, November 11.—N° 2800.

WILLIS, WILLIAM.—" Improvements in processes for copying " or reproducing by the agency of light, drawings, engravings, " lithographs, and photographs, and written and printed docu- " ments."

By this invention a positive photographic copy may be at once obtained.

The principal feature of the invention "consists in obtaining " copies or reproductions by the action of light by preparing a " surface with a solution containing a soluble chromate mixed " with an acid which will combine and form a compound with the " oxide of chromium formed by the action of light, and darken- " ing the impression produced by the unaltered chromate by " means of aniline or other organic substance capable of forming " a dark and insoluble compound with the said unaltered chro- " mate,"

In copying a mechanical drawing on paper, the sensitizing solution contains certain proportions of bichromate of ammonium and phosphoric acid. When the paper has been impressed, in the usual way, in the printing frame, it is laid on the bottom of a wooden box; in this situation it is exposed to the vapour of aniline, that evaporates from bibulous paper fastened to the lid of the box, and thus developed. The fixing takes place by alternate washings in weak acid and in water.

Another sensitizing solution contains chromate of copper, sulphuric acid and phosphoric acid. A third contains phosphate of copper, sulphuric acid, and chromic acid.

The pyrrol bases may be used instead of aniline for developing.

[Printed, 4*d*. No Drawings.]

### A.D. 1864, November 25.—N° 2953.

CROZAT, LEANDRO.—(*Provisional protection only.*)—" Improve- " ments in photographic processes, and in portraits or images " produced thereby."

To produce portraits in " double ground," the sitter is placed before a white cloth, the ground of which is coloured and graduated or shaded. The negative is taken in the ordinary way.

" To produce a shaded bust terminated in white."—A piece of paper is applied to the non-collodionized side of the plate, and the size of the bust traced thereon; this outline is marked on a piece of opaque cardboard, and the said cardboad placed in the pressure frame. The " shading ground " has thus been obtained. To obtain the " general ground," a proof of the bust is stuck on the painted side of the glass. " Then continue applying the white " ground portraits on the glass so that the head perfectly covers " the silhouette, and this in its turn should entirely cover the

" shirt front, and by thus supporting the portrait with a tablet of " the same size as the card," expose it to the sun. The double ground, when obtained, must be toned and fixed as usual.

The eyes and shirt front are then to be covered with gum arabic solution to preserve them from the colouring which is to be applied to the whole face. When coloured, greater brilliancy is given to the portrait by stretching it on prepared glass and allowing it to dry; it is then detached from the glass and stuck on cardboard with a solution of gum arabic.

The colouring contains alum, cream of tartar, cochineal, saffron, and magenta. The preservative varnish for preparing the glass contains sulphuric ether, alcohol, and "photographic cotton;" the second solution for preparing the glass is a solution of gelatine.

[Printed, 4*d*. No Drawings.]

## A.D. 1864, December 6.—N° 3048.

MARTIUS, CARL ALEXANDER.—(*A communication from Johann Baptist Obernetter.*)—" Improvements in the application of photo- " graphy to the ceramic art or to glass."

" This invention consists, firstly, in a combination of processes " whereby photographic pictures are obtained, applied to, and " burnt in upon articles of porcelain or glass; and, secondly, in " a peculiar method of causing the said photographic pictures to " be composed of materials which can be fused or enamelled upon " ceramic articles including glass."

A glass plate is coated, in the ordinary way, with a sensitive solution containing gum, sugar, glycerine and bichromate of ammonium; the plate is then dried, in the dark, in a drying stove. A positive picture is placed on the prepared surface of the plate and the combination is exposed to the action of light, for the time requisite to impress the image. The plate, on which the image is now slightly visible, is next treated with a mixture containing porcelain colour with its flux and some dry powdered soap; this operation is continued until the requisite density is attained, the porcelain colour and its flux being deposited upon those parts of the plate which were protected from the action of light by the superposed positive. To effect the removal of the picture in porcelain colour from the supporting surface, it is coated with collodion, dried and immersed in alkaline water. The washed film is then pasted, picture side undermost, on the ceramic article.

The picture, so adapted, is dried spontaneously, and the collodion film is dissolved therefrom by the action of solvents, "after which " the article with its applied picture is burnt in an ordinary muffle " furnace, as conducted in the usual process of enamelling."

[Printed, 4*d*. No Drawings.]

A.D. 1864, December 14.—N° 3107.

CLAUDET, Antoine François Jean.—" Improvements in " photo-sculpture."

" The original photographic pictures similar to those hereto- " fore used to aid the artist's eye when modelling by hand are " themselves enlarged to the exact size intended to be given to " the model, and are projected on to the block of clay or other " material by a camera or cameras, whilst two pictures at right " angles of like size previously drawn by hand," "are generally " employed at the same time to further aid the artist's eye in " judging of the progress of his work, and to determine the " various planes at which every part of the block is to be " brought during the modelling."

A revolving table, the axis of which is capable of being moved to and fro, is used to place the block of clay upon. Thin blades are pressed into the clay so as to receive the picture and thus to afford a guide to the tool. If the work is pushed back and a piece of ground glass brought into focus, the image received on the glass will guide the artist. If a screen be used, it is stated (in the Final Specification) that, "it is better that it should be on " the tool employed by the artist in modelling;" an apparatus for guiding the tool in this case (called the "plastimonograph") consists of a disc, fixed in the middle of a flat bar, which slides in slots that are made through the upright sides of a frame placed in the plane of the image. " By means of a weight, cord, and " pulleys, the bar is suspended horizontally between the sides of " the slots in the uprights, and remains supported in any position " in which it is left during the various stages of the operation; " and it can be made to move in all directions required by the " least impulse of the hand of the artist."

[Printed, 8*d*. Drawing.]

A.D. 1864, December 16.—N° 3119.

CHEVALLIER, François Auguste.—" Improvements in pano- " ramic apparatus."

"My improved apparatus enable me to obtain photographs "which directly give to a fixed point previously determined the "angle that the vertical points make between them passing "through the fixed point and the different points visible from "this fixed point;" in instruments of this class, the axis of rotation of the "optical system" is vertical, and the sensitive surface is in a horizontal plane. If the axis of rotation be horizontal and the plane of the sensitive surface vertical, photographs can be obtained which give "directly the angles of altitude of the "different points visible from the fixed or station point when "the levels are above or below the horizontal level passing through "the fixed point."

In the first arrangement described in the Specification and shown in the drawings, the dark chamber and sensitive surface are fixed and the "optical system" is movable. The "optical "system" may either consist of a rectangular prism and lens, or of a lenticular prism only. Hairs are employed to indicate the angular direction of the vertical and horizontal planes. A tangent screw, in gear with the disc on which the "optical system" is mounted, moves it as required: converging shutters, a vernier to measure azimuths, and a telescope, are the accessories of the instrument.

In the second arrangement, the sensitized surface is fixed and the dark chamber and "optical system" are movable round a fixed centre.

In the third arrangement, a radial movement of the dark chamber and of the "object glass" is adopted, and the sensitive surface is fixed.

A fourth arrangement, either with a lenticular or conical prism, "is an apparatus comprising a dark chamber, an optic, and a "sensitised immovable surface, and giving the compass of the "horizon at a single stroke."

The part containing the optic may be placed under the sensitized surface.

[Printed, 1*s*. Drawing.]

## A.D. 1864, December 21.—N° 3175.

JOHNSON, John Henry.—(*A communication from Gabriel Jules Bourdin.*)—"Improvements in photographic apparatus."

This invention consists of "the application and use of an "apparatus or camera, wherein the photographic plate is rendered

" sensitive and the image or picture taken, and subsequently " developed without the necessity of removing the plate from " such apparatus or camera."

The plate " is pressed closely against the open end of a vessel " of orange colored glass, such vessel being contained between " two uprights, one at each end thereof, which uprights are held " against the vessel by screws or otherwise." The front upright contains the photographic lens, " an opening being left in the end " of the vessel corresponding therewith, and the back one has an " opening made therein provided with a door." When the plate is introduced through the said opening, and the door closed by the pressure of a spring, the sensitizing liquid is admitted to the vessel; for this purpose a tube is inserted through a small aperture in the upper part of the vessel (protected by a valve), the tube having a hollow air-tight collapsible ball at its apex that contains the sensitizing liquid. On the compression of the ball, the liquid enters the glass vessel, and (by suitably turning the said glass vessel) the liquid is caused to flow over the surface of the plate. The remaining liquid is then sucked out by the expansion of the ball and the picture is taken. The image is developed by means of another tube and ball. A syringe may be employed instead of a tube and collapsible ball.

One of the instruments described and shown consists simply of an orange-coloured glass vessel, having the lens tube attached to one end and the door attached to the other end of the same. This instrument is not expressly mentioned in the Provisional Specification.

[Printed, 8*d*. Drawing.]

---

## 1865.

### A.D. 1865, January 2.—N° 10.

GYE, FREDERICK. — (*A communication from Louis Strelisky.*)— (*Provisional protection only.*) — " Improvements in mounting " photographic, printed, and other pictures."

" This invention has for its object the so mounting photo- " graphic, printed, and other pictures that peculiar effects are " produced by reason of two pictures in each case being combined "with and superimposed one on the other on a sheet of glass. "The two pictures are produced on paper, one on thin paper, "and the other on ordinary photographic or printing paper as the "case may be. The picture on thin paper is made transparent "by means of spermaceti or varnish, or other adhesive matter, "and is made thereby to adhere to the glass. In some cases the "glass is ground or deadened on the side where the transparent "picture is applied. It is preferred in all cases that the two "pictures should be tinted or coloured artistically according to "the effect desired to be obtained, but this is not essential. "Supposing it is desired to mount two photographic pictures "according to this invention, in order to produce a coloured "miniature or colored picture of a person, two pictures are taken "from the same negative such as is used by artists when colored "photographic pictures are to be produced, one picture is on thin "photographic paper and the other on other photographic paper. "The picture on thin paper is tinted or colored and then ren-"dered transparent and caused to adhere to a sheet of glass "by means of spermaceti, or varnish, or sufficiently transparent "cement. The other picture is also tinted or colored and is "placed carefully at the back of the other picture and is retained "correctly in position either by adhesive matter or otherwise, "and the whole so combined are then placed in a frame. In like "manner printed and other pictures are combined together and "with sheets of glass."

[Printed, 4*d*. No Drawings.]

### A.D. 1865, January 3.—N° 12.

HELSBY, William George.—(*Provisional protection only.*)—"Improvements in the manufacture of enamelled glass to render "it more useful in photographic art."

"The practice recently introduced by me of polishing the "surface of opal or other enamel on sheets of transparent and "colourless glass provides surfaces well adapted for photographic "pictures and impressions, especially those seen by reflected light "and known as positives. There is however a want of 'depth' "or 'body' in the 'ground' of pictures or impressions taken on "the said polished enamelled sheets, and to remedy this defect, "or supply the desired 'depth' and 'body,' and to provide a "smooth surface, are the objects of my improvements. The opal

" or other coloured enamel and the glass are united together in " the process of manufacture in the usual way, but instead of the " glass to or with which the enamel is secured or made being " transparent and colourless I make it with a 'body colour' of " any desired tint or quality by placing the proper materials in " the 'glass pot;' rose and cream will be the most desirable " colours when the enamel is opal, but the colours may be varied. " When the enamelled surfaces are polished the sheets are ready " for use."

[Printed, 4*d*. No Drawings.]

### A.D. 1865, January 7.—N° 56.

BENTLEY, Barrowclough Wright, and BAILEY, William Henry.—" Improvements in producing and finishing " photographs and photographic transparencies on paper and " other suitable substances, and in the machinery employed " therein."

1st. A photograph, already mounted on cardboard, is subjected to pressure, under a press which forms the second improvement of this invention; it is then placed, upon a suitable matrix, beneath the die of an embossing press. By this means the most prominent parts of the picture are raised, and a picture in "relievo" obtained. The dies may either be produced in the ordinary manner, " or they may be made by any of the well " known photographic processes of engraving or by the 'bichro- " 'matized gelatine process.'" Thus medallion portraits in relievo may be obtained, or imitation "cameos" may be made by colouring the background according to taste.

2nd. "An improved photographic press."—Certain pedestals, upon a cast-iron framework, receive bearings for the roller. The sliding plate, which travels upon the top of the framework and beneath the roller, is actuated by a link which is attached to a centre or pin that passes through lugs attached to the under side of the sliding plate; the other end of the link is connected to the working lever, which is jointed to a pin that passes through lugs fixed to the framework. A sheet of aluminum bronze is affixed to the upper surface of the sliding plate; this is not so liable to corrode as steel. The roller is pressed upon the sliding plate by means of milled-headed screws. The actuating lever enables the sliding plate, with the photograph upon it, to travel beneath the pressing roller. In the larger sizes of presses a reciprocating motion is imparted to the plate.

[Printed, 8*d*. Drawing.]

### A.D, 1865, January 10.—Nº 72.

PETTITT, Edwin.—(*Provisional protection only.*)—" Improve-" ments in giving permanence to and in ornamenting glass " transparent positive photographs."

" My invention consists, firstly, in the following manner of " giving permanence to glass transparent positive photographs :— " I coat or cover the collodion side of the said photographs, or " the side of the glass on which the picture is situated, with " plaster of Paris, cement, wax, or other opaque or semi-opaque " substance, either white or tinted with colour, and I afterwards " cover the said plaster of Paris, cement, wax, or other substance " with a coating of metal which coating of metal may either be " applied by the electrotype process or be applied in the form of " thin sheets or foil attached by any adhesive matter. By the " treatment described the picture is securely protected from the " injurious effects of air and moisture, and great permanence in " the photograph thereby secured.

" My invention consist, secondly, in ornamenting transparent " glass photographs by coating or covering, either wholly or in " part, that side of the glass on which the picture is situated with " coloured materials, such as velvet, silk, or other fabric, gold " and silver leaf, and coloured bronzes, and afterwards coating " the photograph in the manner constituting the first part of my " invention when permanency is required. By the use of the " coloured materials described any required tint may be given to " the whole or part of the photograph, or it may be tinted of " different colours in different parts."

[Printed, 4*d*. No Drawings.]

### A.D. 1865, January 26.—Nº 218.

GAY, David.—(*Provisional protection only.*)—" Improvements " in photo-sculpture and apparatus to be employed therein."

This invention consists, " first, in cutting the material on which " the portrait is taken or printed, as may be found most desirable, " and in using the marginal or outer portions of the material as " templates." Upon a circular table are arranged a dozen blocks, capable of sliding in radial groves towards and from the centre of the said circular table. An upright rod is fixed in each block; a clip affixed to each rod holds one template. The circular table is pivoted, and the pivot (by means of a loose slotted collar) receives and holds the several templates. The plastic material is placed upon the upright pivot, and roughly shaped by hand.

The templates—obtained as set forth above—having been fixed in their respective upright rods:—" I take first that template con-" taining the most prominent outlines, and pushing it into con-" tact with the plastic material aforesaid, hold this template " steady with one hand, and with the other hand move the plastic " material partly round backwards and forwards until the tem-" plate has scraped the material at this part to the shape of the " template, or partly so, and in like manner I proceed successively " with all the other templates, until I bring their edges close " together in the collar that supports them, after which the sculp-" tured model may be finished by hand; and where several copies " are required, I take a mould thereof by any of the well-known " means, and pour plastic material therein in the manner com-" monly practised of moulding objects in plastic materials."

[Printed, 4*d*. No Drawings.]

## A.D. 1865, March 4.—N° 618.

PETTITT, Edwin.—" A method or process for producing a new " kind of photographic pictures."

One photographic picture is produced " from two photographic " images, the one superimposed upon the other, which picture " possesses great relief. Pictures so produced I designate ' bino-" ' graphs.' If I desire to produce a life-sized portrait, I proceed " in the first place to take a double negative of the sitter by " means of a pair of lenses and a camera obscura in the usual " way of taking ordinary stereoscopic negatives; and having " obtained a satisfactory double negative, I separate the two pic-" tures and transpose them, placing the right-hand picture to the " left and the left to the right; and in this order I cause light to " be passed through them in a dark room or camera obscura, and " receive the rays so passing through the pictures by a pair of " lenses placed at the same distance from them as the pictures " were when taken, and the two pictures now become enlarged, " coalesce, and are superimposed upon each other at or near the " distance from the lenses at which the sitter was originally placed "when the double negative was taken." The distance between the double negative and the lenses is increased for a smaller picture and decreased for a larger picture than life size. "A binographic picture can also be obtained from two pictures without "separation or transposition of the double negative."

To copy a "binograph" "a direct positive or negative" may be taken from the binographic image as thrown upon a screen.

Negative "binographs" are adapted for photographic printing; "a double transparent print on glass is first taken from the "double negative, and the binographic negative is enlarged from "the said transparent print or positive."

By means of sights, the middle of the stereoscopic camera "is "pointed directly at the point of the object to be focussed."

[Printed, 4*d*. No Drawings.]

## A.D. 1865, March 10.—N° 677.

REISSIG, THEODOR.—(*A communication from Wilhelm Reissig.*) —(*Provisional protection only.*)—"Improvements in ascertaining "the presence of 'fixing' agents in photographic productions, in "removing the said fixing agents therefrom, and in apparatus "connected therewith."

"This invention relates, firstly, to a test whereby the operator "is enabled to determine whether hyposulphite of soda or other "such fixing agent has been removed by washing or other process "employed in the practice of photography. With this view he "places the prints after they have been washed or otherwise similarly treated, or the fluid in which they have been washed, in "communication with the poles of a galvanic battery, by which "means the fixing agent becomes decomposed, and sulphur (if "the fixing agent above mentioned has been used) appears at the "negative pole. If silver be employed for this pole the effect is "more apparent. As this part of the invention depends upon "the elimination of the injurious part of the fixing agent, it is "evident that it may be used for the removal of the last traces or "of larger quantities.

"Secondly, the invention relates to a mechanical method of "removing the said fixing agents. For this purpose the photographs are caused to rotate at a considerable velocity, so as to "throw off the liquid they contain by centrifugal force. The "apparatus proposed to be employed is a folding frame, covered "with lattice work, and within which the photographs are placed

" and then confined. This framework or cage is mounted upon a " spindle which is caused to revolve rapidly by a winch handle or " other ordinary means.

" The two parts of the invention may be combined, the test of " the former being applied between successive washings and " operations of the centrifugal apparatus."

[Printed, 4*d*. No Drawings.]

## A.D. 1865, March 14.—N° 712.

BROOMAN, Richard Archibald. --(*A communication from Cyprien Marie Tessié du Motay and Charles Raphael Maréchal.*)— " Improved processes for the production of photographc images " capable of being inked with fatty inks."

A portion of this invention consists in the employment, for the above-named purposes, of " chromic acid salts, more acid and " more complex than protochromates and bichromates," in conjunction with " bodies more capable of reducing the oxygen of the " chromates than gelatinous and albumenous solutions."

Another portion of this invention consists in the use of " soaps " of silver," in conjunction with gelatinous coatings that contain " salts of chromic acids," also for the above-named purposes.

In one process, " alkaline trichromates " are mixed with a solution of gelatine that contains gums, acids, or salts, having great affinity for oxygen. Positive inkable images are yielded.

In a second process, bichromates or trichromates of the alkalies, " and of bichloride of mercury," also " bicarbonate of potass " and of bichloride of mercury," are used either pure or mixed with alkaline protochromates, bichromates, and trichromates, with or without the addition of bodies having affinity for oxygen, and added to gelatinous and other similar solutions, for the said purposes, positive inkable images being yielded upon paper.

In a third process, " the chromo-alkalino-mercurial salts above- " mentioned," in connection with gelatinous solutions, spread upon metal, yield (" after prolonged washings ") images inkable negatively.

In a fourth process, gelatine, chromatized with one of the above-named chromic acid salts, is laid upon paper, or upon metal; the chromatized gelatine is then dessicated and covered with layers of argentic soaps. Two superposed images are produced.

[Printed, 4*d*. No Drawings.]

A.D. 1865, March 31.—N° 915.

SMITH, JOHN HENRY.—(*Provisional protection only.*)—" Im-" proved apparatus for mounting photographs."

" The apparatus consists of a fixed block, on which is mounted " a glass " " plate of the desired shape and dimensions, and the " edges of which project a little beyond the fixed block." On this plate is secured a piece of woven fabric, " which is capable of " being kept damp." " The glass or other plate is of the shape " and dimensions of the pictures which are to be operated upon, " and when the photographs are placed face downwards upon the " damped cloth, and are pressed down a little so as to force out " the air from beneath, the pressure of the atmosphere on the " back will keep them down flat, so that their backs may be " gummed or covered with some adhesive composition without " difficulty. The backs of the photographs having been gummed, " the cardboard on which they are to be mounted is placed upon " them, and upon being slightly pressed down the photograph " will adhere thereto, and upon removing the cardboard the " picture wlll be found to be properly mounted thereon. As the " cardboard is generally somewhat larger than the picture it is " important that the latter should be mounted centrally thereon; " to this end a set of horizontal adjusting or set screws provided " with plates at their ends are placed at one side, and also at one " end of the glass plate, and are mounted in vertical blocks, so as " to be capable of being screwed up and adjusted with nicety " according to the amount of margin that is required. When " placing the cardboard on the picture the edges of the cardboard " should be brought up against the adjusting plates, and thus " with very little trouble the photographic pictures can all be " mounted with margins of exactly the same width."

[Printed, 4*d*. No Drawings.]

A.D. 1865, April 8.—N° 1009.

PROUT, VICTOR ALBERT.—" Improvements in photographic " cameras."

In taking photographic panoramic views, a screen is employed, which is placed a short distance in advance of the sensitive surface. This screen limits the size of the portion of the picture which is taken at each operation, its sides being made of a " zig-

" zag form like saw teeth," so arranged, "that the points at the " one side correspond to the centre of the notches on the other " side."

At present a rigid screen board is used in connection with slides of considerable length; in the arrangement preferred, however, flexible sliding shutters that work in guide grooves on the slide are employed. As the slide is pushed forward into the camera the shutter is pushed back by a stud; the shutter travels round a pulley at the rear end of the slide, and passes to the back. A stud on the other shutter draws it back so that it comes in front of the sensitive surface, and protects it from the light. To obtain an accurate representation of the sky at the same time that a foreground is taken, an adjustable screen is employed, by which the light of the sky may be in great part intercepted; this screen is placed at the lower part of the camera, and may either consist of an arrangement of wires that becomes more opaque as it descends from the edge, or a number of lines (filled in with black varnish) ruled at graduated distances apart upon a thin sheet of glass. A clockwork apparatus determines the time of exposure of the plate. The removal of a peg admits light to the camera, and starts the clockwork; when a disc has revolved a certain part of a revolution, the peg drops, the clockwork is stopped, and the light is again excluded from the camera.

[Printed, 1*s*. Drawing.]

## A.D. 1865, April 26.—N° 1171.

ROWLAND, John Alexander.—*Provisional protection only.*) —" Improvements in photographic cameras."

This invention relates to a mode of constructing cameras " so " that large pictures or panoramic pictures may be taken with " lenses of moderate size."

" In carrying out my invention the lens is mounted in a brass " chamber or slide, to the end of which is adapted a diaphragm " chamber the aperture of which is of a long narrow form placed " vertically. The lens chamber is held in a vertical position by " means of two fixed studs which work in grooves made longi- " tudinally along the top and bottom of the dark chamber. The " lens is also maintained in its proper vertical position by a similar " arrangement. The image from the lens passes through the long " narrow vertical opening of the diaphragm chamber to the sensi-

" tive surface of the plate. By traversing this long narrow opening of the diaphragm chamber along the grooves provided for the purpose, the changing image from the long narrow opening falls on to the sensitive surface, and the photographic image is thereby impressed on the sensitive surface. The traversing motion of the lens and diaphragm is effected by toothed gearing or by a screw working in a nut connected with the diaphragm chamber, and this mechanism may be worked by means of a winch with an uniform slow motion adapted to it, or the diaphragm may be moved by means of a cord or chain actuated by clockwork capable of being driven at a regulated speed according to the state of the atmosphere."

[Printed, 4*d*. No Drawings.]

## A.D. 1865, April 26.—N° 1174.

SMITH, William Henry.—" Improvements in photographing upon wood, and in the preparation of wood, canvas, silk, glass, and other substances for the purpose of receiving and retaining impressions."

" These improvements consist in the application of two substances in succession to the object on which it is desired to produce a photographic impression. The first substance forms a base in the pores and interstises, or on the surface of the said object for the deposition of the second substance, which is sensitized to make it receptive of photographic impressions." " The bases are composed of valata " [balata ?], "india-rubber, gutta percha, or any gum or resin of a like nature dissolved in benzole or in any other hydrocarbonaceous liquid, or it is formed of gelatine and spirits of wine or an oily matter. The receptives are composed of collodion and gelatine, or any gelatinous substance in a state of decomposition or liquid state, of spirits of wine and of nitrate of silver; the nitrate of silver being mixed with a chloride, a bromide, or an iodide, or some salt having similar qualities."

For some kinds of wood a base consisting of an aqueous solution of gelatine may be used; in the receptive, gelatine, honey, chloride of cadmium, tincture of iodine, tartaric acid, and nitrate of silver may be used. After exposure in the camera, a toning solution of sulpho-cyanide of ammonium and chloride of gold is employed.

" In the preparation of papier mâche, china, earthenware," and other substances, the proportion of materials in the receptive is slightly different from that employed in the preparation of wood.

According to another method, the base contains gelatine, whiting, and linseed oil.

In the receptives, " in certain cases iron may be used in combi- " nation with silver in place of the chlorine salts."

[Printed, 4*d*. No Drawings.]

## A.D. 1865, April 27.—N° 1184.

GRAINGER, Alfred, and GIRDLER, Charles Mitchel.—" Improvements in the production of portraits or likenesses on " certain materials by means of photography."

" Our invention relates to the production of photographic " likenesses on porcelain or ceramic ware, and consists in pro- " ducing the photograph or likeness of a person or thing on " the ware itself instead of first taking it on paper or other " material, and then transferring it to the earthenware to be " afterwards burnt or glazed as heretofore. We prepare a plate " or other article of the size and shape required of porcelain " or other ceramic material, which we glaze with a transparent " glaze and burn it on as usual; on this we take the likeness " by means of photography, we then paint or color the photo- " graph taken by hand or by other application of color, then " glaze or enamel it again, and burn and complete the article " bearing such portrait or likeness. Instead of coloring the " photograph before glazing a second time it may be left un- " colored, and with the effects produced by photography only " which is glazed and burnt in as before mentioned. Borax or " other suitable flux may be used for the glazing, which is applied " and rendered fluid by heat, as well understood in glazing " ceramic wares."

The photographic representation may be produced upon " metal " or other material," " which in other respects is treated and " prepared as hereinbefore described, the glazing and baking " rendering the representation enduring and applicable as herein- " before described with reference to plates or pieces entirely of " ceramic material."

[Printed, 4*d*. No Drawings.]

A.D. 1865, June 5.—N° 1541.

NEWTON, William Edward.—(*A communication from William Augustus Leggo and George Edward Desbarats.*)—" An improved " photo-electrotyping process."

A photograph on glass is varnished, allowed to dry, and laid upon a level slab in a dark room; it is then brushed over with " a substance which upon exposure to the light becomes inso- " luble in water," and the coating is permitted to " stand until " it be quite jellied." The sensitive substance employed is a gelatinous solution of bichromate of potash. When coated, the picture is exposed " to the light face downwards " for the requisite time, and the soluble parts of the jelly removed with warm water. Before the remaining part of the jelly is dry, a plaster cast is taken from it.

To produce an electrotype from the said cast, it is dipped into hot water and laid face upwards until all superfluous water disappears; then, while still warm and damp, dipped into melted wax two or three separate times, " allowing the wax to set each " time." The waxed cast being placed face downwards in some hot wax that has been poured upon a metal plate, and allowed to cool, " the wax upon the plate will unite with that upon the cast " and will form a solid mass, from which the plaster may be lifted " away leaving its exact impression in the wax, which when coated " with plumbago will, by the usual process, yield an electrotype " from which perfect copies of the original may be printed."

If desired, a stereotype may be made from the plaster cast as it comes from the gelatinous surface.

[Printed, 4*d*. No Drawings.]

A.D. 1865, June 12.—N° 1588.

BONELLI, Gaetan.—" A new or improved method of obtaining " or producing optical illusions."

" It is a well known fact, based on certain optical conditions, " and by the use of an instrument known as a phenakistoscope, " that an appearance of motion or animation may be given to " pictures or subjects that have been expressly drawn or photo- " graphed for the purpose. The phenakistoscopic apparatus or " instrument being known it will not be necessary to state how " the pictures are arranged to produce the optical illusions, it " will be sufficient to say that an identical result will be obtained

" with images or pictures of small size and even reduced to " a microscopic scale if to each of the images or pictures a lens " is applied. It is only by the application of as many lenses as " there are images on the disc and not by the successive passage " of all the images under a single lens that the effect will be " obtained. Thus, if a disc carries sixteen images, as many lenses " or magnifying apparatus would be necessary, which receive " a rotary motion in the same direction as the images before " which each lens will be respectively placed."

" The pictures or images thus reduced enable me to construct " phenakistoscopic apparatus, whether simple or stereoscopic, of " sizes and shapes not hitherto done, and thereby forming a novel " application of microscopic photography."

In the Provisional Specification, the method of making objects appear animated is stated to be by placing the pictures upon a small rotating disc; " each picture is brought in front of a lens " or magnifying glass, whereby the statue or group appears to " the eye in motion in all the different positions in which it is " represented, and in an apparatus of very small or limited " dimensions."

[Printed, 4*d*. No Drawings.]

### A.D. 1865, June 20.—N° 1653.

CARLEVARIS, PROSPERO. — (*Provisional protection only.*) — " Producing a light applicable to photographic purposes, to light- " houses, and to other illuminations."

One of the scientific principles upon which the obtainment of this light is founded is the property that the oxide of magnesium has " of spreading," on being placed within the oxy-hydrogen flame, " an intense, bright, and constant light, and which is " admirably suitable to photography and general illumination.

" The process is as follows:—A piece of chlorid [chloride ?] of " magnesium, larger or smaller according to the effects of light " required, is placed upon a small prism of gas-retort coal, and " upon it, through a small tube purposely made, the flame of th- " oxyhydrogen gas (the mixture of oxygen and hydrogen) is " directed; or a prism, or even a small and well compressed " cylinder of carbonate of magnesia is placed within the flame " from the same mixed gases. The chlorid of magnesium or " the carbonate of magnesia is directly decomposed and resolved " into oxide of spongy magnesium, from which the intense,

" bright, fixed, and constant light comes forth, causing all the " chemical phenomena of diffused sunlight."

In this light, ordinary gas and atmospheric air may be substituted for hydrogen and oxygen respectively. In large quantities hydrogen "is prepared by causing steam to pass over incan-" descent charcoal;" "oxygen is prepared with manganese and " hypochlorite of lime, with manganese and silicic acid, or with " dry sulphate of iron, the products of whose decomposition are " caused to pass upon platinated pumice stone, turning to profit " the sulphurous acid resulting from the process to the preparation " of sulphites."

[Printed, 4*d*. No Drawings.]

## A.D. 1865, July 6.—N° 1791.

SWAN, JOSEPH WILSON.—" Improvements in the production " of printing surfaces by photographic agency, and in obtaining " prints therefrom."

" Photo-mezzotint printing."—When the image is produced by means of a negative, bichromated gelatine and colouring matter is employed as set forth in No. 503 (A.D. 1864); after the gelatinous tissue has received the actinic impression and previous to its development, the said tissue is mounted upon a surface of glass. The uncoated surface of the glass is placed towards the light. Warm water is used to dissolve the soluble portions of the gelatinous coating, and thus to develop the image. The plate bearing the gelatinous image is surrounded with a rim, hardened by means of a protosalt of iron or of sulphate of alumina, coated with silver whilst wet, and electrotyped in copper. The resulting electrotype is backed up and used for "photo-mezzotint " printing." The said printing is performed as follows :—Warm gelatinous ink is poured upon the greased surface of the electrotype and allowed to cool and solidify; an even pressure is applied to a piece of paper placed over the ink, a press with an elastic tympan being employed for that purpose. The prints thus obtained are fixed by means of a solution of alum. When the image is produced by means of a camera, the surface of the glass to be coated is previously covered with a thin film of caoutchouc or coagulated albumen.

To adapt this invention to copper-plate printing, the sensitive gelatine has an increased quantity of colouring matter, the rim is not used, and thin walls are formed in the recesses of the plate to prevent the removal of the ink in the act of wiping.

To adapt this invention to typographic and lithographic printing, a "crayon photograph" is produced. For this purpose, charcoal, or other opaque substance, is mixed with the sensitized gelatine. When the image is mounted on glass (by means of a caoutchouc solution) the soluble portions of the gelatine are removed by means of warm water.

[Printed, 6*d*. No Drawings.]

A.D. 1865, July 7.—N° 1808.

WILLIS, JAMES.—" Improvements in the construction of port-" able dark tents or chambers for photographers."

A number of steel ribs are attached at their upper ends to the top notch of a tube capable of sliding on a rod. "The lower " ends of these ribs are pin jointed to the outer ends of stretchers, " the inner ends of which are attached to a notched ring fixed " on a tube." When the latter tube is slid upwards, the stretchers and ribs will come towards the stick. At the upper end of the rod a top notch is fixed, from which wires are suspended, as in an umbrella; the said wires are jointed to the ribs at some distance from their upper ends. When the movable tubes are respectively slid down, the ribs will be suspended from the top of the rod, " and will take a curved form like the ribs of an umbrella." The said frame will fold into a small compass, and the tent thus formed should not exceed half a circle. "The stretchers form when " opened out a horizontal frame, which being covered forms a " surface on which a sink and other vessels may be placed. The " frame is covered with a fabric lined with black cotton, velvet, " or other fabric capable of preventing the passage of light into " the interior of the tent or chamber. An apron is formed to the " tent suitable for surrounding the person, as has heretofore been " the case with other forms of portable tents, or dark rooms or " chambers. Light is let into the tent through yellow cloth or " fabric inserted at suitable parts of the tent, as has heretofore " been the case."

[Printed, 10*d*. Drawing.]

A.D. 1865, August 1.—N° 1984.

WELLS, FRANCIS ROSS.—(*Provisional protection only.*) —" A " new or improved method of producing a photographic image " on the surface of copper or other metal plates."

"This invention relates to a new or improved method of pro-
" ducing a photographic image on the surface of a plate of copper
" or other metal, the object being to etch or engrave thereon. I
" propose to sensitive the plate by the vapour of iodine or the
" fumes of bromine, or with iodine combined therewith, when a
" photographic image can be produced thereon and developed by
" heat or by the fumes of mercury. A picture or engraving, or
" any picture or drawing, transparent or semi-transparent, can be
" produced on the surface of the plate to serve as a guide for
" etching or engraving thereon, so that facsimiles can be pro-
" duced. In some cases the solution of iodine or bromine may
" be used in lieu of vapours."

[Printed, 4*d*. No Drawings.]

### A.D. 1865, August 15.—N° 2110.

HENRY, MICHAEL.—(*A communication from Henry Avet.*)—(*Provisional protection only.*)—" Improvements in the production of
" surfaces by means of photography."

This invention relates to the production of printing and other surfaces. On the collodionized side of a glass negative is spread a film of gelatine combined with bichromate of potash, sometimes first forming a border of wax. The plate is then placed horizontal to allow the film to dry, and is exposed to the action of light on the non-collodionized side. "Next, by means of the camera
" obscura, the portions of the film which the black and half tints
" have preserved from the action of light are dissolved (when
" gelatine or gum is used this can be done by hot water). When
" dry the subject will appear in relief." "Any antiphotogenic
" shade resulting from the bichromate is removed. The subject
" being metallized by the nitrate of silver" is then electrotyped, or a gutta percha impression thereof is electrotyped. "Sometimes
" instead of applying the film in solution it may be applied in
" leaf or scales." Leaf gelatine may be sensitized by immersion in a solution of bichromate of potash. "To produce an artificial
" grain a drawing is made in fine white lines on a colored
" antiphotogenic background, or vice versâ." "A photographic
" negative is made by first photographing the grain and then the
" object, or vice versâ." "The negative is then treated according
" to the above-described method. A design of the grain may be
" obtained by making a negative from a reduced drawing in black
" lines on white ground; it may be used as (a transparent) or

" applied on another negative with the collodionized surfaces in " contact."

[Printed, 4*d*. No Drawings.]

### A.D. 1865, October 13.—N° 2648.

BRINCKERHOFF, JOHANNES DE WITT.—" An improvement " in preparing paper and the surfaces of other materials for use " in photography."

The object of this invention is to produce a partially absorbent but porcelain-like surface.

The paper is treated with a thin solution of gelatinous materials and kaolin, then with a solution of alum or tannin.

French gelatine is softened by soaking in water, and dissolved by heat; with this solution a certain proportion of kaolin is mixed, and the paper is dipped into the resulting solution, whilst the said solution is hot. The paper is then treated with a strong solution of alum or tannin.

" The paper prepared in the aforesaid manner is much stronger " than before being prepared, and not being dissolved by the " solutions employed in photography, is much more easily mani- " pulated in the subsequent process of toning and fixing the " picture, and the washing can be performed much more easily " and perfectly than heretofore, because the solutions employed " do not become so much incorporated in the paper."

" A solution of borax may be employed in place of alum to " combine with the gelatine and form an insoluble compound as " aforesaid. In place of employing kaolin I may make use of " sulphate of barytes mixed in with the gelatine to give the white " color, or said sulphate of barytes may be produced by mixing " nitrate of barytes with the gelatine, and then the alum solution " effects a double decomposition, forming sulphate of barytes."

[Printed, 4*d*. No Drawings.]

### A.D. 1865, November 25.—N° 2754.

NEWTON, WILLIAM EDWARD.—(*A communication from Laurent de Montgolfier.*)—(*Provisional protection only.*)—" Improvements " in the preparation of photographic papers."

The object of these improvements is " to simplify the operations " relative to the obtaining of positive proofs, and also to effect " considerable economy in the cost of producing such proofs by

" dispensing with the use of the chloride of gold or platinum " usually employed in the developing process. For this purpose, " a small quantity hypophosphorus acid, phosphorous acid, " phosphoric acid, or any salt having such an acid for its base, such " as phosphate or phosphite of soda, potash, or ammonia is to be " mixed with the albumen prepared for covering the surface of " the paper. The proportion of acid or salt employed in relation " to the albumen may vary from 2 to 20 per cent."

A solution of nitrate of silver sensitizes the paper. "The " proofs will take different shades of brown according to the " length of time they are in the press. After two or three suc- " cessive washings in distilled or rain water, the proof without " undergoing the changing or developing process," is submitted, in the dark, to a bath of hyposulphite of soda containing chloride of silver, " in order to remove all the phosphate of silver and " leave the lights white; the proof is then to be exposed to light " for several hours while immersed in the hyposulphite bath, " and then washed dark shades will thus be produced which " nearly resemble those produced by printing from copper plate " engravings."

[Printed, 4*d*. No Drawings.]

### A.D. 1865, November 9.—N° 2891.

NEWTON, William Edward.—(*A communication from William Gibson.*)—" Improvements in preparing the surfaces of paper, " leather, woven and other fabrics and substances for receiving " photographic pictures, engravings, lithographs, and prints, and " for rendering such substances fire and water proof."

" The chief object of the invention is to provide the above " named substances or materials with a smooth enamel or surface," for the above-mentioned purposes.

" The invention consists in combining with any of the above " named materials," an adhesive mixture, composed of gelatine or albumen, with clay or oxide of zinc. The gelatinous mixture is rendered insoluble in water by means of gallic acid, boracic acid, or tannin, dissolved in water.

The gelatinous mixture may be incorporated " with the paper " pulp in the early stages of the paper manufacture, and the paper " web before delivery from the paper machine may be caused to " pass in contact with the astringent solution, and subsequently " between drying and calendering rollers."

Cloths and other woven fabrics may be treated with the gelatinous mixture, dried, and then submitted to the action of the astringent mixture and of polishing rolls. Wood, leather, and other substances may be similarly treated.

Layers of the said gelatinous mixture may be applied to a sheet of glass, so as ultimately to form a panel; the addition of plaster of Paris to the adhesive mixture facilitates drying. When the mass is dry, it is rendered insoluble by means of the astringent mixture, and an insoluble plate or panel is thus produced.

[Printed, 4*d*. No Drawings.]

## A.D. 1865, November 15.—N° 2937.

BÜNGER, WILLIAM.—(*A communication from Charles Augustus Steinheil.*)—"Improvements in photographic lenses."

The improved photographic lens "consists of two lenses " arranged and combined with a perforated disc, diaphragm, or " stop placed midway between them, the lenses and stop being " mounted or fixed in correct position in relation to each other."

The "periscopic objective lens" "admits of angles of 90°, or " even more, being taken correctly, and the size of this image is " equal to double the length of the focal distance of the objective " lens the latter is calculated from the main point or centre of " the objective lens, which lens consists of two lenses of the same " description of glass and of positive focus. These lenses may " be quite equal or similar to each other, placed in inverted posi- " tion with the concave sides opposite to each other, and at such " a distance apart that all the chief rays cross each other at the " same point in the axis between the two lenses. This is the " only chief point of this combination of parts forming the objec- " tive lens, and the two lenses must be sufficiently curved that the " image shall fall within a plane which is normal to the axis " whereby the lenses take the well known form in which the " raised surfaces project outwards."

With this lens "near and distant objects appear equally clear, " and it is not necessary that the camera should be capable of " adjustment, as its correct dimensions can be determinined when " first made," the image is therefore determined "by simply " observing the image angle above and on the sides of the " camera."

[Printed, 8*d*. Drawings.]

A.D. 1865, November 16.—N° 2949.

SARONY, OLIVER.—" Improvements in the rests or apparatus " employed when taking photographs of the human figure."

" For these purposes the stand or pillar which carries the rests " is arranged to be fixed by screws to the floor in any desired " position. Through the upper surface or tube of the stand a " bar (or it may be two bars) slides, and it is capable of being " raised or lowered, and of being retained in any desired position " in respect to the upper part or table of the stand. At the upper " end of this sliding bar are applied two horizontal plates, which " are capable of being drawn towards each other by a screw bolt " and thumb nut. Between these two plates is a third plate, " through which the screw bolt passes, and about which the third " plate can turn as on an axis. On the third or central plate is a " projection to which is fixed a bent or curved plate, which is " formed to receive a slide to which a chair back is attached, " whilst the seat of the chair is separate and the part where the " back is ordinarily fixed to the seat is masked or hidden by a " fringe or drapery. The projection carries a socket through " which an inclined hollow bar or tube slides." The inclined bar can be fastened in any desired position, and carries at its end a hollow socket through which a cylindrical bar slides and turns; this bar receives the rest for supporting the body. The horizontal bar supports a second slide (in a dovetail groove), which slide carries a hollow socket, through which an upright cylindrical bar slides and turns, and can be retained in any desired position. " At the upper end of this bar is a socket which receives a rod, at " the end of which is applied the rest for the head, and such " rod can be retained in any desired position in its socket by a " thumbscrew."

[Printed, 1*s*. 2*d*. Drawings.]

A.D. 1865, November 17.—N° 2954.

BULLOCK, EDWARD, and BULLOCK, JAMES.—" Improve- " ments in the application of photography to the obtaining of " printed proofs or impressions or engravings."

For the purposes of this invention "reticulated negatives" are produced. One means of obtaining a reticulated or granulated surface upon the negatives is to place such a surface " face to face " with any ordinary negative," and to copy both together through the light, "thus producing a transparency from which a negative " must be taken, a print from which upon paper prepared by any " of the bichromate and ink processes known in the trade will " have the reticulated or granulated appearance afore mentioned."

In another process, a transfer paper, with reticulations already imprinted upon it, is employed. Ordinary or photographic paper is coated with a glutinous substance, and printed with a reticulated pattern. " In this case the specks of ink themselves form a " medium, and by their aid excessive contrasts are avoided and " half tones secured. Such picture when so obtained is passed " to a lithographic stone or zinc plate, and a printed proof pro- " duced therefrom; by the aid of chromolithography coloured " proofs may be produced." The coating the paper with a glutinous substance may " be conducted in connection with " bichromate of potash or bichromate of ammonia." The said transfer paper may be used " whether the impression be a litho- " graph, a zincograph, an impression from an electrotype, or " from an engraved or etched plate."

Other processes for granulating the negative, mainly depending upon the use of the camera, are set forth in the Provisional Specification.

The process in which "the specks of ink themselves form a " medium" is not alluded to in the Provisional Specification.

[Printed, 4*d*. No Drawings.]

## A.D. 1865, November 28.—N° 3053.

NEWTON, Alfred Vincent. — (*A communication from Frederick Von Egloffstein.*)—(*Provisional protection only.*)—" An im- " proved mode of obtaining printing surfaces by photography."

" This invention relates to the use of a heliographic and photo- " graphic spectrum for producing printing surfaces from trans- " parent photographs. The spectrum may be composed of a " single sheet of highly polished level ground plate glass, free " from colours and covered with a good asphaltum etching " ground, heated and smoked over a wax paper in the manner of " the engravers' black etching ground. The plate when cooled " off is ruled over by the diamond or other point of a correct " ruling machine, using light pressure to prevent the chipping off " of the ground and the flaking or breaking of the glass.

" Upon the insertion of an edge guide or pattern, varying in " profile from the mathematical line or straight edge of the ruling

" machine, and guiding the course of the diamond or other point " it will be obvious that any system of lines from the straight " line to the semicircle with equal or varying intervals can be " traced with facility and preserving the perfect parallelism re- " quired for the different spectra."

" The plate prepared to receive the engraving is coated first " with a sensitive heliographic varnish. Upon this varnish the " spectral image is produced by the light falling through the open " spaces of the spectrum made as above described. The spectrum " is thus imprinted upon the varnish previous to its receiving " the photographic image by means of a second exposure to " the light. Both images are thus blended into one, the spec- " trum giving texture to the photographic image. Then may " follow the ordinary heliographic manipulations of developing " the picture."

" The photographic picture " " overpowers the spectral image," the " spectrum," however, remains sufficiently strong to hold the printers' ink.

[Printed, 4*d*. No Drawings.]

A.D. 1865, December 2.—N° 3092.

WRIGHT, Anna Josephine.—" An improved mode of apply- " ing photographic paper pictures to glass."

The face of a paper photograph is covered with a varnish composed of Canada balsam thinned with spirits of turpentine, " taking care that the varnish is spread evenly over the entire " surface of the picture." When thoroughly dry the face of the picture is attached to the glass to be ornamented by means of a thick solution of gum arabic. When the gum is quite dry, the back of the picture is coated with the above-mentioned Canada balsam varnish, also taking care that it is evenly spread. When this varnish coating is dry, if the picture be not perfectly transparent, the Canada balsam varnish is again applied. Then, by careful washing with spirits of turpentine, as much of the varnish is removed as will lay bare the surface of the paper. If it be desired to remove the paper, leaving only the picture attached to the glass, a soft rubber is applied evenly over the entire surface of the paper; this rubber is covered with the finest description of glass paper, and it rapidly removes most of the paper; the remaining portion of the paper is carefully rubbed away with a sponge. The picture may now be painted, and is again varnished with the Canada balsam varnish.

Another plan, not noticed in the Provisional Specification, consists in steeping the paper photograph in the Canada balsam varnish until it becomes perfectly transparent. The clarified picture is at once placed on the glass to be ornamented, and is caused to adhere thereto by the varnish. The transparent and affixed picture may then be treated as set forth above.

[Printed, 4*d*. No Drawings.]

### A.D. 1865, December 11.—N° 3190.

GRISWOLD, Victor Moreau.—(*Provisional protection only.*) —" Improvements in photographic surfaces, and the compositions " and process for preparing the same."

The object of this invention is the production of a photographic picture upon an opal surface in a cheap and easy manner.

To carry out the invention the following solutions and preparations are employed :—

" Solution No. 1.—Opal collodion." This solution contains alcohol, sulphuric ether, acetic ether, gum kowrie, gum shellac, solution of cotton, castor oil, and glycerine.

" Solution No. 2.—Sensitizing." This solution contains water, alcohol, muriatic acid, chloride of calcium, chloride of ammonium, and bichloride of mercury.

" Solution No. 3.—Opal picture enamel." This solution contains shellac, borax, water, linseed oil, sulphuric acid, and albumen.

Solution No. 2 is added to solution No. 1 in certain proportions, to produce a sensitive film, according to the strength of the " ammonia nitrate bath;" a small quantity of the sensitizing solution is only required when the plain silver bath is used with " ammonia fuming." The dried plate is exposed under the negative, then immersed in " solution No. 2," to which water has been added, and gently rubbed with sponge. The picture is then washed, toned, and fixed; " solution No. 3" is used to " finish" the picture.

For printing in the camera " solution No. 1" is sensitized in the same manner as bromo-iodized collodion, and then by nitrate of silver.

A mixture of unslacked lime and ammonia alum is used to produce ammonia fuming for " the plain silver bath."

In fixing with cyanide of potassium or hyposulphite of soda, the picture is converted into a negative; it is reconverted into a positive by the application of solution No. 2 diluted with water.

The enamel (solution No. 3) "preserves the purity of the "white film."

[Printed, 4*d*. No Drawings.]

A.D. 1865, December 15.—N° 3248.

PARKER, THOMAS. — (*A communication from Prospero Carlevaris.*)—" Producing an oxyhydro-magnesian light applicable to "photographic purposes, to lighthouses, and to other illumina-"tions."

This invention "consists in producing a light by placing the "oxide of spongy magnesium in a flame produced by a mixture "of oxygen and hydrogen."

A piece of chloride of magnesium ("same chloride of mag-"nesium that melts into metallic magnesium") is placed between two platinum wires, or otherwise mounted, so as to permit an oxy-hydrogen flame to be directed upon it. Well compressed carbonate of magnesium may be substituted for the chloride of magnesium. In either case, at the high temperature employed, "spongy oxide of magnesium" is produced, and gives an "intense, bright, fixed, and constant light." "The two gases, "let them be pure or ordinary lighting gas, and atmospherical "air, containing more or less oxygen, must come from two "different gasholders through two separate pipes, and must be "mixed only at the end of these pipes in a very small tube, "forming in that manner a real chalumeau (oxy-hydrogen "soldering pipe, or pipe to burn oxygen and hydrogen together.)"

The principles involved in the production of this light are:—

1st. The infusibility of the oxide of magnesium.

2nd. The volatility of the said oxide (in a small degree) at high temperatures.

3rd. The property that the said oxide has, in an oxy-hydrogen flame, of giving an intense and constant light that may be applied to photographic and other purposes.

4th. "The property that the spongy oxide of magnesium has "of being the best of all magnesium oxides to produce the "oxy-hydro-magnesian light."

[Printed, 4*d*. No Drawings.]

A.D. 1865, December 29.—N° 3363.

BAGGS, ISHAM.—(*Provisional protection only.*)—" Improvements " in the construction of stereoscopes and stereoscopic apparatus."

" This invention consists of certain improvements in the con- " struction of apparatus for operating by the reflection, refraction, " and transmision of light in such manner as to cause two or " more stereoscopic views, pictures, or objects to apppear to meet " the eye from the same direction, and capable of dissolving one " into the other, or of being caused to appear or disappear at " pleasure either slowly or quickly.

" A simple form of the apparatus for producing these effects " consists of a square piece of plain plate glass, the same being " placed at such an angle relative to two stereoscopic pictures " contained in the instrument that whenever required one of the " pictures is seen by the direct transmission of the rays of light " therefrom through the substance of the glass, the other picture " being seen by reflection from the surface of the same glass. By " shutting out the light from either picture and illuminating the " other the first is caused to disappear or dissolve, and the other " to take its place in the organ of vision. Two or more mirrors " or other reflecting surfaces may be used in place of one for " producing these effects upon the well known principle of the " camera lucida. Lenses may also be employed in a similar way " to cast the stereoscopic pictures upon a screen or other surface, " or in the air, so as to form dissolving stereoscopic pictures, and " to render the effects more striking the same may be magnified " on the principle of the telescope or microscope, as well " understood."

[Printed, 4*d*. No Drawings.]

---

## 1866.

A.D. 1866, January 9.—N° 66.

SKINNER, JAMES.—(*Provisional protection only.*)—" An im- " proved mode of preparing albumenized paper, and the process " connected therewith."

" The paper is floated on the albumen taken from the eggs of " common poultry; the albumen may be used either with or with-

" out admixture with a portion of some gelatinous substance, " such as arrowroot, tapioca, sago, or any other similarly suitable " substance, and in such proportions as not to injure the albumen. " The paper, after the first part of the process is effected, is then " dried off in a temperature varying from 60° to 90° Fahrenheit, " more or less. The albumenized paper is then rendered insoluble " by being put into a case of metal, wood, or other suitable ma- " terial, and each sheet is separately suspended therein, the case " being perforated all round. Steam at the requisite temperature " is then admitted into the case in such quantity and for such " a length of time as is sufficient to render the albumen with " which the paper is coated quite insoluble. Some time after " being taken from the box or case and being perfectly dried " the paper is again coated with ordinary albumen in the same " manner as in the first instance, and dried in the usual manner."

[Printed, 4*d*. No Drawings.]

### A.D. 1866, January 12.—N° 105.

WOODBURY, Walter Bentley, and DAVIES, George.—" An improved method of and apparatus for finishing impressions " (in colored gelatine or other soluble material) obtained from " metallic or other plates produced by the aid of photography."

The object of this invention is to produce a clear white margin, with a sharply defined edge, upon the said impressions from metallic or other plates (see No. 2338, A.D. 1864). The margin dispenses with the ordinary mounting upon paper.

The invention consists :—

1st. In cutting or scraping off the superfluous gelatine "by " means of four cutters or scrapers arranged in a rectangular " form in such a manner that the area thereby enclosed can be " diminished or increased at pleasure."

2nd. Protecting the central picture, and washing off the composition from the edges or margin with warm water. The impression may be laid upon an india-rubber surface; if a plate, of suitable size and shape, be then firmly pressed upon the said impression, the margin may be washed off by means of a sponge. In a second method, a solution of alum (or other solution that will render gelatine insoluble) is poured into an open frame that is suitably placed upon the impression; the solution is then poured off, the impression dried, and the margin washed off. In a third method, an open frame or "mat" being placed over the central portion of the impression, a solution of india-rubber in benzole is brushed over the impression to render the central portion of the gelatine impervious to warm water.

Another way of producing pictures with clear margins, is to take the original impression on glass that has a waxed surface, and, after removing the margin therefrom, to press thereon a paper already coated with adhesive cement. This paper adheres to the impression, and the paper and impression together may be removed from the glass, the requisite margin being afforded by the paper.

3rd. Printing impressions on paper direct with a clear margin, by means of a dished plate or frame, the edge of which arrests the flow of gelatine and colour. A press is described and shown, in which a "frisket" (hinged to the interior of the lid of the press) carries the dished frame. When the lid of the press is closed and the impression taken (in gelatine), as the superfluous gelatine is squeezed out it is arrested by the edge of the dished frame, passes between it and the edge of the mould, and does not spread over the paper.

[Printed, 8*d*. Drawing.]

A.D. 1866, February 2.—N° 324.

WINSTANLEY, David, junior.—"Improvements in producing "printing surfaces by the aid of photography."

In this invention plane printing surfaces (from reliefs or intaglios produced by the aid of photography) are produced by means of a number of pointed wires.

A layer of bi-chromatised gelatine is exposed to light in the usual manner under a photographic negative; the actinic influence of the light thus admitted to the gelatine renders its parts more or less insoluble in proportion to the extent to which they have been acted upon by that light. In the picture produced, by acting upon the exposed gelatine with warm water, the gradations of light and shade are perfectly represented by corresponding thicknesses of gelatine. Each of the said wires have a like conical or pyramidal termination, and—the surface of points produced by a bundle of such wires being placed upon one of the said protographic reliefs or intaglios—a copy of the respective lights and shades is produced by the relative height of the corresponding points above the plane of the picture. The height of the cone of each wire being equal to the greatest depression of the mould, and the wires being fixed in their containing frame in the position which the said relief has given them, the uneven surface of points thus obtained is planed or otherwise reduced to one general flat surface. The result is a surface in which the lights and shades of the original photograph are represented by dots or marks of a diameter greater or less according to the elevations or depressions of the relief or intaglio from which the uneven surface was produced; this surface can be used in block printing. A cast, in type metal, may be taken from the said wire surface, and, when ground or cut, may be used for printing by the "copper plate" method.

To obtain gradations of light and shade by dots, equal in magnitude but differing in number, a plate (prepared for copper plate printing by the above method), in which the impression has considerable depth, is ground from behind until every point is just represented by a hole. Very fine and equal sized wires are pushed through the said holes, a frame, containing a large number of the said wires, being placed over the plate for that purpose. Upon binding these wires together a surface is obtained "in which the "lights and shades are represented by greater or less numbers of "dots having the same size. This surface may then be planed, "cut, or ground, or casts or impressions in reverse or fac-simile "may be taken from it as in the method before mentioned, and "used for obtaining pictures or impressions either by the 'block "'printing' or by the 'copper plate' method."

[Printed, 6*d*. No Drawings.]

A.D. 1866, February 8.—N° 396.

DALLMEYER, John Henry.—(*Provisional protection only.*)—"An improved photographic lens."

This lens embraces a very large angle of view, say 100° or more. "It is composed of three lenses, two of crown glass and one of "flint glass, making a triple combination," and similar in this respect to the triple achromatic lens already manufactured by the inventor; but, "instead of being composed of three combina-"tions, each acromatic or nearly so in itself," it consists of three very thin single non-achromatic lenses, "which, however, when "acting in unison, form an achromatic whole."

The construction is as follows:—Two crown lenses are "wrought "to a sharp edge of suitable focal length;" they are either identical in this respect or they are not so, but certainly they are of " different or unsymmetrical forms or unequal ratio of radii, i.e. " the anterior crown lens (No. 1) may be a plano-convex, or " nearly so, convex side outside, and the posterior lens (No. 2) " a deep concave meniscus, also convex outside." Between these two lenses, separated at a certain definite distance, a diaphragm is placed; the said diaphragm divides the space between these two crown lenses in the ratio of their foci, and almost immediately in contact with this diaphragm is placed "a flint glass lens, plano- " concave by preference, and of such focal length as with the two " crown lenses above named will form an achromatic whole."

[Printed, 4*d*. No Drawings.]

## A.D. 1866, February 13.—N° 449.

GILPIN, CHARLES.—(*A communication from Francis Kossuth and Louis Theodore Kossuth.*)—" Improvements in the production of " copper or other metallic plates for the purpose of printing " therefrom."

To obtain the said plates, a photographic glass plate, bearing the positive or negative image of the object or design to be printed, and whilst still wet, is treated with a solution containing bi-chromate of potassium and gelatine, so as (when dry) to form a uniform thickness of sensitised gelatine on the glass plate. The layer of gelatine is then exposed to light, washed in warm water, dried, and metallised for electrotyping by means of an alcoholic solution of nitrate of silver acted upon by sulphuretted hydrogen.

The image on the said photographic glass plate, for relievos, is obtained by taking on a plate of glass, a positive image " obtained " by transparency through another negative one (representing the " object to be reproduced in the relievo engraving), on which a " striped plate of glass has closely been applied." For " incision " engravings " a negative image is taken from the object, the said negative image being grained or shadowed with lines or dots.

Various methods of producing the " grain " are set forth. One method consists in the use of a striped glass plate in contact with the sensitive surface; the stripes are made with a diamond point. The image of a grain previously lithographed on paper may be taken on the sensitive surface. To produce the grain at the same time that the image of the object is taken, a camera is employed that has two opposite object glasses, the images respectively of the object and of the grain are, by this means, focussed at the same time upon the same plane.

[Printed 6*d*. No Drawings.]

A.D. 1866, February 14.—N° 469.

HENRY, MICHAEL.—(*A communication from Henry Avet.*)—
" Improvements in photography, and in the process of producing " printing surfaces and other like surfaces by the aid of photo-" graphic agency."

One method of producing the said surfaces consists " first in " producing an artificial grain on plates, then in obtaining a " raised or sunken design or subject on such plates, and, lastly, " in obtaining a metallic surface from the same, the design or " subject thereon being raised or sunken."

Gelatine, or other organic substance, mixed with an alkaline bichromate is applied on a translucent plate already bearing a photographic image. The light is caused to act on the said organic substance through the thickness of the plate; when the parts unaffected by the light are dissolved, a subject in relief will remain on the said substance, from which a metal surface suitable for printing may be obtained by means of electro-metallurgy.

Printing surfaces may thus be formed from engravings or other pictures, and either sunken or in relief. The figuring of porcelain may also be accomplished by this means.

To produce the above-mentioned grain, a representation of very fine white lines on an anti-photogenic coloured ground is produced on the photographic plate, either before or after the image of the subject is taken; or the representation of the subject and of the grain may be photographed simultaneously by using two plates. The grain may consist of dark lines on a white ground.

Negatives with an artificial grain may be employed to obtain superior photographic positives.

[Printed, 4*d*. No Drawings.]

A.D. 1866, February 14.—N° 473.

NEWTON, HENRY EDWARD.—(*A communication from Leon Jaubert.*)—" Improvements in optical instruments."

Amongst other details of instruments, this invention relates to the lenses and other parts of photographic cameras.

The first improvement consists "in the application to all optical "instruments in which it is required to cause one tube to slide "on or in another of a screw guide, whereby great precision and "accuracy of motion for bringing the object into the field is "obtained."

The second improvement relates to microscopes and to verniers.

The third improvement consists in the application of a universal joint to the foot of optical instruments.

The fourth, fifth, and sixth improvements relate to microscopes.

The seventh improvement relates to a new kind of object glass and eye-piece.

The eighth improvement relates to an arrangement of the prisms of the binocular microscope, which "is applicable also to "photographic apparatus employed for producing simply photo-"graphic images having the appearance of sculpture." A number of object glasses may be arranged so as to see the same object through them; "this may also be applied to photography for "obtaining the appearance of relief or intaglio."

The ninth improvement relates to a binocular magnifying glass.

The tenth improvement relates to the stage of a microscope.

The eleventh improvement "consists in the adaptation to "microscopes and other optical apparatus of a new kind of "object glass," which may be applied to photographic apparatus, and which is composed of three lenses; the first "is con-"vergent, the second spheroidal or ellipsoidal, and the third is of "a periscopic form." This compound lens is free from spherical and chromatic aberration and astigmatism. The glasses of optical instruments are composed of concentric layers united in groups. "The forms given to these glasses are obtained by a process "which consists in giving to the glass (when in a soft state) the "form which is required, either by centrifugal force or by means "of a plate which will spread out successively the layers of "glass; or the object may be effected by placing one above the "other thicknesses of glass of different densities softened to "the required point, and then united together by means of heat."

The twelfth improvement relates to an apparatus for producing these glasses with concentric layers. A variable rotary motion is given to the central tube which encloses channels for the supply of air and gas, whereby the fusion of the glass is effected; the mould is placed upon the said central tube. The grinding of these lenses is effected upon a lathe with two transverse motions besides the ordinary circular motion.

The drawings also show a photographic camera for use in microphotography. A prism enables the operator to focus the object upon the rough glass.

A prism "for photographic apparatus for producing black or " dark parts, framing, &c.," is also shown.

A megascope, with the above improvements, may be constructed to serve as an apparatus for enlarging photographs.

[Printed, 3*s*. 6*d*. Drawings.]

A.D. 1866, February 17.—N° 505.

WOODBURY, Walter Bentley.—(*Provisional protection only*.)—" Improvements in the production of ornamental surfaces " for jewellery and other purposes."

This invention relates to the production of the said surfaces by the aid of photography; it " consists in the use of a com- " bination of materials one of which is opaque and the other " transparent or semi-transparent. An ornamental 'relievo' or " 'intaglio' surface being produced on the former the trans- " parent or semi-transparent material is forced therein and united " thereto, and the picture or ornament will be seen through the " upper material more or less, according to the thickness of the " same."

An impression in porcelain from a metal mould may be taken for the opaque material, the said metal mould being produced by the method set forth in No. 2338, A.D. 1864. Glass or enamel, as the transparent material, is then forced into the said impression. These two materials being united by fusion, the surface of the latter is ground level.

As a modification, a gilt metallic surface may be filled in with the semi-transparent material, or the impression may be taken in ebonite, and then filled in with semi-transparent material.

" The only condition necessary to this invention being that the " relief shall be upon the lower or opaque material constituting " the body of the picture or ornament, which must be of a dif- " ferent or opposite color to that of the semi-transparent material " used for filling in the surface."

[Printed, 4*d*. No Drawings.]

A.D. 1866, March 17.—N° 797.

ASHTON, ROBERT HOWE.—" Improvements in pictures obtained " upon paper, glass, porcelain, or other surfaces with transparent " or semi-transparent materials."

The first part of the invention relates to improvements upon the Woodbury photo-relievo process as set forth in No. 2338 (A.D. 1864); it consists in combining therewith another printing or other method of tinting in any desired colours.

In the Woodbury process, an intaglio surface is obtained, the concavities of which vary with the lights or shades of a photograph; a cast is obtained from the said surface in gelatine mixed with a pigment. The coloured gelatine is poured upon the said intaglio surface, and the paper or glass to receive the picture being placed thereon, the superfluous colouring matter is squeezed out, and the picture is formed by varying thicknesses of ink.

The paper to receive the picture is first printed upon with the desired number of tints by chromo-lithography, or the tints are produced by any other known process; it then receives the Woodbury relievo picture above referred to. The registration necessary to ensure the correct position of the several parts of the picture is accomplished by a press like that employed in the Woodbury process, the projecting pins for registration being allowed to pass through or on one side of the glass plate of the press.

Other pictures besides photographs may have the gelatine process applied to them, so as to obtain casts similar to those given by the Woodbury process for printing from.

The second part of the invention is as follows:—If the outline of a figure be painted in black upon the negative, the said figure may be combined with the rest of the picture by printing it with the tint obtained by chromo-lithography. Pictorial effects may thus be introduced.

Modifications of the above processes are set forth.

[Printed, 6*d*. Drawing.]

A.D. 1866, March 20.—N° 820.

LAROCHE, WILLIAM SILVESTER. — (*Provisional protection only.*) — " Improved means and apparatus for producing new " effects in photographic portraiture."

This invention consists in placing the person or persons whose portrait or portraits is or are to be taken in front of a background, "and then placing in front of and close to the sitter or sitters a "screen or frame about six feet high and five feet wide," covered with canvas, and having an opening cut in the centre of it of any required size or shape. The front of this opening is surrounded with an ornamental frame.

The said frame "is made to slide up and down a groove in a "framework, which is made to support it in a perpendicular "position, and can be raised or lowered to suit the height of "any person or persons by means of counterweights with ropes "attached running over wheels fitted to the framework, or by "means of a rack and pinion adjustment fixed at the side of the "aforesaid framework, and worked with a crank handle, or in "any other suitable manner." By means of the ordinary photographic camera and subsequent processes the frame and sitter or sitters are photographed at the same time, "the result being a "complete cabinet picture of any size with portrait and ornamental frame complete."

[Printed, 4*d*. No Drawings.]

A.D. 1866, March 31.—N° 920.

WRAY, WILLIAM.—"Improvements in achromatic object "glasses."

Amongst other uses, this invention is employed in the manufacture of photographic lenses, so as to ensure "absolute coinci"dence of visual and actinic foci" therein.

By means of this invention "perfect achromatism is obtained, "and the secondary and all sub-spectra are destroyed."

According to this invention, the irrationality existing between flint and crown glass is destroyed by the introduction, between the flint and crown lenses, of a fluid or partially fluid meniscus lens, composed of a mixture of oil of cassia with Canada balsam. "The high irrationality of the meniscus lens is (in contradis"tinction to Dr. Blair's plan) opposed to the comparatively low "irrationality of the crown and flint lenses" in the method employed in this invention.

The drawings show "a triple arrangement of glass lenses with "two cement films, one enclosed on each side of the concave "lens," and "a double arrangement of glass lenses and a cement "film between them."

" The power of correcting irrationality and the dispersion of " the cement film increases in direct proportion to the amount of " oil of cassia used." Other oils and balsams may be used, but oil of cassia mixed with Canada balsam is preferred. The borosilicate of lead glass may be used as a thin meniscus corrective lens. " The negative solution of the question of achromatism " may be used when heavy flint glass is employed. In this case " the dense glass is an internal concave lens, and the irrationality " is balanced by two external convex lenses which differ in their " irrationality."

[Printed, 8*d*. Drawing.]

## A.D. 1866, April 5.—N° 980.

COX, Edgar.—(*Provisional protection only.*)—" Improvements " in portable dark chambers or tents for the purposes of photo- " graphic manipulation in the field, and of containing and trans- " porting the necessary apparatus."

An oblong box contains a washing trough, tap and tubing, cistern, chemicals, and bath. The hinged lid of this box carries an additional flap, and to each end of the box is hinged a corresponding flap, so that all actinic light may be excluded from the operator while manipulating, by means of the said flaps (when elevated) and a tent cloth suitably stretched across the tops of the said flaps and secured to the waist of the operator. The tent cloth may be carried in the box, beneath the above mentioned trough.

A four-wheeled carriage, constructed with a tank or reservoir, may be used to transport the box from place to place; it is also provided with suitable arrangements for carrying the requisite tripods, and other similar appliances.

" In order to meet those cases in which it may be necessary to " dispense with the carriage, as in ascending mountain passes " and elsewhere, the upper part of the box with the cover, flaps, " trough or sink, and other necessary details may be made capable " of removal from the box or case, so as to be carried with the " tent cloth and other apparatus by poles or otherwise, and set " upon portable legs capable of being screwed or otherwise " attached to it."

[Printed, 4*d*. No Drawings.]

A.D. 1866, April 6.—N° 987.

GRÜNE, Wilhelm.—(*Provisional protection only.*)—" A new " magic photography."

Photographs are prepared " in such a way that if desired they " may be made invisible upon the paper, and afterwards be per- " manently reproduced in their original condition."

The photographs are taken upon albumenised paper which has been previously washed with a solution of silver and dried. They are fixed with " subsulphite " [hyposulphite ?] of soda, washed, and dried; they are then placed in a solution of chloride of mercury, " and there left until the picture has become totally " invisible; they are then soaked for about six hours in cold " water, and afterwards dried."

The reproduction of the photograph upon the paper is caused by means of " a filtering paper, which is made by soaking in a " solution of subsulphite of soda, and when the same is dry, and " it is desired to reproduce the photograph upon the first men- " tioned paper," the invisible photograph is laid with the glazed side uppermost and a sheet of the prepared filtering paper is placed on the top thereof; both the sheets are then saturated with water, and the invisible photograph will rapidly become visible, " which " effect is caused by the action of the subsulphite of soda in the " filtering paper, which dissolves and ' brings out ' the photo- " graph. The picture is then to be washed several times in clean " water, and when dry may be mounted in any desired manner."

[Printed, 4*d*. No Drawings.]

A.D. 1866, April 17.—N° 1085.

GENGEMBRE, Philippe William.—(*Provisional protection only.*)—" Improvements in stereoscopes, which improvements are " also applicable to instruments for exhibiting single views."

This invention consists :—1st. In the mode of arranging the slides around a roller. 2nd. In the device for securing the slides to the roller. 3rd. In making the roller reversible. 4th. In the mode of adjusting the lenses. 5th. In the manner of regulating the movement of the views or slides.

The box is constructed in two parts, and the bearings of the roller rest in notches formed in the sides of the lower half of the said box. Corresponding notches being in the upper half, the roller is held in journals when the box is shut. When the box is open the roller may be reversed. Knobs on each end of the roller enable it to be turned from the outside. The slides are attached longitudinally to the roller by means of elastic loops fixed to two rings at suitable distances from each other on the roller. "If de-"sirable two views or slides may be placed back to back in each "pair of loops, by which means the number of the views may be "doubled. Both sets of views may be exhibited without chang-"ing their position in the loops, by simply turning the roller end "for end." The lenses are placed in a door in the upper part of the box, "and the reflector is secured upon the inside of the "cover above the lenses." For single pictures only one lens is used. A spring retains the slides in position as they are brought before the lenses. "The roller is prevented from turning back-"ward by means of a small ratchet wheel or other suitable "device."

[Printed, 4*d*. No Drawings.]

A.D. 1866, April 26.—N° 1179.

HEDLER, CHARLES.—(*Provisional protection only.*)—"Improve-"ments in the production of photographic pictures."

The object of this invention is "to produce around the picture "a narrow white border enclosed within a wide black or darkened "border, covering the remaining surface of the paper." Before subjecting the sensitised paper to the action of light, its border is covered, "leaving only sufficient surface exposed for the produc-"tion of the vignette or picture." For this purpose a sheet of black paper is placed on the prepared paper, the sheet being cut out in the centre to the size and form of the picture required. After exposure to light, a narrow border of flat metal is placed round the vignette, and the vignette or picture itself is also covered; "the border and plate may be held in position by a "series of vertical pins, which may be subjected to pressure "in the apparatus and so held in position." "By a further "exposure the whole external border of the paper beyond the "picture will become darkened, and even rendered black; and "between or within this dark margin and the vignette or picture "a narrow white border will result from the covering by the "narrow border preventing the exposure of that portion of the "paper." "The picture is afterwards to be fixed and treated by "any of the well-known methods to ensure its permanency."

[Printed, 4*d*. No Drawings.]

A.D. 1866, May 8.—N° 1315.

WOODBURY, Walter Bentley.—(*Provisional protection only.*)—" Improvements in producing designs upon wood and " other materials by the aid of photography."

A metal reverse is taken from an insoluble gelatine mould as put forward in No. 2338 (A.D. 1864). An electrotype from this metal reverse is then obtained in thick metal, the said electrotype being a mould and a fac-simile of that in gelatine above mentioned.

Having obtained this electrotype mould in relief, the inventor heats it " to a dull red heat (or to a sufficient heat to scorch or " discolor wood or other suitable material)," and he presses it " on " to the surface of a piece of truly planed sycamore or other suit- " able wood, and a picture or design will be the result wherein " the shades are represented by the different amount of burning " or scorching which the wood undergoes, owing to the greater " or less prominence of the various parts of the heated copper " mould, those parts where the heated mould actually touches " the surface being the darkest, and so on in proportion. Instead " of wood other materials may be used, such as velvet, ivory, " cardboard, paper, or other substance capable of being blackened " by heat, and the effect is not necessarily produced by scorching, " as the same effect will be obtained if the wood, paper, or other " material is prepared by any well-known chemical substance " having the property of being blackened or darkened by the " application of heat."

[Printed, 4*d*. No Drawings.]

A.D. 1866, May 9.—N° 1334.

DALLAS, Duncan Campbell.—" Improvements in the pro- " duction of printing and other surfaces in relief or intaglio."

The design is photographed or drawn upon a glass plate in a medium that intercepts the actinic rays. Over the design a solution of bi-chromatised gelatine is poured, and (when the coating thus formed is sufficiently dry) the uncoated side of the plate is exposed to light. The parts not acted upon by light are then softened and caused to swell by treatment with cold water; this is poured off and the design is repeatedly washed with warm water till the design is free from the gelatinous mixture, and allowed to dry.

Increased relief or a granulated surface to the design is obtained by the following processes:—

1. If the original design has been produced in bitumen of Judea, increased relief of the design may be obtained by recoating the surface with sensitive solution, again exposing to light, and then washing and drying. Repetitions of the process give the desired amount of relief.

2. Damping the insoluble design causes increased relief.

3. If the original design has been produced in a soluble material, the said design may be obliterated by the solvent without injury to the insoluble relief, leaving the glass bare at those parts where the coating is dissolved; an increase of depth is thus produced.

4. A granulated surface may be produced on the relief by laying all over it an aquatint ground, or dabbing the surface with a sticky material and sifting powdered rosin thereon.

Before or after either of these four methods, a mould of the surface is taken by electrotyping, casting, or pressure.

"The process is applicable, among other things, to the pro-
" duction of plates, blocks, or other surfaces for embossing or for
" printing."

[Printed, 4*d*. No Drawings.]

A.D. 1866, May 12.—N° 1364.

SOUTHWELL, William Henry, SOUTHWELL, Frederick, and SOUTHWELL, Edwin. — (*Provisional protection only.*)—"Improvements in the production of photographic prints."

"This invention consists in the production of pictures in which
" the aid of lithography is introduced for the purpose of heighten-
" ing the effect of photographs, and is carried out as follows:—
" The photographs are obtained in the usual manner, and the
" prints obtained therefrom also in the usual way. The litho-
" graphic stone intended to be employed in printing is then pre-
" pared by the erasing from it or lowering that part of the surface
" where the photograph is intended to appear, the lithographic
" printing then takes place, and the result is that the ground or
" other portions of the entire picture may be of any desired tint
" or shade."

[Printed, 4*d*. No Drawings.]

A.D. 1866, May 26.—N° 1473.

MC FARLAND, COREY.—(*A communication from John Spooner.*) —" An improved album for exhibiting photographs and other " pictures."

In this album any one of the pictures " can be concealed or " exhibited at pleasure."

The box which forms the album has closed portions and portions provided with glass-covered openings. Certain slides, having projections which extend through the bottom of the case, hold the pictures and enable them to be exposed to view or withdrawn into the closed part of the case as desired, the said case being suitably provided with internal grooves in which the slides move freely. " One end of the case is made to open upon " a hinge or pivot to allow the slides in which the pictures are " carried to be drawn out therefrom. When they are so drawn " out the pictures are inserted into the said slides, which are then " pushed back inside the case." The end of the case is secured by a spring catch or other convenient means.

Porcelain photographs, or others which require a light at the back to show the picture, may be advantageously exhibited by means of this album. When the apparatus is used to exhibit card photographs, two of them " may be placed back to back in " each of the slides;" either picture may be seen " by simply " reversing the box."

The drawings show an album simply consisting of a box as described above, also an album consisting of two parts hinged together " in the form of a book."

[Printed, 8*d.* Drawing].

A.D. 1866, June 18.—N° 1641.

DALLMEYER, JOHN HENRY.—(*Provisional protection only.*)— " Improvements in compound lenses suitable for photographic " uses."

Photographic lenses for taking architectural views consist of " a combination of three simple or uncorrected lenses, two of " which, viz$^{t}$., the first and third, are of crown or plate glass and " both positive, and the second or intermediate lens is of flint " glass and negative;" these three lenses are positioned with intervals between them, and a stop is placed between the intermediate and posterior lenses. The anterior lens is made larger than the posterior lens. The anterior crown glass lens is a deep concave meniscus, convex side outside. The posterior crown glass lens is plano-convex, convex side outside, and the intermediate flint glass lens is plano-concave with the plane side facing the anterior lens.

In photographic portrait lenses which consist of two compound or corrected lenses with a space and stop between them, the anterior compound lens is made larger than the posterior compound lens. To prevent these lenses as well as those for landscapes representing objects in one particular plane with painful sharpness and giving two little detail in other planes, the simple or compound lenses are so mounted that the distance between them can be slightly altered. This has the effect of impairing the correction of the lens to any desired extent, so that then the focus is spread over a space more or less, "and then the definition " of the picture will no where be extremely sharp, nor on the " other hand, will there any where be any marked deficiency of " detail."

[Printed, 4*d* No Drawings.]

### A.D. 1866, July 6.—N° 1786.

FIELD, Lorenzo.—(*Provisional protection only.*)—"Improve- " ments in photographic printing frames."

Two flaps carry respectively the negative and the paper or material on which the positive is to be printed, they "are hinged " together so as to close upon one another, and studs on the " edge of one take into holes in the edge of the other." "The " negative flap consists of a frame carrying an inner frame in " which a sheet of glass is fixed to support the negative. The " inner frame is connected with the principal or outer frame by " means of springs at its corners, so that the negative can take " an exact bearing on the positive surface pressed against it " during printing. When this bearing is obtained the inner " frame can be fixed by means of clamping screws. The positive " flap consists of a frame carrying a panel having a number of " parallel slots in it; an ordinary pneumatic plate holder is fixed " in any required position on the panel by means of the slots, and " the plate of opal glass or a board carrying the positive paper is " held by the pneumatic plate holder. By the facility which the " slots afford for changing the position of the pneumatic plate

" holder the surface to be printed can be brought to any required " part of the flap so as to take a print from any part of a " negative." "The negative being placed in the negative flap " and the opal glass or paper in the positive flap, the latter is " closed or turned down upon the former and secured by hook " fastenings." "The frame is exposed to light in the usual " way, and the progress of the printing may be examined from " time to time by raising the positive flap from the negative " flap."

[Printed, 4*d*. No Drawings.]

A.D. 1866, July 24.—Nº 1918.

WOODBURY, Walter Bentley.—" An improved method of " and apparatus for printing from metal intaglios (obtained by " the aid of photography) in gelatinous or other semi-transparent " ink."

"This invention relates to a method of and apparatus for " rendering continuous and automatic the process of printing " from metallic moulds or intaglios obtained by the aid of " photography," according to the process set forth in No. 2338, or in any other analogous manner.

A cylindrical metal intaglio to print from is obtained by a machine of which the following is a description :—Round the central and non-cylindrical portion of a wrought-iron shaft a cylinder of lead or other soft metal is cast. This roller (when truly turned) is mounted " in a machine similar to a rolling press, " having a sliding table traversing beneath the said roller." The gelatine relief is now placed on the table, and motion is imparted to the roller by means of a winch handle and spur gearing; the plate holding the gelatine relief will be drawn through the machine, and an exact intaglio of the gelatine thereby impressed or indented into the cylindrical surface of the soft metal roller.

In the printing machine, the above-mentioned cylinder is supported by standards, and a continuous length of paper is made to pass round a part of its circumference by means of spur gear, suitably placed rollers keeping the said paper in contact with the cylinder. The ink is fed in between the paper and the cylinder at their point of first contact by means of a steam heated hopper.

Either a gelatinous ink, or a semi-transparent fatty ink, or a water-colour ink may be employed in combination with the said cylindrical metal intaglios.

[Printed, 8*d*. Drawing.]

A.D. 1866, August 28.—N° 2210.

GOULD, WILLIAM.—" A new or improved mode of reflecting " various coloured lights and shades upon stereoscopic and other " objects for producing different effects thereon."

This invention " consists in fixing (either at the front or back " of the instrument) glass slides or shades of any required color " or combination of colors, so that when the stereoscopic slide " is placed in the instrument different colored lights are thrown " thereon." The said slides are held by spring catches attached to revolving bars or pins (horizontal or vertical), which are carried by brackets and rotated by thumbscrews " causing the said " colored slides or shades to rise or fall (turning on their axes) " as desired." A coloured glass slide may also be placed over the top light of the stereoscope instead of at the back. For reflecting on glass stereoscopic slides, the said slides or shades may be placed at the back of the instrument, and also over the top light, " or the upper revolving pin can be made to turn " completely over " the said top light, and thus to serve both for glass and paper objects. " Set screws are also used for regulating " the tightening or loosening of the brackets holding the said " revolving bars or pins."

In the Final Specification the inventor states that he can also apply the invention " to the casting of shades on photographic " and other similar pictures placed in conservatories, windows, " and other like places."

[Printed, 8*d*. Drawing.]

A.D. 1866, September 13.—N° 2355.

BING, LOUIS.—(*Provisional protection only.*)—" An improved " mode of and apparatus for determining the actinic power of " light."

To measure actinic power numerically a graduated transparent medium is used " varying in transparency from the unit to any " required degree of opacity;" to this medium is applied a sensitised strip of paper, " placing it under the transparent

" medium for the purpose of exposing it to the action of the " light, and thereby obtaining an indication of the power of the " actinic rays." "Overlying" plates of talc, glass, or gelatine may form the transparent medium, but the following plan is preferred. A glass vessel having two flat sides is filled with any suitable coloured solution not affected by light. "The upper " plate will have an inclination from the lower plate both in the " direction of its length and breadth, that is, from say the left- " hand corner upwards, and also upwards to the right. Thus " the vessel will be wedge-shaped, tapering in two directions" to the corner which is termed "the unit corner." The whole of the lower transparent surface of the vessel is divided into squares of equal size which are numbered "from left to right with increasing " numbers, and also from the lower and thinner end or edge of " the wedge-shaped vessel to the thicker end."

[Printed, 4*d*. No Drawings.]

### A.D. 1866, September 27.—N° 2502.

DALLMEYER, John Henry. — "Improvements in compound " lenses suitable for photographic uses."

1st. "The production of view lenses or objectives embracing " large angles." These lenses are "composed of two positive " achromatic or actinic combinations of which the higher refract- " ing denser material or flint glass lens occupies the external or " exterior position in each combination." Each combination consists of a deep concavo-convex lens of flint glass, and a deep meniscus of crown glass, each combination being achromatic in itself. The diameter of the back combination is half that of the front combination. This lens equalises the illumination, and ensures a nearly perfect correction of spherical and chromatic aberrations, so that it acts quicker than existing lenses.

2nd. The arranging of double combination portrait lenses "so " that the lenses composing one of the combinations is of such " form and so positioned that by a slight variation of distance or " separation between its elements any desired amount of spherical " aberration can be obtained without at the same time materially " deranging the other necessary corrections of a photographic " objective." Both combinations are of the same diameter; the front combination is composed of a double convex crown lens and a double concave flint lens. The posterior combination is situated at a distance equal to the diameter of the front combination; it is composed of a meniscus crown glass lens, with the concave surface facing the front combination, and of a concavo-convex flint lens, with the convex side outside, " these two lenses " having their adjacent surfaces dissimilar."

[Printed, 8*d*. Drawing.]

### A.D. 1866, September 28.—N° 2513.

CLARK, William.—(*A communication from Joseph de Susini, Jean Marie Onésime Tamin, and Emile Bondonneau.*)—(*Provisional protection only.*) — " Improvements in the means of reproducing " signs, characters, and other marks in the transmission of " messages and signals by electric telegraph apparatus."

This invention relates to the re-transmission of the characters of electro-chemical telegraphs. Either a galvano-plastic or a photographic process may be employed for this purpose. The strip at the receiving station may serve again as a transmitter to another receiving apparatus at a further station, at which a similar reproduction may take place, and so on until the ultimate station is reached.

The strip at the receiving station having been marked by the electro-decomposition of the salts with which it is impregnated, the electric current from the transmitting station having duly acted upon it, a photograph of the message is taken, from which a copy on a conducting ground in non-conducting ink is obtained. The said copy may have a non-conducting ground, the characters being in a conducting material, and it may be produced from the said photograph (either a positive or negative) by photo-lithography or other analogous process. By using this copy in a transmitting apparatus, similar results are obtained to those produced by the original writing; these results may be operated from relay to relay until the ultimate station is reached.

A galvano-plastic process is also given at length.

This process may be applied to the systems of Caselli, Bonelli, Lenoir, Hughes, Gaiffe, and others.

[Printed, 4*d*. No Drawings.]

### A.D. 1866, October 12.—N° 2641.

GRÜNE, Wilhelm.—(*Provisional protection only.*)—This invention is for " developing invisible photographs by the action of the " ammonia vapours evolved from tobacco, and improvements in

" cigar tubes or holders and tobacco-pipes for holding the said " photographs."

In each cigar tube or holder the inventor forms " a transverse " opening, over which the prepared paper is fastened with gum, " starch, or other glutinous matter, so that when the cigar is lit " and the smoke drawn inwards the ammonia vapours will act " upon the paper and cause the photograph to be developed." For tobacco-pipes the inventor adopts " a similar plan, taking " care that the aperture shall be at sufficient distance from the " lighted end of the pipe to prevent the paper from burning;" or he modifies " the arrangement by placing behind the chamber " which holds the tobacco another chamber having an opening in " which is placed a slide holding the paper, there being a space " or perforation around the slide to prevent the draught from " being impeded. The papers are manufactured in the ordinary " manner as follows :—Albumen paper is first silvered and the " desired photograph copied under its corresponding negative, " the paper is then washed and finally placed in a solution of " chloride of mercury until the image becomes invisible, and " after the paper has been well washed with water it is ready for " use."

[Printed, 4*d*. No Drawings.]

A.D. 1866, November 5.—N° 2864.

MEDD, HENRY.—(*Provisional protection not allowed.*) — " Im- " provements in mounts for cartes de visite and playing cards."

" My improvements on the ordinary cartes de visite mounts " and common playing cards consist in printing, impressing, " engraving, or otherwise delineating on the back of such " mounts or playing cards the arms, crests, mottos, monograms, " or other designs representing the armorial bearings or mono- " grammic devices of the person, family, or personage represented " by the likeness in the carte de visite, or to whom the playing " cards belong. Such may be represented in their heraldic " colours, being printed or executed by hand."

[Printed, 4*d*. No Drawings.]

A.D. 1866, November 5.—N° 2869.

DE BRIGES, MALBEC.—(*Provisional protection not allowed.*)— " Improvements in obtaining photographic pictures."

"This invention has for its object to produce directly by "photography grotesque portraits of persons and things. The "process which has been designed with this end consists in "deforming the image of the person or of the object by means "of a mirror, either convex or concave, and photographing the "deformed image produced in the mirror; this deformation may "be regulated by the form given to the surface of the mirror. "Silvered globes of blown glass may be employed, or the "mirrors may be curved in one direction and straight in the "other, and placed either horizontally or vertically and convex "or concave; or concave spherical mirrors or mirrors of any "other form may be employed according as it is desired to "widen, swell out, or shorten the image or deform it in any "other way. The photographic impression is taken directly "from the deforming mirror by the processes and with the "precautions generally used for reflected photographic images. "If desired, parts of pictures thus obtained may be combined "with parts of other pictures obtained in the ordinary manner "so as to limit the grotesque to the extent desired. The pictures "may be copied and enlarged by the ordinary processes."

[Printed, 4*d*. No Drawings.]

### A.D. 1866, November 10.—Nº 2939.

SKAIFE, THOMAS.—"An improved apparatus to be used in "obtaining photograms."

This invention relates to the construction of an apparatus for igniting any composition, the flame of which is sufficiently actinic to produce photographs.

A platform, perforated with one or more touch-holes, is supported by springs so as to permit of its being easily vibrated by a touch of the finger. Connected with the platform is a reflector pierced with a groove, "through which communication is made "with one end of the platform by which it may be touched, or "struck, so that by means of the aforesaid springs it may suddenly vibrate." The deflagrating powder being placed upon the platform is suddenly brought into contact with the light from a spirit lamp placed underneath it, and "at the same instant "the platform being thrown into vibration communicates this "motion to the particles of the powder or other substance to "be ignited, the result of which action is that every particle

" explodes or is ignited simultaneously, and producing a flash " of light, which, acting upon a sensitised plate in an ordinary " camera produces the picture of any object placed before it. " When the picture of a near object is required the powder may " be placed over a touch-hole; if the picture of a distant object " is to be taken then the powder may be distributed over the " platform and over several touch-holes, to all of which the " light may be applied."

[Printed, 4d. No Drawings.]

A.D. 1866, November 15.—N° 2997

BERNIERI, LUIGI.—" Improvements in photography."

This invention " has reference to obtaining photographic back " grounds with inscriptions, monograms, heraldic, and other " devices thereon so as to present to portraits or otherwise a " medallion-like appearance in relievo in place of the plain " unfinished borders as at present, and consists in writing or " printing the said inscriptions, letters, devices, or otherwise upon " the ground glass or semi-translucent surface surrounding the " image upon the negative or collodion plate, so that, by covering " or blocking out the said portrait or other part of the plate not " required to be acted upon by the light, the said inscriptions or " devices as aforesaid may be transferred upon the card, as also " the deeper tints forming the shadows to the outline of the " border in relievo. By this process white inscriptions or other- " wise are left or produced upon the disk or tinted ground of the " medallion as before described."

The drawings represent a medallion portrait at the various steps, as described above, and in its complete form.

" The disposition and arrangement of the different tints, " together with the shades, monograms, or otherwise, may be " greatly varied or receive such other modifications as may be " found necessary."

[Printed, 8d. Drawing.]

A.D. 1866, November 15.—N° 3002.

GRÜNE, WILHELM.—" A chemical decoration on gold, silver, " and other similar metals and colours on porcelain, glass wares, " crystal wares, delft wares, potters' wares, and similar matters " called Grüne's proceeding."

The design on the negative plate is blackened by chloride of platinum, dried, covered with a flux and annealed. The picture film from this negative is obtained by pouring iodised collodion over its picture side, submitting it to the action of nitrate of silver solution, placing it in a frame and exposing it to light transmitted through the negative. The plate is then placed in a solution of sulphate of iron and sulphuric acid; the portions acted upon by light appear in silver, and the plate is washed, fixed by hyposulphite of sodium and again washed. The picture film is then separated from the negative by immersion in water containing glycerine. The film may be treated with chloride of gold or other metallic salts, according to the colour it is to receive. The picture is then bathed in a solution that will colour it by impregnation of the matter of the film; by annealing it afterwards various combinations in different metals may be obtained.

The film is transferred to the object to be decorated by immersing the film and the article together in a large vessel filled with a mixture of water and glycerine. The film is applied to the surface by means of a hair pencil; a coat of flux is then laid on, and the article is annealed, burnished, and polished in the usual way.

[Printed, 4*d*. No Drawings.]

A.D. 1866, November 26.—N° 3113.

COURTENAY, Robert Henelade.—(*Provisional protection only.*)—" Improvements in the preparation of printing surfaces by " the aid of photography."

1st. Metal plates engraved in relief for type printing.—A photographic negative is photographically printed on " to a trans- " fer paper prepared with a solution of gelatine, sugar, albumen, " bichromate of potash or ammonia, or both in combination with " a quantity of bichloride of mercury and hyposulphite of soda." The design is inked with transfer ink, sponged, dried, and " trans- " ferred to the metal plate to be engraved." Either a zinc plate or a copper plate coated with zinc is employed to receive the photographic design. The plate is etched with nutgalls, the transfer ink is cleaned off and the plate inked up. The back is then protected by means of varnish, and (in the case of a coated plate) the zinc that is not covered with the ink is removed, weak acid being used for that purpose. The plate is then electro-etched, and is ready for printing from.

2nd. "Intaglio plates such as are used in the copper plate "press."—A photograph is obtained "on litho transfer paper as "before," and is transferred to the surface of a highly polished copper plate coated with a thin film of zinc; the plate is then etched and inked, the bolder parts of the design being built up thicker with prepared mineral ink. Where there is not any design the zinc is removed with acidulated water, the plate is dried and metallised with plumbago or bronze powder. The plate is then submitted to the action of a small battery in the ordinary copper solution, and thus coated with copper; it is then attached to a powerful battery; the electrotype thus produced is used to print from.

[Printed, 4*d*. No Drawings.]

### A.D. 1866, November 27.—Nº 3116.

FOURNET, AUGUSTE, and NADAUD, OCTAVE.—"A magic "camera."

One of the peculiarities of this invention is that, with a suitable magnifying power, transparent pictures such as photographic positives, may be shown by means of this camera.

"The object of this invention is to construct a magic camera "to show on a ground glass a magnified image of a picture. The "camera is composed of two parts, the first part is a globe or "sphere mode of brass, tin, wood, or any other material with an "opening through the top and bottom to allow of a lamp, candle, "or other kind of light being put inside the globe, the bottom of "the globe being by preference arranged to rest on the socket "of the lamp or other kind of lighting apparatus employed. "Through the side of the globe a round hole is formed, in front "of which is a slide in which the picture is placed, and adapted "to the slide is a tube, inside of which are two lenses of a mag-"nifying power of from four to twenty times, or more.

"The second part is a camera equally made of brass, tin, wood, "or other suitable material; the back end of this camera is con-"nected with the globe by a tube into which the tube projecting "from the globe is inserted. The face or front of the camera is "closed by a ground glass, on to which (when a light is intro-"duced into the globe) is thrown a magnified image of the picture "placed in the slide."

[Printed, 10*d*. Drawing.]

A.D. 1866, December 3.—N° 3177.

WINSTANLEY, David, junior.—(*Provisional protection only.*) —“ Improvements in obtaining basso-relievo or other such “ designs in relief and also in depression by the aid of pho- “ tography.”

This invention relates to the obtainment of the said designs by submitting a layer of chromatised gelatine to the action of light.

In the first place a negative is obtained “ in which the com- “ paratively dense and transparent parts are regulated in their “ gradations according to the respective distances from the lens “ of the several parts which constitute the object to be copied.” “ In order to avoid the extreme effects by photographing various “ tints ” the object is caused “ to assume one tint by the applica- “ tion to its surface of a suitable colour.” To illuminate the object, the inventor avails himself of the fact that the amount of light proceeding in divergent rays from a given point, line, or plane which falls upon a given space in front of that light depends upon the distance of that space from the source of light; therefore the object is submitted only to the diffused light which comes from an opening. A reflector may be used to diffuse the light equally. The rays reflected to the camera will be strongest for the nearest part of the object and weakest for the furthest, with gradations. The photograph so obtained is used for producing a mould in chromatised gelatine, and from this mould a cast in plaster or other material is taken, the said cast will be a basso-relievo. These surfaces may be used for embossing purposes, or the chromatised gelatine itself may “ be so used for some “ purposes.”

[Printed, 4*d*. No Drawings.]

A.D. 1866, December 6.—No. 3221.

LANE, Farindon.— (*Provisional protection only.*)—“ Improve- “ ments in photographic pressure frames.”

“ My invention relates to an improved pressure frame for print- “ ing positive proofs from photographic negatives. It consists “ of a back piece, to which a spring clip is attached, which serves “ to hold the plate and the paper together by one end during the “ printing. To the front edge of the back piece a flap is attached

" by a hinged joint, allowing it to turn back to examine the " print; this flap with a part of the back piece forms a backing " somewhat larger than the plate to be printed from, and this is " covered with black velvet, cloth, or other elastic material to " ensure contact between the plate and the paper. Attached " to the flap are other spring clips fixed to pieces which slide in " sockets so as to allow of their being easily withdrawn and " replaced on the edge of the plate when it is necessary to examine " the print; these clips when on give the necessary pressure to " produce contact between the paper and plate for the purposes " desired."

[Printed, 4*d*. No Drawings.]

A.D. 1866, December 15.—N° 3303.

SWAN, JOSEPH WILSON.—" Improvements in the treatment of " gelatinous tissues of gelatine and gum and of compounds con- " taining such substances."

This invention " consists in the use of salts of the sesquioxide " of chromium, as, for example, sulphate of the sesquioxide of " chromium or the substance known in commerce as chrome alum, " as a means of rendering gelatine or gum (Senegal or Arabic) " or compounds containing those substances insoluble in water."

Amongst the uses to which this invention is applicable are:— The fixing of photographs mounted with gelatine; and the preparation of photographic paper sized with gelatine or gum.

To fix photographic prints mounted with gelatine, they are immersed in a solution of chrome alum, or the prints are mounted on paper wetted with the said solution, or a mixture of gelatine with the solution is employed as the cement wherewith the prints are mounted, the mixture being made immediately previous to using " in consequence of the tendency of the mixture to solidify " after the addition of a solution of chrome alum."

" In rendering gelatine insoluble when used as a glaze, sizing " or varnish for paper, or when formed into a sheet," the inventor applies " a solution containing about five per cent. of chrome " alum " to the gelatinised surface or gelatine sheet, or the surface or sheet is immersed in the above-mentioned solution, or a sheet is formed " from a mixture of a solution of gelatine and the " above-mentioned solution of chrome alum."

[Printed, 4*d*. No Drawings.]

A.D. 1866, December 24.—N° 3393.

ASHTON, Robert Howe.—(*Provisional protection only.*)— "Improvements in producing printing surfaces and carvings from "moulds obtained by the aid of photography."

"According to my invention I use moulds, the different depths "of which correspond to the lights and shades of a photograph, "such moulds being well known and in use for purposes apart "from my invention. From these moulds I obtain printing sur-"faces by passing the tracer of a pentagraph of that construction "known as Jordan's or other similar ruling machine over them, "and which tracer being acted upon by the different degrees of "depression alters the direction in which the etching point moves. "Beneath this etching is placed a sheet of copper, steel, or other "metal coated with a protecting medium, wax, for instance, after "the usual manner adopted by engravers, and the design which "is cut through the said coating may therefore be etched, as is "well understood.

"In producing carvings I employ moulds as above described, "but cause the tracer to regulate the depth to which a rotatory "or other cutting tool acts upon the material to be carved; these "carvings may be ornamental in themselves, cameos, for instance, "or may afford dies for embossing."

[Printed, 4*d*. No Drawings.]

# APPENDIX.

A.D. 1851, August 23.—N° 13,726.

PALMER, James.—"Improvements in delineating objects, and "in apparatus and materials for that purpose."

1st. "Various improvements in delineating objects on sheets of "glass or gelatine placed between the object and the eye of the "operator;" also "various methods of reducing, enlarging, "multiplying, and transferring the drawings thus obtained."

2nd. "Various improvements in materials and apparatus "adapted for the aforesaid purposes." Under this head a mode of preparing insoluble gelatine is described. Amongst other purposes, the insoluble gelatine is applicable to "the obtaining "of photographic images." An aqueous solution is made containing alum, acetate of lead, and borax; upon the subsidence of the white precipitate formed, the clear liquor is drawn off. Into this clear solution either sheets of gelatine, or sheets of glass coated with gelatine, are immersed for several hours. "The "sheets are then taken out and allowed to dry, and are then fit "for use."

To render the gelatine insoluble other salts of alumina may be used. The following are other salts having the same effect:—Bi-chloride of mercury, borax, acetate of copper, sulphate of copper, sulphate of iron, acetate of lead. nitrate of lead, acetate of baryta, fluoride of sodium, fluoride of potassium, cyanide of sodium, cyanide of potassium, sulphate of soda, sulphate of potash, oxalate of soda, oxalate of potash, acetate of soda, and acetate of potash; boracic acid solution has also the same effect.

These sheets of insoluble gelatine, or plates of glass coated with the gelatine which is then rendered insoluble, "may be "employed for photographic purposes, as they admit of being "immersed or covered with the various solutions without the "gelatine being dissolved off the glass."

[Printed, 1*s*. 4*d*. Drawings. See Mechanics' Magazine, vol. 56, p. 197.]

## A.D. 1854, December 14.—N° 2637.

CORNIDES, Louis.—"Certain improved apparatus for coating "or covering surfaces of glass or other material with collodion."

"This invention consists of an apparatus intended to facilitate "and ensure the coating or covering of plates of glass or other "surfaces requiring to be coated with such material, to be em-"ployed for various purposes, both useful and ornamental, by "which means the great loss and inconvenience from the evapo-"ration of æther, hitherto inevitable in such operations, is "entirely obviated."

The apparatus which the inventor employs for the above-mentioned purpose "consists of four principal parts; viz., a "vessel or receptacle capable of being made air-tight, and "intended to receive the objects to be coated; an air-tight vessel "or reservoir to contain the collodion; an air pump in connection "with the above-named vessels, and a worm for condensing the "æther. This apparatus is used by placing the object to be "coated in the receptacle, and after shutting the latter air-tight, "the air is exhausted from it by the air pump, and then the collo- "dion is let into the receptacle from the reservoir by gravity, or "the action of air force pump, or of an ordinary suction and "force pump, placed between the reservoir and the receptacle. "After the object is thus immersed, the collodion is withdrawn "back into the reservoir by the same means, and the receptacle is "then subjected to a temperature sufficient to evaporate the "æther which passes off into the worm." The condensed æther may be drawn off from the worm by means of a cock below. The dry plates are removed by withdrawing the upper part of the receptacle, and a fresh supply is substituted. "The operation may "then be repeated as before."

[Printed, 8*d*. Drawing.]

A.D. 1855, February 6.—N° 275.

GEDGE, JOHN.—(*A communication from J. E. Pointeau.*)—(*Provisional protection only.*)—"Improvements in frames suitable for "photographic or stereoscopic proofs or portraits."

"Instead of the frames at present in use for receiving and pro- "tecting stereoscopic or photographic proofs or portraits," the inventor proposes "to make frames by which a certain distance "shall be preserved between the prepared glass and the proof or "portrait, making the back framing in such a manner as to "secure the said object of exact distance. It is known that "of stereoscopic proofs two are taken from the same subject at "different angles; these are superposed in frames, and only one "is visible, but that one in great relief." This effect however will be much increased by the use of this invention, "which, "from the peculiarity of the parts, forming in the whole a frame, "secures the spectator a view of the object, portrait, or proof "in its highest developement, and produces an effect hitherto "unknown in photographic or stereoscopic objects."

[Printed, 4*d*. No Drawings.]

A.D. 1855, February 9.—N° 309.

PONT, BARTHÉLEMY. — (*Provisional protection only.*) — "A "process of autographic engraving."

This invention "entirely supersedes the object glass, and it is

" the plate itself on which the artist has made his engraving " which gives the most faithful impressions."

"Mode of operating.—Take a plate of glass, prepared with " collodion, and sensitived by nitrate of silver, then expose it " in the dark chamber to receive the impression." The plate " then undergoes the ordinary manipulation of photography until " it presents a completely smooth surface, and of an intensity in " harmony with the tint for the positive proof." The plate is then well washed, fixed with "hyposulphate" [hyposulphite ?] of soda, and treated with a solution of dextrine; the plate then remains twenty-four hours before it is used.

"Execution of the autographic engraving on the plate.—The " plate on a black ground, the part prepared with collodion " uppermost, the artist, with points, pens, or any other instru- " ment fit for taking off cleanly the collodion, is to draw all the " tracings which are to serve as a model for his engraving; he " will easily be able to judge of the effect of his work, for each " part removed presents black tracings, as is the case in a pen " and ink or pencil drawing. Should the black masses appear " to him too heavy, he can lighten them by using silver white " diluted in water, and applied with a very fine brush. If he " wishes to obtain spots completely white he has only to lay on " a thicker coat. The striking off impressions is produced by " the same means as in photography, but only more rapidly."

[Printed, 4*d*. No Drawings.]

## A.D. 1857, June 22.—Nº 1744.

SEROPYAN, Christopher Dicran.—" A mode of preparing " bank notes, bills of exchange, and other papers to prevent " counterfeiting by photography and its kindred processes, and " a mode of preparing an ink for the same."

This invention " consists in using two or more colours, which " do not reflect nor transmit but absorb the chemical rays of " light; one of which shall be applied to the paper either by " printing, staining, or other mode, so as to cover the surface " with a tint or ground of a red, orange, or yellow shade or color, " while an ink of a different color or shade from the tint or sur- " face color shall be used for printing the other parts of the " note, that is the obligatory and ornamental parts upon the said " surface; the said ink consisting of such ingredients as will

" render it equally or more fugitive than the color forming the " tint or surface color of the paper." The inventor has " for this " purpose invented an ink, which is composed of a black pre- " cipitate, produced by the action of an alkali upon sulphate of " iron in aqueous solution; this is washed with water and mixed " with gallic acid or extract of logwood. This mixture well " dried is then ground in boiled linseed oil or painters' varnish. " If not found sufficiently black a little blue may be added. " Other organic or metallic substances may be also employed, pro- " vided they possess the required relative properties of absorption " and fugitiveness to the other shade or color."

[Printed, 4*d*. No Drawings.]

A.D. 1858, April 13.—N° 793.

SPILLER, Thomas.—(*Provisional protection only.*)—" Exhibit- " ing slides in the stereoscope and preserving them from injury, " to enable each slide to be conveyed to the point of view, and " then after use deposit them each in its place in the box without " handling, or exposing the slides to the chance of being soiled, " keeping them always under cover in safety, a box 18 in. by " 8 in. square will hold and exhibit near 1,000 slides."

" The box is made in two compartments, with a square drum " on the top over the parting. Inside the box are placed guide " lines or wires, the slides are then fixed in a chain made ex- " pressly " " the length required, one end of such chain is then " fixed to hooks on the guide lines (which slide on them, and " traverse from top to bottom in each compartment of the box), " the chain is then properly folded and laid in one compartment " of the box, and the other end of the chain carried over the " drum, and fixed in like manner to the hooks on the guide " lines in the other compartment of the box. By turning the " drum, over which this chain of views passes, the slides will be " brought before the lens or sight of the stereoscope (which is " fixed on a line with the axis of the drum), and exhibited. " When done with, the drum is turned one-fourth round, and " another view exhibited in the same manner, untill all the " pictures are brought in rotation from the full compartment of " the box exhibited in the stereoscope, and carried into the other " compartment and folded carefully up in their place, when by " turning the drum the reverse way, the views repass through

" the stereoscope again into the first compartment, exhibiting " them both ways." The name of each picture may be told by an indicator; the stereoscope may be fixed to the box with a swivel " and the box may be made to elevate or depress to the " hight required."

[Printed, 4*d*. No Drawings.]

### A.D. 1858, April 22.—N° 887.

MAUGEY, Pierre.—(*Partly a communication.*)—" Improve- " ments in diaphragms for optical instruments."

This invention " relates to a mode of constructing diaphragms, " whereby the size of the aperture may be increased or diminished, " or regulated at pleasure to the greatest nicety.

" This object is effected by making the diaphragm of some " elastic material, which is capable of being expanded or " stretched. The material which is preferred is a disc or sheet " of india-rubber, either vulcanized or in its natural state, or " some other analogous elastic material. A small central " circular opening is made in this disc, the edges of which are " securely held or clamped, and by forcing the disc outwards by " means of any convenient instrument, so as to stretch the " material, the central opening or aperture made therein will " be expanded; or by drawing back the stretching instrument the " aperture will be contracted.

" The stretching instrument may be actuated by means of a " rack and pinion motion, which is the most convenient con- " trivance for the purpose, but other means may be employed " for effecting the object, if preferred."

[Printed, 6*d*. Drawing.]

### A.D. 1859, November 8.—N° 2541.

JOHNSON, John Henry.—(*A communication from Auguste Pinard.*)—(*Provisional protection only.*)—" Improvements in " 'passe-partouts' or mounts for pictures."

The said mounts may be used " for photographic and other " pictures."

" According to this invention in place of making the 'passe- " 'partout' or mount out of one piece of cardboard stamped to " the desired pattern, it is proposed to construct it of three " distinct parts suitably colored or tinted, and disposed in

" relation to each other to give the effect of a solid mount with " a considerable degree of relief. The inner rim of the mount " next to the picture consists of a suitably stamped or moulded " frame of cardboard or thin metal, and may be tinted, bronzed, " or left plain as desired. Upon this rim is laid a flat piece of " cardboard, having an aperture of such a size as to enclose the " moulding of the rim, and this second piece may either be plain " or ornamented, or tinted according to taste; over this again is " placed a third sheet of cardboard, the aperture in which is large " enough to show a considerable portion of the second piece last " referred to. It is obvious that the apertures may be of any " desired form, whether round, square, or oval, and that the " tints and ornamentation of each of the three component parts " may be varied according to taste."

[Printed, 4*d*. No Drawings.]

# INDEX OF SUBJECT MATTER.

[The numbers refer to the pages in which the Abridgments commence. The names printed in *Italic* are those of the persons by whom the inventions have been communicated to the Applicants for Letters Patent. The Roman numerals refer to the Introduction.]

## ERRATUM IN PART I.

Since the publication of the first volume of the present series of Abridgments the following erratum has been discovered therein:—

Page 8, line 26, *for* "Walcott" *read* "Wolcott."

## ERRATA IN PART II.

Page 72, line 15, *for* "Lefarge" *read* "Lafarge."
Page 148, line 19, *for* "No. 2338" *read* "No. 2338, A.D. 1864."

LONDON:
Printed by GEORGE E. EYRE and WILLIAM SPOTTISWOODE,
Printers to the Queen's most Excellent Majesty.

PATENTS FOR INVENTIONS.

# ABRIDGMENTS

OF

# Specifications

RELATING TO

# PHOTOGRAPHY.

PART III.—A.D. 1867–1876.

PRINTED BY ORDER OF THE COMMISSIONERS OF PATENTS.

LONDON:
PUBLISHED AND SOLD AT
THE COMMISSIONERS OF PATENTS' SALE DEPARTMENT,
38, CURSITOR STREET, CHANCERY LANE, E.C

1878.

# PREFACE.

The Abridgments of the Specifications of Patented Inventions are prepared and published in classes, each confined to one subject or group of subjects, in order to facilitate the searches of inventors who are desirous of ascertaining whether their inventions are novel. Without the Abridgments an inventor can, by the aid of the general Subject-matter Indexes of Patents, refer to the Specifications included under any particular subject or head in those Indexes, but this course involves the perusal of the whole of the Specifications so included.

Each volume of the classified Abridgments is furnished with an elaborate subject-matter or analytical index (as well as an alphabetical index of the names of patentees and inventors) to guide the reader to the abstracts of the inventions belonging to any particular section of the class. On reading these, he will be able to determine which Specifications he ought to consult for more minute information.

The Abridgments are arranged in the volume in chronological order, and at the foot of each is stated the price at which the printed copy of the Specification can be purchased at the Commissioners of Patents' Sale Department (38, Cursitor Street, Chancery Lane, E.C.).

This volume forms Part III. of the series of Abridgments relating to "Photography," and embraces the years 1867 to 1876; the first and second parts contain the Abridgments of the Specifications from the earliest date to the year 1866.

The number of Specifications from the earliest period to the end of the year 1876 amounts to 100,042. A large proportion of the Specifications enrolled under the old law, previous to 1852, embrace several distinct inventions, and many of those filed under the new law of 1852 indicate various applications of the single invention to which the Patent is limited. Considering therefore the large number of inventions and applications of inventions to be separately dealt with, it cannot be doubted that several properly belonging to the group which forms the subject of the present series have been overlooked. In the progress of the whole work such omissions will from time to time become apparent, and be supplied in future editions.

To afford a more precise idea of the scope of the present series of Abridgments than can be conveyed by the short title, the following definition is offered of the word Photography,—" the art of copying designs, figures, or images by " the chemical or actinic action of light upon surfaces pre- " pared to receive that action."

With respect to the printing of pictures obtained partly or wholly by photographic means, it has been found advisable to include also all successive processes for transferring or copying such pictures, and all subsequent printing of copies of the same in permanent inks or colours.

Optical and philosophical instruments not used directly in the process of photography are excluded from the present series, and should be looked for in that entitled "Optical, " Mathematical, and other Philosophical Instruments."

It should be particularly observed in connection with the preceding paragraph that all stereoscopes are omitted from the present volume.

Improvements in books or albums for containing photographs or pictures should be looked for in the series relating to " Books, Portfolios, Card-cases, &c.," whilst improvements in mounting, holding, and framing photographs, &c. will be found in the series devoted to " Artists' Instruments and " Materials."

Inventions relating to the application of photographic pictures to such ornamental objects as jewellery, or to the embellishment of boxes, furniture, other articles, and surfaces of all kinds are not contained in the present work unless some novelty in producing the picture is described or claimed.

H. READER LACK.

*October*, 1878.

# INDEX OF NAMES.

[The names printed in *Italic* are those of the persons by whom the inventions have been communicated to the Applicants for Letters Patent.]

# PHOTOGRAPHY.

## 1867.

A.D. 1867, January 3.—No. 20.

HELSBY, WILLIAM GEORGE.—A "system of and apparatus "for manufacturing photographic pictures."

A large number of impressions are made closely together on one sheet of sensitised paper. The pictures can be cut out of the sheet and they are produced simultaneously and of uniform depth of shadow all over the sheet.

The apparatus consists of a camera in which there is mounted the same number of lenses as of impressions to be made; these lenses touch or are near to each other, and are fitted either to produce like pictures, or to produce pictures of different sizes and shadows. That an equal quantity of light may pass to every picture, a sliding shutter in the object piece of the camera is fitted so that it can be opened and closed in the same direction. Thus the extra quantity of light allowed to enter through at one side of the object piece aperture, during the opening, is compensated by being shut off earlier in the closing movement. One arrangement of the sliding shutter is to fit it closely on a pin and to operate quickly by a lever or toggle joint.

The operation of focussing may be facilitated by means of a plate with apertures of the shape of the pictures and at the proper distance from each other.

[*Printed*, 10*d*. *Drawing*.]

A.D. 1867, January 7.—No. 45.

ASHTON, ROBERT HOWE.—*Provisional protection only.*)—"Im- "provements in the production of printing and embossing "surfaces from photographs."

In the first place a mould is obtained by exposing a photograph to light in combination with sensitised gelatine; the gelatine mould is then placed in contact with a slab of soft or fusible metal, and the whole is submitted to pressure. From the metallic mould or impression this formed an electrotype is obtained, which consitutes the printing or embossing surface. Whether or not this electrotype be taken direct will depend upon the photograph having been negative or positive. If it be necessary to form an intermediate mould, a cast is obtained in the following manner:—A sheet of wax is taken and coated with plumbago so that its surface conducts electricity; the fusible metal mould is then pressed upon the conducting surface, and the wax mould resulting from this operation is electrotyped. In a modification of the process, the gelatine mould is pressed upon the slab of wax direct.

Printing surfaces obtained by these means may be used after the manner of copper-plate printing, or may be employed for producing photo-relievo printing.

[*Printed, 4d. No Drawings.*]

## A.D. 1867, January 11.—No. 71.

MORVAN, Arthur Guy. — "Improvements in the art of "making photographic trsnsfers adapted to the purpose of "engraving or lithography, or of making plates or blocks "from which to print the design or other matter trans-"ferred."

Paper (albuminised or not), of suitable texture for photographic printing, is dipped into a bath of sour milk, whey, or lactine to enable it to resist the chemical substances afterwards to be applied; it is then dried.

To coat the paper with a material to receive the action of light, and to be soluble in water where it has not been acted upon by light, dissolve French glue in water, and, while it is boiling, add thereto a solution of permanganate of potash this preparation is applied to the surface of the paper and dried in a dark room. The paper may then receive the photographic impression in the ordinary way.

Immediately after the impression is taken, and before it is developed, the paper is thinly and uniformly coated with a composition containing Judea bitumen, white wax and Burgundy pitch dissolved in a sufficient quantity of essence of lavender. This coating is on the sensitised side of the paper and it is allowed to dry in the dark. By this means colour is given to the picture and a medium is formed to transfer the picture to the wood, stone, or metal surface.

The paper is placed, with its black side upwards, in a bath of cold water. The spaces not acted upon by light are thus removed, together with the superincumbent black coating. The proof is cleansed by a sponge, dried, and transferred by contact and pressure to the stone, metal, or wood surface to be printed from if upon stone, or etched or engraved if upon metal or wood.

[*Printed, 6d. No Drawings.*]

A.D. 1867, January 17.—No. 121.

NEWTON, William Edward.—(*A communication from George Streat.*)—(*Provisional protection only.*)—"Improvements in "tinting or colouring the surface of paper or other materials "or fabrics."

This invention consists in tinting or straining the surface of these materials by submitting them to the action of smoke. The sheets of material are stretched upon tables within a chamber into which smoke is introduced. The smoke may be obtained by the burning of tobacco, hay, &c. The smoke subsides and condenses upon the surface of the material, and produces a permanent tint proportioned to the density of the smoke. By placing upon the surface patterns of any desired configuration, so as to entirely or partially prevent the action of the smoke on the parts covered up by them, figures can be left of the original colour or nearly so of the surface, while the uncovered surface is tinted to any desired depth of colour.

This improvement is adapted to the tinting of paper for photographers, &c.

[*Printed, 4d. No Drawings.*]

A.D. 1867, January 23.—No. 174.

ROSS, Thomas.—"Improvements in the manufacture of "lenses for photographic cameras and various optical instru- "ments."

This invention consists in the employment of different

descriptions or qualities of flint glass or of crown glass for both the convex and concave glasses of the above lenses, the convex being made of light flint or light crown glass (having a low refractive and dispersive power), and the concave being made of heavy or dense flint, or heavy or dense crown glass, which has a high refractive and dispersive power. The different degrees of refrangibility of the two descriptions or qualities of crown glass, or of the two descriptions or qualities of flint glass counteract or neutralise each other.

[*Printed*, 4*d*. *No Drawings.*]

A.D. 1867, February 13.—No. 407.

NEWTON, William Edward.—(*A communication from Adolphe Guillaume Larauza.*)—(*Provisional protection only.*)—A " mode " of and apparatus for taking photographic portraits, pictures " or representations."

This invention consists in the production of a " photo-" graphic statuette," or a portrait showing the back and the front of the person photographed. The portrait is taken from two sides, and the pictures produced on paper are to be stuck together at the edges, but leaving the bottom open.

The person to be photographed is placed on a turntable which has two stops, one exactly opposite the other. The turntable is thus capable of turning exactly through half a circle, so that the person is first seen from the front and afterwards from behind. The corresponding photographic pictures form two exact halves, which may be placed back to back. After the edges of the two portraits have been struck together, the whole is expanded by introducing a piece of wood or other material into the interior.

The photographic statuette is placed in a kiosque, which is composed of four, five or more looking glasses placed in connection with each other, and may be ornamented with gilding, sculpture, or otherwise.

[*Printed*, 4*d*. *No Drawings.*]

A.D. 1867, March 8.—No. 659.

LAKE, William Robert. — (*A communication from John Cushing Crosman.*) — A " mode of coating paper and other " materials with fluid substances, solutions, and compounds " for photographic and other purposes."

The sheet of paper is thoroughly wetted by being immersed in suitable fluid; it is then deposited on a level table, so that its under side comes into contact with the upper surface of the table, by expressing air and any free fluid from between the sheet and the table and by using bibulous paper to remove superfluous fluid from the upper surface of the sheet. When in this condition, the coating fluid or material is applied to the upper surface of the sheet, by pouring the fluid upon the surface near the centre of the sheet. A deckle may be used if a considerable depth of fluid is required to make the coating.

When a solution of gelatine mixed with carbon is used to coat paper for photographic purposes, the mixture is employed at from 100° to 200° Faht. When albumen is used, either alone or in connection with gelatine, the temperature is kept below the point at which albumen coagulates. The coated sheets are not removed from the level table till the coating has sufficiently set.

[*Printed*, 4*d*. *No Drawings*.]

A.D. 1867, March 13.—No. 732.

LEE, EDWARD.—A "mode of and apparatus for producing "pictures, designs, letters, and other characters or figures "upon glass, metal, wood, and other materials."

To transfer photographs from photographic paper to glass, the surface of the photograph is first covered with varnish. Powdered resin or other powder (which, when applied to the surface will render the same waterproof), is placed on the surface and fixed by passing the paper over a wet stone or between a pair of rollers. Other modes of effecting the fixing are by heat, or by allowing the resin to become incorporated with the colour applied to the photograph, so as to form a varnish. Instead of the powdered resin, gold, bronze, or other metallic leaf or powder may be employed, or the latter may be mixed with the resin to spread over the surface. When metallic leaf or powder is employed, the application of the wet stone is not required.

To produce a coloured impression of the photograph, the various colours are applied to the surface of the same instead of the varnish, the said colours being covered with the powdered resin.

In transferring the design to the glass or other surface, if resin alone has been used, the surfaces are immersed in water, and in that condition the design is laid upon the glass, with the coloured or varnished side in contact with the glass. Pressure is then applied to the back of the paper by means of a burnisher or a pair of india-rubber rollers. The photograph being allowed to dry, the paper is wetted at the back and is then removed, leaving the photographic image on the glass.

[*Printed*, 4*d*. *No Drawings*.]

A.D. 1867, March 19.—No. 793.

CLEMENTS, James Moore. — "Improvements in ironing, " pressing, finishing, and damping machinery."

Two or more hollow cylindrical rollers are mounted horizontally between standards, and revolve either in the same or contrary directions. They may be capable of moving to and fro by means of levers or screws at equal or unequal speeds, motion being given to them in any suitable manner. The rollers may be heated; pressure is applied to them by means of an adjusting screw and spring, fitted in the framing above the upper roller.

For photographs, publications, or objects of similar nature and large size, rollers are employed having a space between them sufficiently large to admit of the passage of pad blocks together with the article to be pressed.

A board or table is fixed in such a position that a portion of the lower pressing cylinder or roller may pass through it, which table receives the work to be passed through the rollers.

[*Printed*, 1*s*. *Drawing*.]

A.D. 1867, March 21.—No. 825.

HART, Herbert William.—(*Provisional protection only.*)—A " process for producing indestructible photographic pictures."

According to this invention, pictures are produced on the surface of glass enamel with which a metallic plate is coated. The picture is burnt in by heating the plate until the enamel is softened and partially melted.

The photograph may be taken direct on the enamelled surface; or a collodion film with the photographic image thereon may be transferred from its glass plate to the enamelled surface; or a negative, taken in the ordinary manner, may be printed on paper, transferred from thence to enamelled surface, and burnt in.

[*Printed*, 4*d*. *No Drawings*.]

A.D. 1867, March 23.—No. 849.

EDWARDS, Ernest. — (*Provisional protection only*.)—" Im-" provements in photographic pictures and in apparatus for " producing them."

A lens of very short focus and large aperture is fitted to a camera arranged to produce one or more different photographic pictures successively upon the same sensitive plate. Different parts of the plate are exposed to the action of the light from the lens by means of an instantaneous shutter placed in front of the lens or behind it. For these negatives the sensitising bath is a silver bath of full strength and neutral or slightly acid. The developing solution is sulphate of iron to which is added loaf sugar and alcohol. The fixing solution consists of hyposulphite of soda. After fixing, the pictures are varnished.

To print enlarged copies from the above negatives, the above mentioned camera is attached to the front of another camera. The distance between the negative and the lens and between the lens and the picture to be formed determines the size of the copy. A mixture of collodion with a solution of white wax in ether is poured on to the glass plate and rubbed off again to give place to the film sensitised in an acid bath. The picture is then taken in the second camera and developed with a solution containing excess of acetic acid. The picture is fixed with cyanide of potassium, and toned with a solution of bichloride of mercury containing salts of gold or uranium. The film is washed with hyposulphite of soda solution and then dried. The dried picture is transferred to paper by immersion in water and by pressure.

[*Printed*, 4*d*. *No Drawings*.]

A.D. 1867, March 30.—No. 947.

WOODBURY, Walter Bentley, and ASHTON, Robert Howe.—" Means of producing designs upon paper."

By means of this invention transparent designs are produced upon paper, such as water-marks, by pressure from reliefs or intaglios obtained by photography. Also, by pressure from the said reliefs, and with the interposition of thin paper charged with a greasy substance, a design is produced which may be transferred to stone for printing by the ordinary lithographic process.

The relief is first obtained in gelatine by the process described in No. 2338, A.D. 1864; it is mounted on a hard metal plate, and the paper to be impressed is rolled or passed through a press in contact therewith. The paper is rendered transparent at those parts which have been in contact with the raised parts of the design. This gives a positive picture by reflected light. If a positive picture by transmitted light be required, a transparent positive is used, in place of the ordinary negative, to produce the gelatine relief. Instead of the above method, a metal impression or electrotype from the gelatine may be used with the ordinary paper machinery to produce the water-mark.

The greasy coloured paper used in the second part of the invention may be similar to the carbonised paper used for copying letters.

[*Printed, 4d. No Drawings.*]

A.D. 1867, March 30.—No. 955.

HIGGINSON, George.—(*Provisional protection only.*)—"Improvements in producing opalline pictures into china, glass, "porcelain, or any other enamelled surface."

By applying certain chemicals, in various proportions, according to the nature of the article, to any surface of the above description, photographic pictures may be struck into the surface in such a manner as to form permanent photographs. The chemicals used are salts of uranium, pyrogallic, gallic, nitric and citric acids, "cyanide, silver, iron, and "zinc."

[*Printed, 4d. No Drawings.*]

A.D. 1867, April 1.—No. 968.

BROOMAN, Clinton Edgcumbe. —(*A communication from André Adolphe Disderi.*) — (*Provisional protection only.*) —

" Means of producing surfaces in relief or in intaglio for " printing and other purposes."

1st. The transformation of a photographic image into an electrotype mould.

2nd. The production, with this mould, either of impressions with inks of any colour upon paper, textile fabric, glass, ceramic ware, wood, stone, and metal, or of " countertypes " in relief or in intaglio on any metal, or of engraved blocks for printing.

The photograph in relief is obtained by means of an albuminous or gelatinous coating containing nitrates, chlorides, bichromates, prussiates, and the like, double salts of gold, silver, copper, and the like. After being exposed to light, the parts that have not been acted upon by the light are dissolved out by water or other solvent.

The electrotype mould is obtained by varnishing the contour of the glass on which the photograph is made with gutta percha, and depositing a layer of metal upon the photograph.

The mould is covered with ink at the ordinary temperature and applied to the surface to be printed; desiccation then takes place, and, by heating the mould gently, the image is left adhering to the surface.

The matrix or mould may be subjected to a second galvanic operation to obtain relief designs in gold, silver, copper or other metals; these designs may serve for printing purposes.

[*Printed*, 4*d*. *No Drawings*.]

A.D. 1867, May 22.—No. 1526.

NEWTON, William Edward.—(*A communication from Henry Cook.*)—(*Provisional protection only.*)—" Improvements in portable photographic apparatus."

1st. A cylindrical box, constructed to contain fifty or other convenient number of prepared and sensitised glass plates, which are brought, by means of the simple rotation of the case, one after another, directly over a door, which opens and shuts automatically.

2nd. The photographic apparatus or camera.—This apparatus is like an ordinary opera glass. Its lenses are exactly similar in focus; one projects the image to be photographed on to the focussing glass, which occupies the large part of the tube; the other produces the image on the sensitised plate, which occupies a small dark chamber formed in the opposite tube. The above-mentioned cylindrical box is pushed over an opening in the dark chamber; upon the doors of the box and of the chamber being opened simultaneously, a sensitised plate falls into the chamber. When the view has been taken, the apparatus is turned upside down, and the plate which has been exposed falls back into the box, which is then removed. The door of the box closes, and a small movement to the right or left brings another sensitised plate over the door and the opening in the camera.

A stand for the camera consists of a socket and upright post provided with a universal joint.

[*Printed, 4d. No Drawings.*]

### A.D. 1867, May 25.—No. 1553.

SIMPSON, James.—(*Provisional protection only.*)—An "ar-" "rangement of apparatus to be employêd in producing pho-" "tographic pictures."

By means of this invention, enlarged photographic pictures may be obtained from small negatives; also from such negatives positive prints may be produced that are of better quality than those yielded by the ordinary methods.

An ordinary photographic camera and lens is used in connection with an elongated conical dark tube, arranged telescopically in order that it may be diminished or elongated in length. The camera is connected with the dark tube, so as to have the lens enclosed and embraced by the narrow end of the cone, the base of the cone being provided with a slide constructed like the ordinary dark slide of a camera to receive sensitised paper. The ordinary dark slide of the camera at the reverse end and outside the conical box receives the negative to be printed from. The rays of light, entering the box, penetrate the negative as a medium, and, in their passage through the lens, become gradually enlarged until they are received by the sensitised surface at the base of the cone, the result of which is a direct enlarged positive print.

[*Printed, 4d. No Drawings.*]

A.D. 1867, May 29.—No. 1593.

GAGE, FRANKLIN BENJAMIN.—A "means or method of illu-"minating the shadows and harmonising the lights and "shades in photographic impressions."

According to this invention, in developing the photographic image, the inventor employs a partial development in the camera by the aid of diffused light. The development is afterwards completed in the usual way, taking the precaution to reduce the strength of the developing fluid with water.

The photographic impression is taken in the usual way, then some plain dark dead surface is placed in front of the camera, the sensitive surface still remaining in the camera. The covering is removed from the lens tube and the sensitive surface is exposed to the light reflected from the dark surface, while the dark surface is kept in gentle motion. For a dark dead surface, a piece of thick black woollen cloth is attached, by one edge, to a stick. This is held horizontally and gently moved in front of the camera with the left hand, while the lens tube is uncovered with the right hand.

[*Printed, 4d. No Drawings.*]

A.D. 1867, June 7.—No. 1676.

PETRZYWALSKI, JEAN.—A "camera obscura."

This apparatus consists of a square box formed of two parts one sliding within the other, which facilitates the adjustment of the focus and forms a dark chamber, at one end of which is a sheet of ground glass, whilst the object to be reflected is placed at the opposite extremity. On each side of the object a lamp is placed, a similar one being in front thereof; the rays of the three lights are concentrated on the object by reflectors. An enlarged image is produced on the ground glass by lenses fixed in a sliding panel in front of the lamps and object.

The above described apparatus may be used to reproduce photographic pictures upon a magnified scale, so that a magnified view of the picture may be seen upon the sheet of ground glass. Another use of the arrangement is to enlarge photographs, by making use of the ordinary photographic process; positives or negatives can be equally well taken at day or night. The magnifying power is "from 50 to 500

" degrees;" the process can therefore be used to take a facsimile of the smallest insects on an enlarged scale. "As a " photographic album it admits of the successive reproduction " of carte de visite or other portraits in a magnified form, so " that a number of people can see the picture at the same " time represented on the ground glass at the end of the " apparatus."

[*Printed*, 10*d*. *Drawing*.]

A.D. 1867, July 8.—No. 1994.

LOEWE, JULIUS MARIA.—(*Provisional protection only.*)—" Improvements in the preparation of canvas and other similar materials for enlarging, printing, and developing objects " from negatives or positives obtained by means of photography."

Linen, cotton, silk, &c., are prepared by means of an aqueous solution of gelatine; these materials are then treated in the manner herein-after described for canvas.

Canvas is prepared in the ordinary way for photographic purposes, that is by using oxide of zinc or lead, ivory instruments being employed to spread the coating; virgin wax dissolved in essence of lavender and turpentine is added; three coatings are given with this mixture. When the canvas is dry, it is rubbed with an aqueous solution made with iodide of zinc, potato and pumicestone, and dried with a flannel till the surface becomes brilliant.

For developing the picture on canvas by sunlight an aqueous solution is used that contains albumen, chloride of sodium, chloride of ammonium, tartaric acid, carbonate of soda, iodide of potassium, iodide of ammonium, bromide of potassium, bromide of ammonium, iodine, nitric acid, chloride of gold, and sulphuric ether saturated with virgin wax.

For enlarging by artificial light, more chloride of ammonium and iodine are added to the above solution, together with magnesium, iodide of lithium and iodide of cadmium.

For printing in a pressure frame by contact, the first solution is weakened and chloride of sodium and gold are added thereto.

[*Printed*, 4*d*. *No Drawings*.]

A.D. 1867, July 26.—No. 2170.

SILVY, Camille. — "Improvements in photographic apparatus."

By combining this apparatus with a camera, the ordinary frames and dark tents may be dispensed with, and panoramic views may be taken on all points of the horizon. It consists of a metal tube, in the centre of which is a little movable wooden roller, on which the sheet of sensitive paper is wound. One of the ends of this sheet is stuck on to the roller, whilst the other end is fixed on another similar little roller placed out of the tube, and which serves to unwind the sheet through a slot cut in the whole length of the tube. The tube enclosing the sheet of paper is then placed in another tube fixed vertically on the left side of the camera.

In order to operate, the sheet of paper is gradually unwound at the rear part of the camera, which is purposely of a circular form, by means of a second little roller which is put into another tube fixed vertically on the right side of the camera. When a panoramic view has been taken, the paper is wound up anew in the first tube, answering the purpose of a sheath, and contains the latent picture, which may afterwards be developed by the ordinary means.

[*Printed, 8d. Drawing.*]

A.D. 1867, October 3.—No. 2784.

GEDGE, William Edward.—(*A communication from Hippolyte Rochat.*—(*Provisional protection only.*)—" A portable laboratory " specially intended for photographic excursions."

This apparatus consists of a rectangular wooden case closely joined, the dimensions of which are determined by those of the basins and frame intended for the operations. The larger lid opens from the front backwards, and a smaller lid opens from the back forwards; between these lids there is a middle part which remains fixed. The openings are provided with yellow glass. A movable covering or shade is joined by hinges to the fixed part of the lid. The sides of the shade slide on the exterior of the sides of the case; they are fixed at a certain height by an adjusting screw. The two sides and the front are pierced with a round opening to which is fitted a sleeve which terminates in an elastic wristband. A groove at the bottom of the case at the end, and at the height of the basins, receives the frame and gives to the frame the necessary inclination to prevent the fall of the glass plates into the interior of the apparatus. A spring keeps the frame in position.

Two basins, of equal dimensions, being placed in the interior of the apparatus, one containing the sensitising bath, the other the developing bath, the collodionised glass plate is placed in the frame. The latter is introduced into the dark chamber by the back opening, which is reclosed, and the glass plate in the frame is taken by the hands in the sleeves and immersed in the silver bath, drained on blotting paper and replaced in the frame; it is brought back into the apparatus after the pose and the development is made in the second basin. The plate is afterwards withdrawn by the large opening in front, and the image is washed, fixed, and strengthened.

[*Printed*, 6*d*. *Drawing*.]

A.D. 1867, October 17.—No. 2915.

WHITE, Otis Converse.—" Improvements in chairs for the " use of dentists, barbers, photographers, or other persons, " such invention also being applicable to other useful pur- " poses."

From the lower part of the back of a chair a fixed arm extends which has a socket to receive a pivot on the lower end of a tubular rod. A sliding clamp encompasses the rod, which not only receives the rod but also a curved bail, the two ends of which are hinged to the back of the chair. The clamp has a set screw and a saddle or bearer, which rests against the bail. A similar saddle in the rear of the clamp revolves on a journal; this saddle not only extends into a longitudinal slot in the tubular rod but rests against a solid rod extending into and projecting from the bore of the tubular rod. An arm extending from the head rest is connected to the solid rod by means of ball-and-socket clamp. A set screw in the clamp contracts its jaws upon the ball. The above-mentioned sliding clamp fixes the tubular rod to the bail and the solid rod to the tubular rod.

In another head rest, the upper part of the sliding solid rod is curved, so as to carry the ball very close to the head rest and between a curved plate and a clamp plate; the curved plate is fastened to the back of the head rest.

The head rest is formed or upholstered with a cushion, so raised on one side and two edges as to leave a recess or hollow within the cushion.

[*Printed*, 8*d*. *Drawing*.]

A.D. 1867, October 30.—No. 3057.

PIERCY, FREDERICK. — "Improvements in the method of "tinting or coloring photographs," &c.

An outline of the photograph is traced on the back thereof or on a sheet of paper. The picture and paper are placed together on a lithographic stone or plate having on its surface the colour which is to be transferred to the picture. The picture is placed face downwards with the sheet of paper between it and the stone, then by rubbing those parts in the tracing that are to be left light in the picture, the colour at those parts on the stone is transferred to the sheet of paper. If the traced outline has been made on the sheet of paper, the rubbing must be applied to it instead of to the back of the picture. More colour may be removed from the stone by means of bread. The picture is then laid face downwards on the stone and passed through the press.

A finely ground or ruled glass plate may be used in the above process instead of the lithographic stone.

This invention also consists in transmitting rays through a lens from the photographic negative on to a tint or on to the paper covering the tint so as to place in proper position the colours or to remove colour as in the above method. Oil colour may be applied to photographs to increase their permanence. A ground may be printed in oil colour on the paper, before or after printing the photograph, which can be worked upon with pencil, chalk, crayons, or oil or water colours. Albumenised paper may be charged with printing ink, and the colour may be transferred to the photograph by pressure.

[*Printed*, 4*d*. *No Drawings*.]

A.D. 1867, December 13.—No. 3542.

SINTZENICH, EDWARD REED.—(*A communication from Daniel Reed.*)—(*Provisional protection only.*)—"Improvements in the

" treatment of gutta percha, india rubber, Honduras gum, " and the other allied gums for the production of a prepara- " tion or preparations applicable as a varnish, cement, paint " stain, water and air proof material, solid or elastic material, " and for other purposes."

Any one of these gums is dissolved in benzine, and alcohol is added in sufficient quantity, thereby causing the purer portions of the gum to collect in the form of a card which is redissolved in benzine for the finer uses.

When used as a substitute for collodion for photographic purposes, the treatment consists in the use of the solution of the proper consistency in the manner in which collodion is now employed. For photographic printing purposes, and for the production of sun pictures, the treatment consists in producing a surface of the solution in the finer state on any convenient material, or using sheets of the same on which may be conveniently printed the photograph.

[*Printed, 4d. No Drawings.*]

---

## 1868.

A.D. 1868, February 3.—No. 363.

DOMENECH, Jose Maria, and JONTE, Frederic Pierre.—" Improvements in combined photographic apparatus for out- " door operations."

This apparatus forms a package which can be carried on the back by straps; it contains the camera with all the requisites for taking a view or portrait, as well as chemicals and materials for preparing, developing, and fixing the plates. A flexible cloth covers the entire apparatus, as well as the operator, and serves the purpose of a tent. Sufficient light is admitted to the interior of the apparatus through glass of a non-actinic character. The box is so constructed that, on fixing it upon the tripod stand, it at once opens out into a large and convenient tent, with the lens and camera attached in front. The operator takes up his position inside the tent, and before him is the camera, which he can open or shut; his stock of glass plates are over the camera, as well as a water cistern; on his right hand are the shelves containing the chemicals, and at his left the nitrate bath, and before him the tray to receive the waste in washing.

One of the apertures in the cloth is to enclose the operator's waist. A peculiar water supply valve is fitted to the flexible supply tube for washing the plates; it consists of a clip with an elastic spring, which is released by the pressure of the finger and thumb when the flow of water is required.

When the camera is used as a separate apparatus from the dark room, an additional tripod is used, with three sliding pieces.

[*Printed*, 1*s*. 2*d*. *Drawings*.]

A.D. 1868, April 11.—No. 1206.

BROOMAN, Clinton Edgcumbe. — (*A communication from Augustin Marion.*)—(*Provisional protection only.*)—" Improvements in photography."

The collodion film used in this invention is prepared with collodion and castor oil.

The image from a negative picture on glass is transferred to a collodion film. This is done by spreading the film on the glass, allowing the excess to run off and drying. The film together with the image which it bears is then raised. When the picture is varnished, some time is necessary for the removal of the image and water is used. A positive picture on the collodion film is thus obtained.

Gelatinised paper prepared with carbon is made by spreading upon paper a regular layer of gelatine coloured black. This is sensitised by means of solution of bichromate of potash and the coloured side is applied to the collodion film. When the paper has been exposed to light for the requisite time it is cleansed by means of water; the water also causes the gelatine to swell and the image to appear in relief upon the black layer of gelatine. This image is then transferred to albumenised paper.

The albumenised paper is floated on its back on water, and the layer of black gelatine is placed upon the albumen. The two sheets are then lifted together and pressed between plates of glass for an hour or two. The two papers are hung up to dry and submitted to the action of steam in a closed box. The steam coagulates the albumen and leaves the gelatine soluble. The two papers are put into hot water and separated. The albumenised paper retains the image, which is visible and doubly unalterable.

[*Printed*, 4*d*. *No Drawings*.]

A.D. 1868, April 21.—No. 1297.

BING, Louis.—A "mode of and apparatus for ascertaining " the actinic power of light."

For this purpose, a surface is provided on which the light falls, so as to produce a graduated series of shades which are defined by a graduated figured scale. A sheet of photographically sensitive paper is applied to the surface thus lighted for a determinate time; a graduated representation of the progressive change of tone from light to darkness is thereby obtained. By this arrangement the photographer is enabled to expose his plate or printing paper for the time requisite to produce determinate results.

The apparatus consists of a dark box, in which is mounted, on a vertical axis, a cylinder to receive a sheet of sensitised paper. The paper is secured to the surface of the cylinder by springs. One side of the box is movable and has a vertical slit cut in it. In the slit is mounted a glass tube, of square cross-section, having three sides of yellow glass, and the fourth side of transparent glass with the above mentioned scale marked thereon. The movable side is pressed against the cylinder by springs. The top of the tube has a horizontal slide for shutting off or admitting the light. The light admitted to the tube acts on that portion of the sensitised paper which bears against the transparent side of the tube. The cylinder can be rotated, so as to receive successive impressions and the impressions may be compared with one another.

The graduated scale is produced by geometrical projection, upon the supposition that the light falling on the horizontal section of the top of the tube is of equal intensity in every part.

Other modifications of the apparatus are alluded to.

[*Printed*, 10*d*. *Drawing*.]

A.D. 1868, May 2.—No. 1443.

JOHNSON, JOHN HENRY.—(*A communication from Langlois and Company.*)—" Improvements in apparatus for exhibiting " minute photographic pictures, and in pictures or slides to " be used therewith."

The apparatus for exhibiting the pictures does not give this invention a title to be included in the present volume, but the improvements in the pictures or slides are stated to be independent novel features.

In one case two or more successive images of the same object in different positions are placed on Stanhope lenses, one image to each lens. In another case microscopic views are placed on movable strips of glass, thereby admitting of the disposal of a considerable number of different subjects on the same strip. By these slides "the production of " apparently animated microscopic photographic pictures" is accomplished.

[*Printed, 8d. Drawing.*]

A.D. 1868, May 7.—No. 1499.

HENDERSON, ARTHUR CHARLES. — (*A communication from Octave Victor Fournier.*)—" Improvements in the manufacture " of plates for heliographic engraving."

The object of this invention is to obtain, on the above plates, a grain or hatching capable of rètaining the printing ink.

The process which forms this invention consists in first taking a photographic proof on glass of the hatching required to be produced and which has been previously printed or traced on any surface. The hatching may be square, lozenge, dotted, or otherwise. Then a positive photograph is printed on glass of the object to be reproduced. Thirdly, these two photographs are placed one on the other so as to touch, and from them is printed a negative picture. From the latter picture is obtained, by the ordinary means, a heliographic plate.

" A similar but less satisfactory result may be obtained by " taking in two different positions the image of the grain or " hatching and the object itself on the same plate and vice " versa."

[*Printed, 4d. No Drawings.*]

A.D. 1868, May 11.—No. 1532.

WEBSTER, William, and BARNES, Robert William.—(*Provisional protection only.*) —"Improvements in head and "waist rests for the use of photographers."

On a horizontal bed of considerable length is placed a sliding block whereto the rising stem is affixed. To traverse the block on the bed, a toothed wheel is mounted on it and gears into a rack formed on the bed; the wheel is provided with an operating handle. Or a screw on the bed may pass through the block. That part of the rising frame which is affixed to the sliding block is formed with a groove, and a bar is fitted so as to be easily moved up and down therein, set screws being employed to secure it at the required height. At or near the top of the bar, the caliber supports are attached, and, to admit of their being moved backwards and forwards at a right angle to the length of the bed, they are borne in a sliding piece; when desired, a hinge or joint is provided, so that they can be moved round sidewards.

The invention also consists "in affixing the rising stem of "head and waist rests already in use, or of the ordinary con-"struction to the said sliding block when it is supported on "the said bed or carrier, and operated as described."

[*Printed*, 4*d*. *No Drawings.*]

A.D. 1868, July 1.—No. 2104.

SMITH, James Alexander.—(*Provisional protection not allowed.*)—" Improvements in photographic apparatus."

Inaccuracy is said to occur in photographic images by the sensitised plate "being a flat transverse surface intercepting "a cone of diverging rays at all the inequalities of distance "from the focus of the camera lens as now in use at which "the various parts of the flat base of a cone arė unequally "distant from its apex."

To correct this inaccuracy, a lens or lenses is or are inserted between the focus of the camera lens, as now in use, and the sensitised plate, and thereby, " or by other optical means and "appliances suitable for that purpose," redirecting the actinic rays into parallel rays, or into such other direction as after the rays have passed through and diverged from the focus will enable the image to be projected on a flat sensitised surface with its parts in due proportion to each other.

The inventor also proposes to insert into the camera, a diaphragm with a small aperture in it, to intercept those actinic rays which do not converge to or pass through the focus; thereby securing the actinic efficiency and homogeneous purity of the colours proceeding from the object photographed, with a view of producing a coloured image of the said object.

[*Printed, 4d. No Drawings.*]

A.D. 1868, July 13.—No. 2201.

EDWARDS, Ernest.—" Improvements in photography."

A coating of bichromatised gelatine colouring matter is applied to a surface of paper, &c. and an image is obtained upon it by means of light under a photographic negative. To the surface bearing the image is applied a paper prepared with gelatine, albumen, &c. which is rendered insoluble. The parts of the sensitised gelatine and pigment not acted on by light and therefore remaining soluble are washed away with warm water, together with the paper that originally carried the sensitive film. The picture is in this case reversed.

When a direct picture is required, the negative is transferred from its glass plate by producing on its surface alternate films of india rubber and collodion till a sufficient strength has been obtained. These films, together with the negative, are then removed and the prepared gelatine is printed upon from their reversed side. Modifications are described in which a mirror, or lenses or direct sun light is used. Or the sensitised gelatine together with the alternate films may be mounted on the insoluble prepared paper.

In enlarging photographs, the position of the negative may be reversed in the above process; the printing may take place through the glass.

In the preparation of the pigmented paper, carbon, earths, and aniline, or other dyes may be used, also finely divided glass of various colours.

Collodio-chloride of silver paper is prepared, by pouring the collodio-chloride on to a glass plate, applying the transfer paper to it, and removing them together from their support.

Iodides or chlorides may be added in the preparation of the transfer paper, which is used in place of albumenised paper.

[*Printed*, 4*d*. *No Drawings.*]

A.D. 1868, July 20.—No. 2280.

RAINE, Joseph.—(*Provisional protection only.*)—A "method " of and apparatus for obtaining panoramic pictures by " means of photography."

The camera is divided into two or more equal parts by a division constructed to produce a picture with a soft or vignetted margin; by this means the margin of the second picture taken unites with that of the first so as to yield a panoramic view by the junction of the several pictures obtained by moving the camera through the requisite degree of space.

The division or divisional partition is a band with serrated edges and passing over two pulleys, one at the top, the other at the bottom of the camera. The width of the band is adjusted so that when a lens is fitted into each division of the camera and a sensitive plate is exposed when the camera is pointed at any uniform body, the serrated edge of the band (which must be moved by means of a handle attached to the axis of one of the pulleys over which it travels) shall so soften the edges of each picture as to produce a continuous effect of lighting.

To apply this for taking a panorama, the camera is made to rotate upon a table in such a manner that after a view has been taken on one portion of the plate, by means of one lens, the camera is rotated a certain previously determined distance and the second picture is taken. By these arrangements, the pictures form a continuous panoramic view.

When portraits are to be taken, the division may be removed, and instead of having a separate lens to each division, one lense only (fitted on a sliding front) may be used.

[*Printed*, 4*d*. *No Drawings.*]

A.D. 1868, July 30.—No. 2391.

DAVIES, George.—(*A communication from Hyacinthe de Font Réaulx and Amédée Barret.*)—(*Provisional protection only.*)—A "process for obtaining heliographic plates for printing."

Behind a positive or negative of the object to be reproduced is placed a sheet of glass marked with a design which gives, by transmitted light, the appearance of hatching or grains. A photograph is then taken from this double plate by transmitted light. Or a photograph of the hatching may be placed behind the photographic plate to be reproduced.

The photograph with the grain and with the picture on it, obtained as above, is covered with a very thin coating of bichromatised gelatine, then with a second and thicker coating of the same; the thickness of the latter coating is according to the amount of relief to be obtained. After desiccation, it is exposed to light and then immersed in water. The picture is represented by reliefs and depressions. The gelatine is then to be removed and transferred to another flat surface; to facilitate this operation, the gelatinised plate is immersed for some time in tepid water to which glycerine or sugar is added. The gelatinous surface is then metallised and an electrotype obtained therefrom; the electrotype forms the printing surface.

[*Printed*, 4*d*. *No Drawings*.]

A.D. 1868, August 10.—No. 2498.

FRUWIRTH, Daniel, and HAWKINS, Abdiel.—A "process "for producing raised printing surfaces."

From a negative, a positive picture is produced on a copper plate, the plate being sensitised by a paste containing bichromate of potash, hyposulphite of soda and chloride of ammonium, together with albumen, gelatine, and Nuremberger or French glue. The unaltered bichromate is dissolved out by a mixture of methylated spirit with acetic acid. A faint reversed impression is left behind, which is capable of being transformed into a surface in relief by means of a process resembling the glyphographic process.

Instead of printing the negative on a copper plate, it may produce a positive on a plate of glass or ebonite. In this case, the sensitising solution contains gum, gelatine, Nuremberger glue, bichromate of potash and oxalic acid. The unaltered bichromate is dissolved out with cold water.

In making a cast from the reversed impression, the high lights may be built up with wax or varnish. The plate is then allowed to swell in water and dry. The surface is brushed over with a solution of ox gall and coated with a composition containing sulphur and tar to which may be added india-rubber or gutta percha. From the plastic mould thus obtained, a stereotype cast in intaglio in metal may be taken. Or a cast may be taken in plaster of Paris.

In obtaining a surface for printing by a letter press machine, the fine work is built up by means of a roller covered with varnish. More varnish may be applied with a brush. An electrotype from the intaglio is thickened with type metal mounted on a wooden block.

[*Printed*, 4*d*. *No Drawings*.]

A.D. 1868, August 11.—No. 2507.

ARGAMAKOFF, ALEXANDER.—(*Partly a communication from Basile Bauler.*)—" Improvements in the production by means " of photography of printing surfaces adapted to lithography, " typography, and chromotype."

For printing surfaces adapted to typography, the following processes are employed:—Strong ammonia is added to an aqueous solution of gelatine and a thin layer is poured on to glass; the layer is dried at the ordinary temperature and then washed in a bichromate solution containing alum. The dried prepared surface is then exposed to light under the original. A warm solution of gelatine is poured on the prepared surface till it forms a thick layer; this layer can be taken off the prepared layer, when a design is obtained, either in relief or hollow according to which portion of the gelatine has swelled. A compound containing wax, gutta percha, and oil is poured on the glass, and relief is given to the resulting gutta percha plate by hollowing out the large white spaces by hand. The gutta percha plate is then coated with black lead and an electrotype matrix is taken therefrom. The matrix is covered with turpentine, which is allowed to evaporate, then with black lead and the electrotype taken therefrom is adapted to typography when the relief is increased by corrosion. The design can be transferred from the copper plate to a lithographic stone.

Application to the chromotype:—An outline relief (one for each colour) is obtained from the coloured original by means of an outline drawing. From the outline relief an impression is taken on paper in blue; originals (consisting of spots) are thus obtained, from which clichés are made separately for each simple colour.

A copper plate, prepared from a photographic negative may be treated with collodion and gelatinous ink to produce, by pressure, an image resembling a photograph.

[*Printed, 6d. No Drawings.*]

A.D. 1868, October 10.—No. 3118.

HART, FREDERIC WILLIAM.—A "material or substance to "be used for varnishing or protecting the surfaces of litho-"graphs, photo-lithographs, and other like printed surfaces."

The print should be on a well-sized hand paper. When this is not the case the print is laid on a level surface and both sides are sponged with a warm solution of starch; it is then hung up to dry and calendered. Albumen of white of egg, prepared with a whisk, is used to coat the printed surface of the paper. During the operation of whisking, matter of a deliquescent nature, such as chloride of calcium or nitrate of ammonium (in small quantity) is introduced into the albumen, to prevent the albumen on the paper from becoming thoroughly desiccated, as in this state it is conveniently rendered insoluble. The paper having become dry, in the ordinary sense of the term, the surface coating is rendered insoluble by means of steam or boiling water.

[*Printed, 4d. No Drawings.*]

A.D. 1868, December 4.—No. 3695.

MARSDEN, HENRY LOUIS DELABENE.—(*Provisional protection only.*)—"Improvements in machinery or apparatus for toning, "fixing, and washing photographic prints, which improve-"ments are also applicable to rocking machinery or apparatus "for other purposes."

One or more vessels, containing the toning, fixing or washing solutions, are so arranged, in connection with machinery, as to receive a rocking motion. The rocking motion is produced by turning on its axis a rod or bar, with or without joints, fitted with one or more bends, pins, or other projections, or with crank or cranks and connecting rod, or rods fixed or movable, which may be tnrned by hand or other motive power. The intention of the rocking motion is to thoroughly separate the prints and keep them in motion while immersed in the solution.

The vessels may be filled with liquid to the required height, and emptied, either by means of jointed tubes of india-rubber connected with an aperture or apertures in each receptacle, or by tubes of other material with or without joints.

[*Printed*, 4*d*. *No Drawings*.]

A.D. 1868, December 7.—No. 3722.

LAKE, William Robert. — (*A communication from George Kendall Proctor*.)—"Improvements in rooms and apparatus "for photographing by artificial light."

This invention consists chiefly in so constructing these rooms that the rays of light from a lamp placed therein will be reflected and concentrated upon the person or object to be photographed. The interior of the room is curved above and at the front and sides of the point whereon the light is to be concentrated, and a magnesium lamp is arranged in such a position that its light is reflected to the object. An opening is made in the curved end of the room for the camera, and a suitable background is placed at the other end. A kerosene lamp is employed for focussing.

[*Printed*, 6*d*. *Drawing*.]

A.D. 1868, December 17.—No. 3849.

POUNCY, John.—"Improvements in the production of plain "or coloured photographic and other pictures."

1st. To produce a picture on the reverse or obverse side before transferring the same to the surface of the material where it is to remain permanently.—A photograph is produced in carbon, &c. mixed and prepared in oil, spirit or water, and sensitised by methods described in No. 780, A.D. 1858, or in No. 267, A.D. 1863, or by other methods. The colours are laid on the reverse or obverse side of the picture before transferring the same.

The monochrome colour (carbon for instance) is applied all over the surface of the material on which the picture is to be produced and the negative is applied to the prepared or non-prepared surface, when the light hardens the colouring matter, leaving the other parts to be dissolved off by spirit or water.

2nd. To produce a picture in any transparent substance, the transparent substance to be afterwards restored to its original opacity.—The solution for giving transparency to the paper, &c., consists of oils, wax, and glycerine, or like substance. To restore the opacity and develop the picture boiled water is employed; the picture is finished by immersing it in turpentine, naptha or benzol.

[*Printed, 4d. No Drawings.*]

---

# 1869.

A.D. 1869, February 3.—No. 336.

JOHNSON, John Robert.—" Improvements in the manufac-" ture or production of photographic pictures."

By this invention photographic pictures in carbon or other pigments are obtained. This invention relates to those pictures which are obtained on a film or tissue of sensitised gelatine and retained, in a reverse condition, upon the paper or support upon which they are produced; or transferred, by india-rubber paper to another surface, so as to be in a non-reversed position.

1st improvement.—Instead of using moist pigment preparation, a sheet of prepared pigment is employed, the sheet being dissolved and mixed with a definite quantity of colourless gelatine. The pigment preparation contains the finely ground colour together with gelatine and sugar; the mixture is poured upon a slab to dry.

2nd improvement.—Making tissue in continuous lengths and using the gelatine compound, without forming the ordinary tissue. The support of porous paper is supplied from a roller in a continuous sheet and is brought into contact with the surface of the pigmented gelatine solution. Instead of porous paper a plate of metal is employed together with a film of wax; the developed picture may be transferred to its final support of paper, &c.

3rd improvement.—The use of soluble resins, &c., which become insoluble when dry, for cementing the pigment picture to its permanent support.

4th improvement.—The mounting of the gelatine image upon an impermeable surface without cement, for models for engraved plates. Also transferring the gelatine image from a glass or metal surface to the surface of paper, by means of an intervening film of wax.

5th improvement.—Transferring the picture from its temporary support of wax paper to its permanent support, by means of varnish and pressure.

[*Printed*, 6*d*. *No Drawings.*]

A.D. 1869, February 8.—No. 385.

SARONY, Oliver.—"Improvements in photographs to give "them artistic effect, and to produce enlarged and permanent "pictures from small negatives."

A small negative portrait is taken and enlarged to any suitable extent or variation upon glass, either by sun or artificial light. The enlarged portrait is backed up with rough drawing paper on which has been lithographed hatched shading or free hand lines.

The hatching or toning, instead of being lithographed on paper, may be reprinted on the same glass containing the transparent picture, and in that case the portrait or picture may be backed up with plain paper.

[*Printed*, 4*d*. *No Drawings.*]

A.D. 1869, February 10.—No. 417.

FISCHER, William Henry.—(*Provisional protection only.*)—A "photographing printing process."

Gelatine and gum arabic are dissolved in water, and lump sugar is added thereto. After filtering this mixture, bichromate of potash or of ammonia in solution is added thereto. The mixture is then ready to pour on the glass, slate, or metal which is to form the positive picture. After drying, the plate so prepared is placed under the negative, so that the film side of the negative shall be in contact with the gelatine film. After a proper exposure, the picture on glass, &c. is ready for the printer, and is printed in an ordinary lithographic press in the usual way employed by lithographic printers, and with their usual inks.

[*Printed*, 4*d*. *No Drawings.*]

A.D. 1869, March 20.—No. 853.

ROBINSON, John Vincent.—(*Provisional protection only.*)— " Improvements in the production of negatives applicable for " photographic printing, and for obtaining raised printing " surfaces."

1st. The production of negatives for photographic printing. —A glass plate is coated with silver, metallic alloy, or preparation of aniline. A drawing being made, with pen or pencil, upon this surface, the lines thereof are removed by means of a graver; or the drawing may be at once etched in without the aid of the pen or pencil sketch. The lines thus etched admit light through the plate, and the result is an etched negative.

A print may be obtained from the negative on photographic paper, in the usual manner.

2nd. The production of raised surfaces for printing purposes by means of the above-mentioned etched negative.—A plate of glass, &c., is coated with a layer of gelatine to which has been added an alkaline bichromate. The negative being placed in contact with the prepared surface of gelatine and the combination exposed to light, the parts of the gelatine film acted on by the light are rendered insoluble in water. The gelatine plate is then placed in warm water, and the result is a relief picture corresponding to the lines of the etched negative. From this raised surface a stereotype or electrotype cast may be taken and printed from in the usual manner.

[*Printed, 4d. No Drawings.*]

A.D. 1869, April 28.—No. 1312.

ISAAC, Lewis. —" Improvements in folding or collapsible " tables, stands, and stools."

This invention is designed to enable the above articles to be closed up within a small compass when not in use.

In a tripod stand for a photographic camera, a sliding collar, on a vertical guide bar, carries hinged tie rods, the ends of which are formed with racks that may be placed on pins projecting from the legs, so that the several legs may stand at varying angles and yet be held firmly in their respective positions. The legs are jointed half way by means of hinge joints; and the upper part of each leg has pivoted to it a flat plate or catch, having a slot in its side, which receives a pin fixed to the lower part of the leg, so that the two parts of the leg are held rigidly together, until the plate is drawn back, when the legs can be folded up.

[*Printed*, 3*s*. *Drawings*.]

A.D. 1869, June 10.—No. 1796.

COOK, Walter.—"Improvements in apparatus for taking "photographic pictures by which the operator can take a "number of prepared sensitive plates and expose them succes-"sively without the aid of extra changing box or bag."

A box is divided into two compartments. In each compartment a number of plates can be placed with their prepared sides towards the lens. In front of the box (or on the side nearest the lens) there is a shutter, which, on being drawn out, opens one of the compartments to the lens, and the front plate receives the image, being pushed forward in its place by means of a spring from the back. On the shutter being closed, it carries down the front plate (which has just been exposed) with it, leaving the next plate ready for exposure. A slide is then drawn up at the back of the box; this action draws a plate from the bottom compartment to the top, and a spring holds the plate at the top while the slide is pushed down again; thus the plates are kept in rotation till all are exposed.

"In front of the box a camera is attached by hinges or "other means and being easily detached allows the operator "to focus the picture. The plates are kept from rubbing "each other by a covering of paper, cardboard or other sub-"stance round the edges and at the back or not as may be "thought desirable."

[*Printed*, 8*d*. *Drawing*.]

A.D. 1869, June 30.—No. 1965.

COURTENAY, Robert Henelade.—"Processes for producing "printing surfaces."

1st. A collodion, or other like solution, is prepared with

precipitated silica. A positive picture is taken upon a glass plate covered with the silicated collodion, and is printed on to a film composed of parchment size, sugar candy, gum, and chromic acid, the film being mounted upon glass. The film is freed from sensitive salts by washing, drained, immersed in chloride of gold, and drained again. It is then coated with a solution of nitrate of silver and grape sugar and immersed in a solution containing proto-sulphate of iron and grape sugar, prior to being placed in a bath for electro-depositing copper upon it. The electrotype is removed from the bath, treated with boiling water to remove the matrix therefrom, and burnished at the edges, and at the high lights; it is then fit for printing from in an ordinary printing press.

2nd. Producing raised surfaces for printing from such as type printing from a negative ingrain. The design is printed from the negative direct upon a matrix similar in composition to that set forth in the first improvement, which is similarly treated. An electrotype from this matrix is suitable for printing from.

A positive may be printed by printer's ink from the negative ingrain on to a paper prepared with the substance before-named and transferred to stone for photo-lithographic printing.

In heliographic engraving a conducting surface may be produced on the photograph by precipitating silver powder from the nitrate by copper and spreading it over the matrix.

[*Printed, 4d. No Drawings.*]

### A.D. 1869, July 5.—No. 2022.

GRÜNE, Friedrich Wilhelm.—(*Provisional protection only.*)— "Improvements in transferring photographs to wood, metal, "ivory, and other surfaces."

A photograph, obtained by the ordinary collodion process, is applied, face downwards, on the surface to which it is to be transferred, and, is fixed thereto by means of gum, albumen, or other adhesive substance. Alcohol is applied on the back of the photograph to remove the water or humidity; then ether is applied, whereby the collodion is removed and the metallic picture alone remains on the surface. Thus the transfer is so made as not to leave any layer or trace of foreign substance on the surface likely to prejudicially affect the surface for engraving or ornamentation. The surface can then be engraved ; or the image may be fixed by varnish for ornamental purposes.

[*Printed, 4d. No Drawings.*]

A.D. 1869, July 20.—No. 2193.

TREVOR, DAVID.—(*Provisional protection only.*)—" Apparatus " for facilitating the production of photographic pictures ; " also improvements " in the mounts or cases and fittings for " the same."

1st part of the invention.—Taking several views of one sitter from different positions, or of several sitters, on the same sensitised plate, with one sensitising. The apparatus for holding the sensitised plate to accomplish this result consists of a circular disc of wood, working in a corresponding sized hole formed in an outer rim, which is square for the sake of easy application to the camera. In the disc is a square opening, which is fitted with a door for securing the sensitised plate therein. The front of the apparatus is obscured to the extent of the eccentric space requisite for taking the intended picture. If the disc be divided into six parts, one picture is taken, the cover is put over the lens, and the disc is revolved one-sixth of a revolution ; another picture is taken, and so on until the whole series of six is complete.

2nd part.—In mounting these pictures, the whole series may be exposed on a revolving disc, so as to be turned upwards in succession ; or openings may be formed in the upper part of an additional disc, the compound picture being revolved underneath.

3rd part.—The leaves of the albums for the portraits comprised in this invention are so arranged as to permit of the whole to be seen at once, by turning the album, or by revolving the picture so as to bring the portrait into view, the album or case being stationary.

[*Printed, 4d. No Drawings.*]

A.D. 1869, September 18.—No. 2718.

DENNE, THOMAS JAMES, and BILLINGSLEY, HENRY. — " Means of producing transfers and blocks for lithographic " and other printing."

1st. Producing a printing block from a photographic negative.—An ordinary negative on glass is coated with a mixture of gum resin, bichromate of potash and water, and, when dry, is exposed to the action of light so as to print a picture on the coating. The non-solarised portion of the picture is dissolved out by hot water, leaving a raised picture on the glass, from which a mould or electrotype can be taken from which other printing blocks can be produced.

2nd. A mixture, composed of bichromate of potash, gum arabic and distilled water is spread upon paper. The paper is allowed to dry, placed under a negative and printed with a copy. The surface of the picture thus taken is covered with a groundwork of litho-tint produced on a stone; it is then floated on warm water to remove the free gum and ink, and the picture can then be transferred on to a stone for printing from.

3rd. A process for producing negative copies of pictures without the use of a camera and lens.—A sheet of thin paper is coated with chloride of sodium and allowed to dry; it is then floated on a bath of nitrate of silver, dried, and exposed under an original in a frame. This negative picture is fixed by hyposulphite of soda, washed, and dried; it may be used to print positive pictures in the photographic printing frame. Or transfers can be printed to throw down on stone and print from by the above photo-lithographic process.

[*Printed, 4d. No Drawings.*]

A.D. 1869, October 19.—No. 3049.

WINDOW, Frederick Richard.—A "process for producing "photographs in pigments."

This invention consists in applying, for the above purposes, the chemical action of chromic salts on soluble organic bodies together with the treatment with powdered pigments of surfaces prepared with hygroscopic compounds combined with a soluble chromate ofter insolation through a cliché.

A sensitive solution is prepared which contains honey, glucose, albumen, dextrine and bichromate of ammonia. This solution is spread upon the surface to be sensitised, or poured upon it, dried, and exposed, under a negative, to light. Powdered pigment is then brushed over the surface until the image is developed and then fixed by alcohol and dried. The soluble portion of the coating is then dissolved out with acidulated water. The print is washed and finally passed through water rendered slightly alkaline with ammonia. If the print is to have a glazed surface, collodion is the fixing agent instead of alcohol.

For pigment prints upon paper with a glazed surface the above sensitive solution minus the albumen is employed. A plate is coated with this solution, insolated under a negative and developed with powdered pigment. Collodion is then poured over the coated surface and left to dry; it is then transferred to gelatinised or albumenised paper.

The developed print may be transferred to other surfaces having a glutinous coating, by means of pressure and moisture.

The developed print may be transferred to stone or zinc for direct printing therefrom. For this purpose the print is developed with a powder of a fatty or resinous nature. The developed print is laid upon the heated stone and pressed. On damping the back of the print the transfer is effected.

[*Printed, 4d. No Drawings.*]

A.D. 1869, October 30.—No. 3151.

MEWBURN, John Clayton.—(*A communication from E. Rye.*) —A "photographic process for preparing printing sur-" faces."

The object of this invention is to allow of impressions, equal to those obtained by lithography, being taken on common paper with printing black or lithographic printing ink from photographic negatives.

A mixture is poured over a very thick piece of plate glass, highly polished or plain. The mixture is composed of gelatine, albumen and water. The glass bearing the film is dried in an oven and is then exposed to light on the reverse side. The glass is then washed, dried, and another mixture is poured over it; this mixture contains gelatine, albumen, bichromate of potash, bichromate of ammonia, gum myrrh, gum benzoin, nitrate of silver, iodide of cadmium, and bromide of potassium. The glass is again dried; it is then ready for copying; this is effected by placing a photographic negative on the glass.

When the negative has been copied on the glass, the latter is washed under strong pressure of water, so as to extract all the chromate. The glass is again dried and is ready for printing.

[*Printed*, 4*d*. *No Drawings*.]

A.D. 1869, December 8.—No. 3543.

EDWARDS, Ernest.—"Improvements in photo-mechanical "printing, and the reproduction of designs."

This invention relates to printing from a surface coated with bichromated gelatine which has been acted upon by light.

The film is hardened by means of alum, sensitised by a chromate, and then acted upon by light. It is washed in water, coated with a second layer of insoluble sensitised gelatine, and the operation is repeated till, after exposure under a negative, all the details appear. The plate is freed from bichromate by washing and is then ready for printing.

In printing, the plate is wetted and placed in an ordinary printing press in which the pressure is vertical. To obtain pressure in the shadows of the gelatine, a mould is made by warming a sheet of gutta percha, placing it on the plate and pulling the press, so that the mould may fall, each time, exactly in its place.

In inking, to avoid tearing up the gelatine film by the adherence of the ink to the plate, rollers or dabbers of india-rubber are used, and the lithographic ink is thinned with tallow and oil. To procure prints of more than one tint from one plate, the inks are prepared of different degrees of stiffness and of different colours, the stiffest ink being first applied, the next stiffest (which adheres only to the half tones) is then rolled in and so on. To obtain prints with margins a mask is employed.

Sheets of transparent material are coated with a substance impermeable to the actinic rays. By the removal, in places, of the non-actinic layer, a design is made upon the sheet or plate. Prints from this plate are obtained by photography or by the above-described greasy ink process.

[*Printed*, 4*d*. *No Drawings*.]

## 1870.

A.D. 1870, January 12.—No. 89.

HEMERY, Thomas George.—(*Provisional protection only.*)—
" Improvements in photography."

This invention relates to the production of one negative from a combination of several negatives, in order to photograph two or more sitters, one in front of the other.

1st method.—The first figure, or that nearest to the camera, is photographed and developed in the usual way, as a negative, with a perfectly transparent background. The figure being covered with an opaque varnish, the back of the plate is coated and sensitised and receives the image of the second sitter through the front or used portion of the glass, thereby leaving a counterpart of the first figure on the back of the plate, perfect in outline but transparent after development. Lastly the opaque substance is removed from the front of the plate and (by exposure to light) a transparency is taken of the double negative.

2nd method.—One of two similar impressions of the first figure is rendered opaque, and is placed in the camera in front of a second plate so as to print a transparent figure thereon. The second plate and the unused impression of the first figure are combined together to afford a single negative. More negatives than two may be combined by this method.

3rd method.—The two similar impressions of the first figure differ by their backgrounds, one being transparent, the other opaque. The first is superposed on the photograph of the second figure (having rendered the first figure opaque) and a transparency is taken of the combination. By combining two transparencies a perfect negative is obtained.

[*Printed*, 4*d*. *No Drawings.*]

A.D. 1870, January 22.—No. 201.

JOHNSON, John Robert.—" Improvements in the prepara-
" tion of paper or other surfaces for the production or manu-
" facture of pictures by photography."

This invention refers to those photographic pictures in

which the colouring matter is carbon (lampblack) or other pigment.

1st improvement.—The compound spread upon paper, for these purposes, usually contains gelatine (or its congeners) sugar and bichromate of potash, the use of the sugar being to give flexibility to the compound, and to increase its sensitiveness and its solubility. This part of the invention consists in using soft soap (oleate of potash) instead of sugar in the above compound.

2nd improvement.—Instead of using a pigment ground in oil, mixed with bitumen or asphalt and soap, with or without a bichromate, and acting upon the compound with oil of turpentine to dissolve out the portions not acted upon by light, this portion of the invention consists in substituting gelatine or its congeners for bitumen or asphalt, thus enabling warm water to be used to develop the picture.

3rd improvement.—Replacing the gelatine, in the above compounds, wholly or in part by curd or caseine, &c., which is insoluble in warm water but which is rendered soluble by ammonia or other alkali. The caseine is precipitated, with rennet or by an acid, from skimmed milk. The resulting curd is collected on a filter, dried by pressure and dissolved in dilute ammonia. This solution is used instead of a portion of the gelatine solution.

[*Printed*, 4*d*. *No Drawings*.]

A.D. 1870, February 7.—No. 346.

WOODBURY, WALTER BENTLEY.—(*Provisional protection only*.) —A "method of producing surfaces by the aid of photo-"graphy."

1st. A gelatine relief is produced photographically, from a negative, of a design in black and white. From the relief a reverse is obtained in any soft metal, by hydraulic or other pressure. The reverse may be printed from, as an engraved plate, in fatty opaque inks. An electrotype may be made from the soft metal plate, if desired.

2nd. Obtaining a granular surface on the gelatine relief, by substituting for the colour (in the carbon process) in the gelatine and bichromate, a coarse granular powder. This produces, on the relief, an ink-holding ground. In some cases, emery powder is dusted over the partially dry gelatine relief; the emery adheres only to the parts that represent the dark portions of the design.

3rd. Producing designs on cardboard, leather, &c. direct from the gelatine, by pressure.—A film of gelatine colour and bichromate is exposed under a photograph, and the exposed side is brought into contact with a sheet of copper under water, the two surfaces being pressed into close contact. The combination is then placed in hot water, which removes the greatest portion of the gelatine, and dried. The plate when passed in a press, will reproduce the design on the cardboard, &c.

[*Printed, 4d. No Drawings.*]

A.D. 1870, February 9.—No. 383.

FOLEY, ARTHUR, and FOLEY, HENRY.—"Improvements in " the production of photographic pictures or designs on " veneers and solid wood, and in the application thereof to " cabinet work, furniture, fancy wood articles, and other " decorative purposes."

To produce the photograph, a piece of even grained veneer is planed, on both sides, with a toothed iron plane. One side is left rough, the other is smoothed by scraping and rubbing with glass paper. The smooth side is damped, so as to cause the grain of the wood to rise; the wood is then allowed to dry, and the alternate damping and drying is repeated until the wood grain ceases to rise. The prepared surface is treated with chloride of sodium, dried, and sensitised with nitrate of silver. Printing is then performed, on the sensitised wood, from a photographic negative.

To lay the veneer thus printed upon, it is cut to the shape required, and its rough surface is prepared with glue. The veneer is then caused to adhere to the article to be decorated, by heat and pressure. When the veneer is dry and hard, the surroundings of the table top, &c., are filled up by veneers, and, when these are dry, they are planed down to the surface of the photograph. The whole surface is then French polished.

According to another plan, the veneer carrying the photographic picture is placed upon another veneer that is sufficiently large to cover the whole surface of the article to be ornamented. Both veneers are then cut to the figure or shape required, by means of a fine saw. The photographed veneer is then allowed to drop into the aperture in the larger veneer, and the combined sheet is ready for decorative purposes.

Solid woods are treated in a similar manner to the above.

[*Printed, 4d. No Drawings.*]

A.D. 1870, February 9.—No. 385.

BAYLIS, Charles.—(*Provisional protection only.*)—A "means "of preventing impersonation."

The object of this invention is to identify the presenter of a ticket as the person to whom that document was issued. For this purpose, the photograph of the real owner of the document is made an integral part of the document itself.

The paper which forms the face of the ticket is made sensitive to light, so that a photographic image may be obtained directly thereon. In railway tickets a portion of the ticket is cut out, and the photograph is inserted between it and the mount; by embossing or printing a fugitive border round the portrait, any attempt to change the picture will be detected. Paper may be prepared with a silk or coloured thread or threads woven into the same, and the photograph may be printed over the threads.

This invention may also be applied to passes for exhibitions, in the preparation of letters of credit, and for documents that will identify annuitants, &c.

[*Printed, 4d. No Drawings.*]

A.D. 1870, March 18.—No. 802.

JANICOT, Charles.—"Improvements in producing photo-"graphic pictures on tissues or fabrics."

1st. The tissue is prepared and sensitised, and the photographic image is produced directly upon it, by the usual processes.

2nd. The photographic picture, on the fabric, is rendered transparent by paraffin or other suitable material.

3rd. The reverse side of the fabric, thus made transparent, is painted, by following with a brush all the lines of the photographic image which is fixed thereon.

4th. When the painting is finished, the fabric is rendered

opaque by covering the dry painted side with opaque colour, or by pasting the painted fabric on paper or other opaque surface, so as to intercept the transparency of the coloured image.

5th. The fabric bearing the photograph is then mounted in a frame.

There is thus obtained a painted cloth, or surface, on which the painting, applied on the reverse side to the photographic image, appears on the obverse side by the direct transparency of the fabric.

[*Printed*, 4*d*. *No Drawings*.]

A.D. 1870, March 31.—No. 941.

MC CRAW, WILLIAM.—" Improvements in photography by " processes which combine printing, painting, and trans- " ferring, for the production of chromo-photographs."

A positive is produced on glass from an ordinary negative. The glass is coated with photographic collodion, so as to form a structureless or transparent film. The film, containing the picture in light and shade, is painted with appropriate local colours and the painted surface is cemented to canvas, or other final support; the glass is then removed, leaving an untouched photograph supported by colours transferred to canvas. The picture may be varnished or worked upon by the artist's hand.

Instead of the film being painted whilst upon the glass, it may be supported on paper for that purpose. This gives the picture in its true relative position, if a direct print on glass from a negative be employed.

This invention is applicable to photographs produced by the carbon process, enlargements by the camera, by the Woodbury process, &c., by which a photograph can be obtained on glass having a similar appearance on either side. The dusting-on process may be used, also vitreous or enamel colours may be employed. The process is available for the decoration of glass, enamel, or ceramic surfaces, and the colours may be burned in. The colours may be applied immediately to the canvas and the transparent photograph may be fixed thereon. When the picture is transferred to a rigid surface, the colours can then be applied by chromo-lithography or other means.

[*Printed*, 4*d*. *No Drawings*.]

A.D. 1870, May 10.—No. 1333.

EDWARDS, Benjamin Joseph.—" Improvements in apparatus " for photographic printing."

This apparatus relates and is applicable to the production of prints in which more than one negative is used.

A frame or frames to hold the negative or negatives from which the prints are to be made.—The negatives are held firmly in their places by a grooved sliding bar pressed against the end of the plate by a screw.

A frame or frames to which the sensitive paper for printing is firmly attached.—These frames or boards are made by means of guiding pegs to register exactly to the frames which carry the negatives.

Masks may be used to protect certain portions of the print from the action of light during the printing of other parts of the picture, these masks being cut by means of impressions from negatives. A print having been taken from the negative upon sensitive paper fixed to the frame, this frame is separated from the frame carrying the negative and the mask is adjusted on the print and fixed thereon by india-rubber solution. The frame carrying the print and mask is then brought into contact with the negative from which the next portion of the picture is to be made, and the whole is pressed together. The print is then lifted away and the mask may be removed by means of benzole.

[*Printed*, 4*d*. *No Drawings*.]

A.D. 1870, June 15.—No. 1718.

HOPKINS, William Jeffrey.—(*Letters Patent void for want of Final Specification.*)—" Improvements in the construction " of roofs and sides of houses and other structures."

Photographic operating rooms are included among the structures above referred to.

The object of the invention is to dispense with the use of putty to fasten the glass and to allow of the rain water leaving the roof with freedom. The **U** or **V**-shaped plates or bars that form the styles and sash bars are held in position by rivets; these plates form the channel to receive the rain water from the glass and lead it to the rain water pipe. Studs, at various portions of the length of the sash bars, hold clips which press upon the glass to keep it in place. The edges of the plates are covered with prepared rubber, &c., for the glass to rest upon.

The glass plates are capable of being removed, in the case of portable houses.

[*Printed, 4d. No Drawings.*]

A.D. 1870, August 4.—No. 2171.

WOODBURY, Walter Bentley.—A "method of producing "surfaces by the aid of photography."

1st. The preparation of a gelatine relief produced photographically from a negative that has no half tints.—The relief is dried, and a reverse is obtained therefrom in any soft metal, by hydraulic or other pressure. The metal plate is used as an engraved plate for printing from in fatty opaque inks. An electrotype may be made from the soft metal plate, when a number of prints are required.

2nd. A method of obtaining a granular surface to give an ink-holding ground, when the photographic negative possesses gradations of light and shade.—A coarse granular substance is mixed with the bichromated gelatine; when acted upon by light transmitted through the negative, this produces a relief having the required granular surface. In some cases, emery powder is dusted over the partially dried gelatine relief; the powder adheres only to the parts representing the dark portions of the design, thus giving to the soft metal plate produced by pressure therefrom a corresponding granular effect. According to another method, two coatings of the granular gelatine mixture may be given, the first holding a finer and the second a coarser powder.

3rd. A method of producing direct from the gelatine relief, by pressure in contact with cardboard, paper, &c., designs on these materials.—A film of gelatine colour and bichromate is exposed under a photograph; its exposed side is brought into contact with a sheet of metal under water, and the two surfaces are pressed into close contact. The placing of the plate in hot water removes the greater portion of the gelatine, leaving the design in relief on the surface of the plate; it is then dried and passed through a press in contact with the cardboard, paper, &c. A design similar to a water-mark is the result.

[*Printed*, 4*d*. *No Drawings*.]

A.D. 1870, August 16.—No. 2266.

WINDOW, Frederick Richard.—" Improvements in obtain- " ing proofs or impressions from printing surfaces produced " by photography, and also in obtaining such printing sur- " faces."

1st. A mode of obtaining impressions from any surface obtained by photographic agency.—Two or more plates, instead of one, are produced from the same negative, each plate receiving a sufficient degree of exposure. The subject is thus divided into a series of flat tints. If the process be one yielding a surface giving a flat tint, each plate represents a tint and has a register mark obtained directly from the negative. These plates are printed in succession in a more or less transparent ink.

If the process be one yielding surfaces with a certain degree of half tone, a smaller number of plates is necessary.

By stopping out, an analogous process is applicable to the production of proofs in colour.

2nd. A mode of obtaining printing surfaces on stone or zinc.—By means of a negative, a positive image of the subject to be reproduced is obtained in a compound which forms a resist upon the stone and will absorb moisture from the atmosphere, such as a sensitised mixture of albumen and honey. The parts not exposed to light adhere to the stone, and the whole is pressed. Or, a negative image is laid down upon the stone, after wetting it and scraping out the superfluous air and moisture. The non-actinised parts are washed away. The stone is washed in alum, dried, inked up, and the gelatine image sponged out. The remaining fatty image is treated for lithographic printing.

Another way of forming a printing surface upon a metallic surface is by a gelatine film spread upon sugared paper. The gelatine is rendered insoluble by glacial acetic acid, printed, and washed. The film is then mounted upon a well grained copper plate and dried; it is then ready to receive the ink.

[*Printed*, 4*d*. *No Drawings*.]

A.D. 1870, September 13.—No. 2464.

BALDOCK, GEORGE, DENNE, THOMAS JAMES, and HENTSCHEL, AUGUST.—"Improvements in the manufacture of "casting blocks and printing blocks used for heliographic and "other printing."

Preparing caseine or curd of milk for the above purposes.—Milk which has become sour and set by keeping is separated from grease and other extractive matters by skimming it, putting it into a bag, and submitting it to the action of water and pressure. When dried, the substance is suitable for use.

To prepare a plain block for casting and printing.—The foundation is made by dissolving the caseine in slightly ammoniacal water and adding China clay and glycerine; paper or cardboard is coated with this composition. The coating is made insoluble by means of weak acetic acid and allowed to dry.

This block may be sensitised with bichromate of potash, exposed under a negative and washed with soap and soda solution. The result is a relief picture from which a printing or casting block can be made.

To prepare a paper negative for working on a plain block.—A drawing is made on paper in a compound containing caseine and Indian ink or lampblack. The whole surface of the paper is covered with lithographic ink, and the paper is placed in slightly ammoniacal water; the drawing, with its covering of ink, washes away, leaving the undrawn parts black.

A method of making a printing block to print half tone printing, to obtain a stereotype or electrotype directly from a negative.—The block is composed of caseine, soap, and bichromate of potash. The composition is printed through a negative, which is coated with it. The non-actinised portion is removed as above, and from this electrotypes can be taken.

Photolithography.—Paper is coated with a mixture of caseine, soap, and bichromate of potash. The dried paper is printed under a negative and the print is rolled in with lithographic ink. The picture being plunged into slightly ammoniacal water, the non-actinised parts are dissolved away, leaving the picture, in ink, upon the paper; this is transferred down upon a lithographic stone. A similar process is used in photozincography, or "zinco-photography;" in this case the zinc is coated directly with the composition and the paper is dispensed with.

A method of preparing for printing, the surfaces of wood, metal, paper, &c., which is especially applicable to the preparation of a printing block on wood.—The wood is coated with a composition containing caseine, zinc, white or white lead, and chloride of ammonium. This coating is dried, covered with nitrate of silver, exposed under a negative and fixed with hyposulphite of soda; the block is then ready for the wood engraver. The composition may be modified by substituting for the zinc white or white lead, China clay, pipe clay, flake white, &c.

[*Printed*, 6*d*. *No Drawings*.]

A.D. 1870, September 15.—No. 2485.

EDWARDS, Ernest.—"Improvements in photo-mechanical "printing."

This invention relates to further improvements upon that set forth in No. 3543, A.D. 1869, to improvements upon other methods of photo-mechanical printing, and to new methods of accomplishing the same operation.

1st. "Methods of removing the printing films from the "surfaces on which they have been prepared to those on "which they are to be placed during printing at the print- "ing press."—The level surface to receive the gelatine film is cleaned and rubbed with a solution of wax in ether. Adhesion of the film to the support for printing from is ensured by coating the support with Canada balsam, collodion, india-rubber, or other cement or varnish. The film may be coated with india-rubber and pressed into contact with the support by heavy pressure; it may be made flexible by the addition of glycerine or a non-drying oil.

2nd. When the printing surface is to be level, a plate of soft metal is used as the support, and heavy pressure applied, which imbeds the gelatine film into the metallic surface.

3rd. In printing proofs, great intensity of the shadows is obtained by dusting over them, after printing, powder colours such as are used in lithography. When great intensity is required, the copper plate method of inking with a dabber is employed and the surplus ink is cleared off the plate with a roller. When prints are required with margins, metal masks are used, instead of the paper ones described in No. 3543, A.D. 1869.

4th. To produce reversed or non-reversed pictures from ordinary negatives.— A very deep actinic impression is obtained on the printing film, which is inked on the reverse side to that which has been in contact with the negative. Or, a gelatine print is transferred for development to the surface of a film of insoluble gelatine, the surplus gelatine being washed away with hot water; the plate is dried, damped, and printed. Printing films may be prepared from ordinary negatives; or, the back of the negative may be placed in contact with the printing film and the printing may take place through the glass. Plates may be prepared for printing by rendering a gelatine film insoluble by the addition of a solution of perchloride of iron and tartaric acid; those parts that have been actinised become again absorbent. Transparent positives are necessary for printing by this method.

5th. When lithographic or other stone is used, it is thoroughly dried and immersed in a solution containing gum or gelatine, bichromate of potash, and alum; the stone is dried, actinised, washed, and printed from. A printing block may be made by the admixture of plaster of Paris with sensitised gelatine.

6th. To produce prints in transfer ink on transfer paper from the printing films, these prints are transferred to wood blocks, or to stone or zinc for cutting or engraving. Or, the plates on which the transfers are to be made are placed in contact with the printing film.

[*Printed, 6d. No Drawings.*]

A.D. 1870, November 9.—No. 2948.

WALLACE, William Vincent.—(*Provisional protection only.*) —" Improvements in the preparation of surfaces adapted for " painting, photographic, and other printing."

The object of this invention is to provide a substitute for paper, &c. used for the above purposes, the surface having great delicacy and evenness.

The invention consists in the preparation of metal or other surface, impervious or nearly so to the mixture herein-after stated, by the application of wax or its compounds. Sun bleached wax is mixed by heat with a white body colour, such as flake white, &c. The mixture is applied to the surface by heat or by other convenient means.

[*Printed, 4d. No Drawings.*]

A.D. 1870, November 19.—No. 3036.

SAWYER, JOHN ROBERT.—(*Provisional protection only.*)—"Improvements in the preparation of printing surfaces in photo-mechanical printing."

A layer of bi-chromated gelatine is placed upon a metal or other hard surface, dried by hot air and exposed under a negative; the plate is then subjected to the action of water and allowed to dry and harden at about 100° Faht. After a lapse of forty-eight hours, the plate is placed on the carriage of a lithographic press and fixed to a piece of soft wood. The plate is then allowed to imbibe as much water as possible and the surface is dried; transfer ink is rolled in and a lithographic impression is taken on transfer paper. If the transfer be perfect, the picture (cut to the required size) is damped and is placed on a polished lithographic stone heated to 100° Faht. The print is covered with soft paper and the stone is passed several times through the press. The back of the print being sponged with water, the stone is turned and passed through the press in the opposite direction. The back of the print is softened with water and gently rubbed until the paper can be easily removed from the stone. The impression is then gummed and inked up like an ordinary lithographic transfer, and any writing is added to the stone. The whole is then printed as a lithograph.

In another method, paper, prepared with gelatine and lampblack, is floated on a solution of bichromate of potash; when dry, it is exposed under a negative, immersed in water, and, when limp, laid face downwards, in the proper position, on a polished lithographic stone, making it adhere firmly by means of a squeegee. The stone is immersed, face downwards, in warm water, which removes the paper and the surplus gelatine and carbon, leaving the picture on the stone;

any writing is added and the whole is rolled in with lithographic printing ink and worked off in an ordinary press. Sometimes the negative is coated with india-rubber solution and transferred to paper, which transfers the picture to the stone.

When only a few copies are required, they may be printed from the original gelatine film. To render the film sufficiently hard for this purpose, it is treated with a spirituous solution of a hard gum or lac.

[*Printed*, 4*d*. *No Drawings*.]

---

## 1871.

A.D. 1871, February 16.—No. 401.

BROWN, JOHN.—"Improvements in the manufacture of "collars, cuffs, and shirt fronts."

Printed surfaces, prepared by photolithography or photozincography, or other photographic process, are employed for ornamenting the above articles when they are made of paper and paper cloth.

In the first instance a photographic picture is taken of a piece of lace, for instance. A print from the picture is taken on to a surface of gelatine prepared with bi-chromate of potash. The gelatine surface is coated with lithographic ink, moistened with water and sponged over. Those portions of the surface which have been rendered insoluble are thus removed, together with their coating of lithographic ink, leaving an image, in lithographic ink, of the lace; this image is subsequently transferred by pressure to the surface of a lithographic stone, and the paper from which the articles are to be manufactured is printed therefrom, and is subsequently cut up and manufactured into collars, cuffs, and shirt fronts, in the ordinary manner.

[*Printed*, 4*d*. *No Drawings*.]

A.D. 1871, February 28.—No. 540.

HASELTINE, GEORGE. — (*A communication from William Augustus Leggo.*)—A "photo-electrotyping process and stop-

" ping-out plate to be used therefor, and for other similar " purposes."

This invention relates to further improvements in "Leggo- " typing" as set forth in No. 1541, A.D. 1865.

The photograph, being varnished and dried, is rubbed carefully with a piece of fine cotton wool, so as to give it a fine polish or gloss. The gelatine is applied cold, the exposure to light is made, and the gelatine coating is washed in hot and then in cold water; it is soaked in hyposulphite of soda solution, washed, soaked in protosulphate of iron solution, and again washed. The plaster cast is now taken and the depth required is obtained by further applications of sensitive jelly with exposures to light and further manipulation.

The further manipulation consists in soaking the plate for a short time in dilute nitric acid, washing and drying, prior to a second application of the bichromated gelatine; the sensitised compound is applied warm. A third application of jelly may be made in the same manner.

Another part of the invention consists of a stopping-out plate. A collodion negative, taken from the positive photograph to be Leggo-typed, is exposed with a second sensitive collodion film and the latter, which is a soft blurred looking copy of the original, is placed over the jellied positive during exposure to prevent the light from acting too strongly upon the design itself or upon certain portions of it. Other stopping-out plates made by hand may be used.

[*Printed*, 4*d*. *No Drawings*.]

### A.D. 1871, March 8.—No. 622.

EDWARDS, BENJAMIN JOSEPH.—" Improvements in apparatus " for taking photographic pictures."

This invention relates especially to apparatus for taking photographs upon glass plates by means of collodion sensitised by silver salts, the plates being used in a wet state.

The object of the invention is to prepare the plates and to develop the pictures upon them without the use of a dark chamber.

The dark slide used in this invention, instead of a hinged back, has a shutter sliding in grooves, similar to that in front of the slide. The sensitising dish has grooves upon its upper edge fitted to those upon the back of the dark slide. When the dish and dark slide are temporarily attached, the collodionised plate having been previously placed in the dish with the collodion side uppermost, the solution of silver is admitted to the dish at a slit formed in one side or end thereof. There is a reservoir outside the dish at the side of the slit; the reservoir may be either closed or open, if open there is a hanging screen for the fluid to flow under into the dish and to exclude the light from the interior of the dish. When the dish is inclined diagonally, the fluid passes through the slit back into the reservoir.

When the sensitising fluid is to remain in the reservoir, so as to be carried about in it, the top of the reservoir is closed and there is a movable partition between it and the dish. The slit is deeper than that described above and there is a movable door and fixings to press the door against the slit.

The plate being sensitised by means of the silver solution, the solution is returned into the trough and the plate is removed from the dish to the interior of the dark slide. For this purpose the dish has a flexible bottom; the shutter having been withdrawn, the plate is pressed into its position by hand. Or the dark slide and dish may be inverted and the plate allowed to fall into its place. In another plan cranked wires or bars may be operated to raise the plate by means of an external knob. In a fourth method a pneumatic plate holder is used for the purpose. After the introduction into the dark slide, the shutter is closed behind the plate and the dark slide may be removed from the dish, and exposed to light in a camera.

To develop the picture, a dish similar to the sensitising dish is employed, but it has its bottom part made of yellow glass, in order to watch the process; the dish also has a plate of yellow glass working in grooves in its upper part, as well as the grooves for the dark slide.

When the picture is sufficiently developed, the developing fluid is poured away, and water is admitted to the dish, so as to wash the plate repeatedly. The plate may then be finished and fixed in the usual way.

[*Printed*, 1*s*. 6*d*. *Drawings.*]

A.D. 1871, March 18.—No. 730.

DENNE, THOMAS JAMES, and HENTSCHEL, AUGUST.—" Im-" provements in the preparation of surfaces for prints, print-" ing, and other purposes."

The naturally made caseine, described in No. 2464, A.D. 1870, may be advantageously used for the purposes of the present invention.

By means of this invention paper photographs are made practically permanent, being thereby protected from the action of damp and having a smooth and glass like surface.

Casein is dissolved in soap and water or in ammonia water, and the photograph is covered therewith, either by floating, or by brushing on the solution ; the paper is then dried. To render the surface smooth, a sheet of glass is wiped over with gall and the paper is softened and prepared by means of an aqueous solution contianing acetic acid and gall ; the casein face is laid downwards on to the glass and the back of the photograph is wiped over with a " squeezee " in order to press the paper perfectly flat upon the glass. When dry, the paper leaves the glass with a polished appearance.

Colouring ingredients may be added to the casein, and flexibility may be given to the casein surface by means of glycerine.

[*Printed, 4d. No Drawings.*]

A.D. 1871, April 20.—No. 1038.

LAKE, WILLIAM ROBERT.—(*A communication from Ebenezer Ashton Goodes.*) — " Permanent photographs on glass and " process for producing the same."

The photograph is taken upon glass, in the usual way by coating the plate with bromo-iodised collodion. The coating is then excited in the silver bath and submitted to the action of light. The picture is developed by pyrogallic acid or proto-sulphate of iron, fixed by cyanide of potassium or " hyposulphate " [hypo-sulphite ?] of soda, washed, and dried.

To stain the picture permanently on glass, " the glass plate " is placed in an air-tight kiln, or otherwise subjected to a " strong heat, until every part of the photograph becomes a " deep brown, care being taken that every part of the picture " is equally heated. It is then gradually cooled, after which

" it may be placed in water, and a dark brown powder formed " on its surface will easily wash off, leaving a perfect copy of " the photograph permanently produced upon the glass plate " in tints varying from a pale yellow to a deep red or " brown."

[*Printed*, 4*d*. *No Drawings*.]

A.D. 1871, May 25.—No. 1409.

GILBEE, WILLIAM ARMAND.—(*A communication from William Augustus Leggo, and George E. Desbarats.*)—(*Provisional protection only.*)—" An improvement in photography."

The leggotypes mentioned in this invention are those described in No. 1541, A.D. 1865.

This invention relates to the production of a negative or positive photograph in which the gradations of light and shade are shown in granulations. This style of photograph is employed in the manufacture of leggotypes and of transfers for stone or zinc.

To make a granulated positive, a copy is made of a negative photograph in contact with a granulated plate (of known construction), the two together forming one object. From the positive so made a surface type is produced by leggotyping, " capable of being printed from on any type printing press."

A granulated negative is made by copying the above-mentioned positive, or an ordinary positive in combination with a granulated plate, the two together forming one object. From this negative, the leggotyping process produces intaglio plates " capable of being printed from in the ordinary copper plate " printing press." Negatives so made are also employed in the production of transfers.

[*Printed*, 4*d*. *No Drawings*.]

A.D. 1871, July 27.—No. 1968.

GRUBB, HOWARD.—(*Provisional protection only.*)—" Improve- " ments in photographic lenses."

Ordinary single landscape combinations or patent aplanatic view lenses are not suitable for architectural views on account of the curvilinear distortion of the image.

This invention consists of a combination of lenses to be applied to the primary lens (this lens being either of the two forms mentioned above), in the vicinity of its diaphragm, which corrects the distortion while it does not alter the focus.

The combination consists of two simple lenses, which may be of similar or dissimilar glass, one placed a short distance anterior to the diaphragm of the primary and of positive power, the other of negative power and posterior to the diaphragm. The anterior lens is plano-convex with its convex surface to the front; the posterior lens is plano-concave with its concave surface to the front. Their respective powers are such that, when placed in their correct positions, their combined power is about equal to zero. Displacement of the rays is thus produced without such refraction as effects the focus of the primary lens.

The power of this combination in correcting the distortion of the primary may be varied at pleasure by alteration of distances between the various components, thus giving the photographer the power of producing at will perfect correction or a slight amount of either kind of distortion.

[*Printed*, 4*d*. *No Drawings.*]

## A.D. 1871, August 17.—No. 2166.

HASELTINE, George.—(*A communication from Horace M. Hedden and Charles A. Hill.*)—"Improvements in the manu- " facture of plates for sun pictures."

This invention consists chiefly in covering iron and other metal plates with a composition best suited for giving the desired tint to such pictures.

Linseed oil is boiled with "India red," until a paste-like substance is produced. When this is cold, it is applied to the plates with a quantity of spirits of turpentine or naptha. The composition is laid on to the surface of the metallic plates with a brush or other means, and afterwards the plates are baked in a suitable oven.

India red yields a chocolate or reddish brown tint. There are other forms of "oxide of iron" "known in commerce by " a variety of names" that may be used. The inventor sometimes prepares the common black ferrotype plates, and then puts a covering of the aforesaid coloured compound over the black coating and so obtains different shades and surfaces.

[*Printed*, 4*d*. *No Drawings.*]

A.D. 1871, September 30.—No. 2590.

STEVENS, HENRY.—(*Provisional protection only.*)—" Improve-" ments in cataloguing, calendering, or registering the con-" tents of libraries and museums, and in the apparatus or " means employed therein."

This invention is not included in the present series of abridgments on account of the improvements in cataloguing, but solely because it treats of a method of arranging surfaces for producing photographs therefrom.

For instance to collect the titles, sample pages, or illustrations of books, the books are combined behind plate glass frames, and, with fixed camera and frame, fifty or more titles, &c. are photographed at a time on one plate. The frames " consist of plate glass sliding in grooves in the upper and " lower sides of a frame left open at the ends." The books when arranged on the glass are pressed down flat by means of padding, &c.

In another method revolving frames are employed; these are " composed of two or more plates (by preference four) and " two or more cameras, according to the amount of work to " be performed in a given time."

[*Printed, 4d. No Drawings.*]

A.D. 1871, October 20.—No. 2799.

EDWARDS, ERNEST.—" Improvements in photo-mechanical " printing, and in apparatus to be used in such printing, " parts of which apparatus are also applicable to other " purposes."

This invention relates to improvements in the methods described in No. 3543, A.D. 1869, and No. 2485, A.D. 1870; also to materials and appliances for carrying out these and other methods, and to methods of photo-mechanical printing in two or more colours.

1st. Rollers for spreading the ink upon the prepared gelatinous surface, or " forme " from which impressions are to be taken.—Oil, or oil combined with an alkali, is added to the composition of treacle and gelatine ordinarily used for making printing rollers. A substance tending to produce insolubility of gelatine, also india-rubber, may be used in the composition.

2nd. The addition of dyes or colours to the water used in damping the prepared gelatinous surfaces.

3rd. One method of obtaining coloured pictures is to make the same number of negatives as of colours. In each negative, all is stopped out except the part to be printed in the given colour. A prepared gelatinous surface is then made from each negative, and each surface is inked of a given colour. Each surface printed in succession produces the picture. Instead of a number of negatives, one negative with a series of masks may be used.

Another method consists in producing a number of prints from a photographic negative and a prepared gelatinous surface therefrom. The actinic effect of each print is the same as that of the paper on which it is printed; for instance light blue upon white paper. The number of prints corresponds to the number of colours, and that portion of each print which is to be of a given colour is worked up in black and white by hand. A negative is obtained from each print, only the parts worked up in black and white being produced. From these negatives separate prepared gelatinous surfaces are obtained, and each surface is used in succession, with a different coloured ink, to obtain the picture.

A third mode consists " in 'masking' out in the gelatine " 'forme' all those parts of the picture except those to be " printed in the particular coloured ink then being used."

[*Printed, 6d. No Drawings.*]

A.D. 1871, October 30.—No. 2909.

WENDEROTH, Frederick Augustus.—(*Provisional protection only.*)—" Improvements in photographic pictures."

Photographic pictures heretofore taken on metallic plates have been negatives, in which the shadows have been produced by the exposed surface of the metal plate.

According to this invention, the shadows are produced by the coating applied to the plate and the lights arise from the light reflected from the metallic surface.

A polished plate composed of Britannia metal plated with silver has its surface rubbed with very fine sand. The rubbing takes place from top to bottom of the picture, so as to cover the surface with a multitude of fine scratches that give brilliancy to the picture in consequence of the light reflected from the surface. Upon the plate thus prepared the picture is printed from a glass negative in any of the known modes, such, for instance, as the collodion dry or wet process, or the collodion chloride process. Chromatised gelatine with a colouring pigment may be employed, so as to vary the colour of the developed picture. In any case the positive picture printed upon the sensitised surface is intensified in its lights by the light reflected from the surface of the metal plate.

When the picture is finished, it is rendered more durable by sealing it to a piece of plate glass the same size as the metal plate, and for this purpose white beeswax or similar material may be employed.

[*Printed*, 4*d*. *No Drawings.*]

A.D. 1871, November 14.—No. 3069.

STILLMAN, William James.—(*Provisional protection only.*)—
"Improvements in cameras for taking photographic pic-
"tures."

The base board of the camera is made in three portions, so jointed and hinged together that they enclose the works of the camera when closed up, and, when open, they form a base for extending the gussets of the camera, in order to obtain the proper focus on the sensitive plate.

Two gussets are in connection with a central partition in the camera; the forward gusset is connected with the frame and slide of the lens, and the hinder gusset with the back which carries the dark slide. The sliding board, carrying the lens frame, moves in grooves in the base board and is so jointed, in connection with the joints of the base board, that the whole can be closed together. The slide or sliding board is adjusted by a screw and nut arrangement. The back is also moved in grooves in the base board, and is clamped in a suitable position by screws.

The back of the camera is so constructed that the under side of it forms a longitudinal section of a cylinder and is fitted to a shoe, so that the back and the shoe may be clamped by a screw, and so that the dark slide may be fixed at any angle of inclination to the base board by this form of swing back.

[*Printed*, 4*d*. *No Drawings.*]

# 1872.

A.D. 1872, January 9.—No. 73.

EDWARDS, Ernest. — "Improvements in apparatus for "photo-mechanical printing, parts of which apparatus are also "applicable to other purposes."

This invention relates to improvements in the apparatus and materials used in carrying out methods described in the following Specifications:—No. 3543, A.D. 1869; No. 2485, A.D. 1870; and No. 2799, A.D. 1871. Also to improvements in carrying out the said methods, part of which are applicable to other processes.

In order to print pictures upon enamelled paper, hard gums or resins are added to the enamel composition with which the surface of the paper is prepared. Oily or waxy substances, or substances tending to make gelatine insoluble, or a combination of two or more of these materials, may be added to the enamel composition.

Machinery for printing from formes, and for damping, inking and masking them.—The prepared gelatinous surface or forme is bent and attached round a portion of the circumference of the main cylinder, the remainder of the circumference forming the inking table. The cylinder by which the paper is pressed upon the forme carries a flexible mask. The forme is damped by a damping roller which revolves in a cistern containing the damping fluid.

To keep the formes saturated with moisture, the water for damping them contains deliquescent salts, or glycerine, treacle, or analagous non-drying substance.

A forme may be prepared by means of a solution of a hard gum in water containing borax or ammonia. Sensitised gelatine may be added to the composition.

Designs may be added to a forme by the application of a solution of alum to the parts where they are required to appear. Indistinct designs may be printed on the back, by a transparent negative, the printing being continued until the light has sufficiently penetrated the forme.

[*Printed*, 8*d*. *Drawing*.]

A.D. 1872, January 17.—No. 149.

CLARK, ALEXANDER MELVILLE.—(*A communication from Henry Vander Weyde.*)—" Improvements in applying colors or tints " to photographs, albumenized and other surfaces."

This invention consists in a method of combining and applying dry colours and crayons with other substances, for the production of mezzotint effects or atmospheric stipples, stippled effects or appearances, in any tint or colour, on albumenised, glutenised, and gelatinised paper surfaces suitable for photographic prints.

The method consists " of rubbing with the aid of a soft pad " dry colors or tints of any kind, condition, or form, and fine " pulverized pumice stone, or its equivalent, using them " together or separate on the albumenized or other surface " until it adheres by penetrating into the grain of the same."

In this invention, the powdered pumice stone is employed in a manner analogous to the oil or water used to thin or dilute the colours in oil or water-colour painting.

" Other materials may be substituted for pumice stone such " as flour of glass, pulverized cuttle fish, and emery powder " when a brown color is required."

[*Printed*, 4*d*. *No Drawings.*]

A.D. 1872, March 1.—No. 641.

SMITH, GEORGE REEVES.—(*A communication from Henry Vander Weyde and Oliver Sarony.*)—(*Provisional protection only.*)—A " method for applying colour or tints to the surfaces of " photographs or other surfaces."

The purpose of this invention is twofold :—1st. To produce, by a rapid and easy process, stippled tints of any shade or colour over the backgrounds and other parts of photographs. 2nd. To fix dry colours or tints on albumenised surfaces.

This method " consists of rubbing with the aid of a soft pad " dry colours or tints in any kind, condition, or form and fine " pulverized pumice stone or its equivalent, using them " together or separate on the albumenized or other surface " until it adheres by penetrating into the grain of the same."

[*Printed*, 4*d*. *No Drawings.*]

A.D. 1872, March 19.—No. 826.

GAREY, William.—" Improvements in preparing paper for " photographic purposes."

This invention has for its object to obtain improved effects when photographic printing is combined with ordinary printing.

The border or other device is printed on the paper that is to be subjected to the photographic process, previously to the albumenisation of the paper. The paper is then albumenised so that the albumenous coating covers the printed part as well as any imprinted portion. " The photograph is then " produced in the usual manner, care being taken to have it " arranged as regards position so as to combine and agree in " the desired manner with the previously printed device. In " this way when the whole is completed the entire surface, " both printed and photographed, will have a uniform glaze " or finish."

[*Printed*, 4*d*. *No Drawings*.]

A.D. 1872, March 20.—No. 840.

WINDOW, Frederick Richard.—(*Provisional protection only*.) —A " method of taking silhouette or outline pictures."

This invention consists in an application of photography to the above purposes."

This invention consists in an application of photography to the above purposes.

A light coloured background is lighted strongly, either from the back like a transparency or from the front. The subject to be silhouetted is placed in front of the lighted background, in a position where little or no light can fall upon it from the front. A picture is then taken, by photography, of the lighted background and of the unlighted subject intervening. By this means is obtained a negative cliché of the subject in the form of a silhouette or outline. This is printed from on photographic paper, or it may be printed by means of photo-lithography on any suitable material.

[*Printed*, 4*d*. *No Drawings*.]

A.D. 1872, May 7.—No. 1395.

SLATER, ALFRED.—(*Provisional protection only.*)—"Improve-"ments in printing from photography."

This invention relates to reproducing copies of photographs in the printing press from an ordinary lithographic stone.

1st. process.—A picture in gelatine leaf is obtained from a photographic negative. Two inking rollers, with lithographic printing inks of different densities, are passed over the leaf, then an impression is taken, from the leaf, on lithographic transfer paper. The impression is transferred from the transfer paper to a lithographic stone from which prints can be taken.

2nd process.—The picture obtained in leaf is embedded into type metal by pressure, thereby leaving a print. This print is then inked, and an impression is taken on lithographic transfer paper, from which it is transferred to a lithographic stone. Prints are obtained from the stone in the ordinary way.

3rd. process.—An undeveloped picture is obtained on a film of gelatine mixed with alum. The picture is absorbent of greasy ink, and the ground is non-absorbent of greasy ink, but absorbent of water. This film is inked with two rollers having inks of different density, and an impression is taken therefrom on lithographic transfer paper. The impression is transferred to a lithographic stone from which prints can be taken.

[*Printed*, 4*d*. *No Drawings.*]

A.D. 1872, May 30.—No. 1642.

FAULKNER, ROBERT.—"Improvements in the preparation "of photographs and other prints for colouring."

The object of this invention is to obtain roughened or granulated surfaces on the above articles, by indentation. These surfaces enable the paper to take colour or shading readily.

According to one method, a sheet of emery, glass, or pumice paper, or of linen is placed in contact with the picture and the whole is pressed between rollers or otherwise. The print is laid with its back to the roughened surface and blotting paper is placed on the face of the print. The print may be damped, and only parts of the picture may be submitted to this process.

In another method, the powder is dusted on to the picture and subjected to pressure against paper or cardboard. The powder is then dusted off and the colour applied.

In a third method, the powder, made into a paste, is laid with a brush, over the parts of the picture to be acted upon. When dry, the pressure is applied. Colouring matter may be mixed with the paste. The paste may be made with water or any liquid that is not gummy or otherwise injurious to the photograph. The powder is removed before applying the colour.

For some effects the powder may remain on the surface; the picture is treated with gelatine.

Engraved plates on rollers may be employed to press the picture when a line effect is desired.

[*Printed, 4d. No Drawings.*]

A.D. 1872, June 5.—No. 1700.

FAULKNER, Robert.—(*Provisional protection only.*)—"Im-"provements in the production of photographic pictures "from negatives, and in the means employed therefor."

The object of this invention is to give the above pictures a lined, dotted, or otherwise varied ground. The ground work may be drawn on paper and a negative prepared from it. A protective film of gum, gelatine, or india rubber is given to the negative by means of a solution. When the negative is dry, transfer collodion is poured over it, this is allowed to dry and the whole is soaked in water. By soaking in water the triple film can be removed from the glass. This film is interposed between the negative of the picture and the surface on which it is to be printed by light.

To present the appearance of a line engraving, the groundwork negative is produced from wire gauze. Instead of wire gauze, natural objects may be employed. A dotted effect may be produced from a pigment held in mechanical suspension in liquid; a negative from this effect may be combined with that taken from gauze. Other effects may be produced on the groundwork by taking various objects and by photographing them more or less out of focus.

These negative films may also be used for superposing objects with more or less faintness over the objects of the principal picture. The superposed objects may be used in combination with the groundwork.

[*Printed*, 4*d*. *No Drawings*.]

A.D. 1872, June 29.—No. 1970.

SPAHN, Emil Peter. — "An improvement in camera " stands."

This invention consists "in the arrangement of a table " which supports the camera on suitable guides working in " the frame of the stand, and connected with a suitable band " or strap passing over a roller; said band or strap containing " weights in its other end to counterpoise the weight of the " table, guides, and camera placed upon the table. And " further, in the arrangement of a slide working in suitable " ways in the frame of the stand at right angles to the guides " supporting the table, and operated by a suitable screw so as " to act against the surface of said guides to hold the same, " and consequently the table and camera in any position the " same may be moved."

[*Printed*, 6*d*. *Drawing*.]

A.D. 1872, July 27.—No. 2246.

SMITH, Charles.—(*A communication from Elijah Alliger*.)—A "mode of lettering and ornamenting mirrors for adver-" tising, decorative, and other similar purposes."

The lettering and ornamenting mirrors do not of themselves entitle this invention to be described in this series of abridgments.

In the Provisional Specification, the inventor states that he produces " on the glass itself, by photography or other suit-" able and efficient means, the pictures, devices, letters, " figures and lines intended to appear or be represented," and silvers the ground or so much of it as may be desired, and then adds " the paint or such other ground as may be " necessary."

[*Printed*, 4*d*. *No Drawings*.]

A.D. 1872, August 24.—No. 2519.

TONGUE, JOHN GARRETT.—(*A communication from Théodore Givry.*)—(*Provisional protection only.*)—" Improvements in the " manufacture of needle and perfumery boxes or cases and " other useful or fancy articles."

The application of this invention does not come within the scope of the present series of abridgments, but the process, in which photography is involved, entitles it to a place here.

A photograph is taken on the platinised glass or " glace " platinée" which is used in this invention; or the photograph is applied thereto. In another method the photographic image is enclosed between two glasses or is covered with collodion, varnish, or transparent silicate.

When the back of the platinised glass is screened from the light, it serves as a mirror, but, when placed before the light, it is transparent and the photograph thereon is distinctly seen.

[*Printed*, 4*d*. *No Drawings*.]

A.D. 1872, September 10.—No. 2684.

LAKE, WILLIAM ROBERT.—(*A communication from Emile Rienzi Weston and Timothy McDonald.*)—A " machine for " burnishing photographs, cards, and other like articles."

A durable polished surface is given to photographs by the application of heat and friction to the article under pressure.

The apparatus is attached to a bed piece which can be secured to a bench or table. The photograph is laid face upward upon a convex bed of wood which carries standards having slots at the top, through which passes a bar; the cast iron burnisher is attached to the bar. The lower edge of the burnisher is turned up and concave, hardened and highly polished. Set screws in the top of the standards are used to force the burnisher down upon its bed. A to-and-fro motion is given to the burnisher by a lever attached to the bar and having its fulcrum in the bed piece. The burnisher is heated by a lamp. One or two strokes of the burnisher impart a brilliant finish to the photograph.

When a stationary burnisher is used, it has a convex roller revolving upon a shaft in standards placed over it. The burnisher is horizontal and a pressure screw is placed under it. A lamp underneath the burnisher supplies heat thereto. The picture being inserted between the roller and burnisher, the former is revolved by a crank, and the picture receives its polish as it passes through the machine.

[*Printed*, 6*d*. *Drawing*.]

A.D. 1872, September 16.—No. 2743.

BECKETT, James. — A "system for 'washing,' 'toning,' "'washing,' 'fixing,' and 'washing' photographic prints, "and apparatus therefor."

Three small troughs, with circular bottoms are used, in which to place the toning, fixing, and washing fluids. Bearings at the ends of each trough receive the axle of a small cylinder which is capable of rotating in close proximity with the bottom of the trough. On removing the prints from the printing frame, they are cut to size, and spread over the surface of the cylinder (picture surface outside), where they are secured in position by elastic bands. By rotating the cylinder in the washing trough, the nitrate of silver is removed from the prints. After the discharge of the remaining water, the toning solution is introduced into the trough and the cylinder is rotated therein until the prints are toned; the remains of the toning solution are then withdrawn from the trough. Several changes of water are then introduced to remove the toning solution from the surface of the prints, and a similar rotation in the fixing trough and subsequent washing, finishes the prints.

Flat surface plates may be used to carry the prints instead of cylinders.

The surface dipping which forms the principle of this invention is said to be superior to the total immersion and saturation of the print which is usually adopted.

The bearings on which the cylinder rotates are, in some cases, attached to a stand which is arranged to receive a loose trough.

[*Printed*, 4*d*. *No Drawings*.]

A.D. 1872, September 27.—No. 2848.

COURTENAY, Robert Henelade.—A "method of producing "and using surface blocks for line and photographic printing "from nature."

The object of this invention is the preparation of printing surfaces by photography and hydraulic pressure.

To prepare a granular negative, collodion containing oleate of silver and farinaceous powder or powdered glass is used.

A soft metal plate is prepared with a grain, and a relief photograph is transferred thereto. A gutta percha cast from the plate is electrotyped; the electrotype is ready for the press.

By means of a collodion without the powder, a positive transparency is taken from the granulated negative. Or a grained plate can be used. Or a smooth plate can be used, with a grain in the composition used for the negative relief. The composition is a mixture of gelatine, sugar, tragacanth, and starch, together with carbon in fine powder. To use the gelatine composition, a paper support is placed on a level slab of glass, and the compound solution is poured on to it. It is then dried, rolled, placed under a photograph, transferred to a metal plate, and the soluble parts dissolved out by warm water. When the plate is dried, the relief design is ready to be engraved by pressure, very great hydraulic pressure being used. The photograph for the overlay is printed from a negative.

To reproduce plates from printed matter, a dense negative is taken from the subject; from the negative, a transparent but dense positive is taken. After the design is forced into the plate by pressure, the boldest parts of the work are treated with bi-chromate of potash and allowed to dry, then coated with gelatine and silex, and dried in the light, rendering the bi-chromate insoluble. After washing away the upper part of the gelatine, the whole is dried and pressed. The plate is washed in alkaline water to remove the photograph, electro-coated with zinc, inked up, and treated with solution of nut galls, phosphoric acid, and gum, leaving the plate in a condition to be used in the typo-lithographic press, or in the lithographic press, or, when mounted, in the type press.

[*Printed*, 4*d*. *No Drawings.*]

A.D. 1872, October 17.—No. 3069.

FERRANTI, Cæsar, and TURNER, Edward James.—"Im-"provements in the artistic treatment of photographic "portraits."

The object of this invention is to obtain a specially dull appearance in the background of the picture and to throw the portrait in relief therefrom, giving to the figure and face a miniature-like effect of finish.

The albumenised or gelatinised paper bearing the photograph is washed with a mixture of prepared ox gall and water. The background of the picture is then treated with dry colour, pastil or crayons, giving a stippled and aerial effect. Lastly the albumen obtained from crate eggs is applied to the figure and drapery.

[*Printed, 4d. No Drawings.*]

A.D. 1872, November 13.—No. 3379.

PALMER, JOHN EASTMAN.—"Improvements in the production of copies of photographic pictures, and in shading and colouring the same."

A transparent picture of the negative, on glass, is coated with a matt varnish and is placed against the light with a sheet of tracing paper covering the same; the outlines of the picture are traced on the paper with a lead pencil. The tracing paper is then placed with the pencil marks against the glass, and the pencil marks retraced with the lead pencil on the other side of the paper. Gray crayon is then rubbed on the outlines on the side first traced, and the whole placed on drawing paper, slight pressure being used over the outlines until the whole of the outlines are marked on the paper. The shading and colouring are then commenced; "crayons, oil, or water colours, powder colours, prepared chalk, or other suitable materials being employed on the said paper as a packing, and placed against the photographic transparency, thus forming by its combination a fine artistic effect."

If clear varnish be used instead of the matt varnish, Reigate silver sand, or similar substance is used to assist in the distribution of the crayon or powder colours.

The photograph used to back the transparency may be taken from an artistic design, on albumenised paper, leaving a white centre to receive the photographic transparency. The photographic design is blended into the edges of the transparency by means of the aforesaid crayon, or other colours. If albumenised paper be used, it is floated with the matt varnish; if plain paper be designed to receive oil colour, it is prepared with size.

[*Printed*, 4*d*. *No Drawings*.]

A.D. 1872, November 25.—No. 3531.

FAULKNER, ROBERT.—(*Provisional protection only.*)—" Im-" provements in applying and fixing dry colours on paper, " canvas, and other porous substances, and in the means " employed for this purpose."

This invention may be employed to produce backgrounds such as are used for photographic purposes.

To carry out this invention, powdered dextrine is mixed with the powdered pigment, and the mixture is applied by rubbing, in the manner usually employed for powdered pigments. The back of the porous surface is moistened with water, which soaking through it, dissolves the dextrine, and thereby fixes the colour.

Instead of applying the pigment in the form of powder, it may be mixed with dextrine, and the mixture made up into crayons, which can be used for drawing on the surface. The surface is moistened before drawing thereon; when dry, the drawing will be fixed.

When large backgrounds for photographic purposes are to be produced, the paper is first treated with a weak solution of dextrine, which is allowed to dry, then the pigments are applied.

Other substances than dextrine, gum, or gelatine may be used, together with a suitable solvent, such as turpentine, benzole, &c.

[*Printed*, 4*d*. *No Drawings*.]

A.D. 1872, December 4.—No. 3654.

WOODBURY, WALTER BENTLEY.—(*Provisional protection only.*) —" Improvements in photo-mechanical printing, and in appa-" ratus and appliances therefor."

1st. A method of obtaining by photography, type printing blocks, either from a subject in line or from a photograph containing only half tones.—When the subject is in line, a photograph of it, that is positive by transmitted light, is made to yield a relief in gelatine by the ordinary method;

the relief is pressed into metal by hydraulic pressure, or by the method stated in the second part of this invention, producing (in either case), a uniform surface in the resulting mould. When the subject is in half tone, as in a photograph from nature, in printing on the gelatine film, a transparent photograph of musquito netting or Brussels net, is interposed betwen the gelatine film and the negative. The soft metal block may be electrotyped for printing.

2nd. Improvements in Woodburytype.—In place of using a collodion film to hold the gelatine of the relief, when the latter is dried and ready for use, the side next the glass is exposed for a few seconds to light, before exposing it under the negative. When the gelatine relief is dried, a sheet of tin foil is laid on it and attached to it at the edges; plate paper being laid on the back of this, the whole is passed through a rolling press. To print from it, the tin foil is mounted on a composition of shellac and asphalt. Instead of fixing the proofs by alum, they are varnished with shellac varnish.

3rd. Printing Woodburytype by machinery.—A cylindrical hole in a solid block of iron receives (loosely) a cylinder of soft metal, which has a central and taper hole. Gelatine reliefs, inserted between the interior of the block and the soft metal, are impressed on the soft metal by pressure of a taper spindle driven into the taper hole. The cylinder and spindle are used in the process of printing.

[*Printed*, 4*d*. *No Drawings*.]

## 1873.

A.D. 1873, February 19.—No. 615.

ABNEY, William de Wiveleslie.—" Improvements in photo-" mechanical printing and in apparatus to be used in such " printing."

A portion of the apparatus used to carry out this invention, consists of a soft roller described in No. 2799, A.D. 1871, and used in producing heliotypes. Heliotypes are also described in No. 3543, A.D. 1869, No. 2485, A.D. 1870, and No. 73, A.D. 1872.

An ordinary negative is placed under paper prepared with bi-chromatised gelatine, and the combination is exposed to light. The picture thus printed is washed in cold water containing alum. A good deal of the bichromate which has remained unaltered is washed out; the paper is then placed on a support of zinc, &c., and the moisture is expelled by pressure. A soft roller charged with transfer ink, causes the ink to adhere only in the actinised parts of the picture. The inked up picture may be dried, again actinised, and transferred in the ordinary manner to stone or metal plates.

When a half tone negative is employed, the back as well as the front of the paper is coated with gelatine to cause it to adhere to the support.

A modification of the invention consists in using a sheet of hardened but not sensitised gelatine. Designs are then formed upon the prepared sheet, with a pen carrying an ink composed of a gelatinous solution that has been sensitised. Aniline dye may be added to this ink. When the writing is dry and has been actinised, the sheet is damped and inked up, and the design can then be transferred to stone for printing from in a printing press.

[*Printed*, 4*d*. *No Drawings*.]

### A.D. 1873, March 4.—No. 776.

LONG, John, FORSYTH, Alexander, and GRIFFITHS, James Robert.—(*Provisional protection only.*)—A "method "of obtaining artistic results in finishing photographic "portraits."

1st. A vignette head, or part, or whole figure is photographically printed on paper.

2nd. A stippled surface is prepared, on a plate of glass, paper, or other transparent medium, in a coloured pigment, and is superposed on the print in the photographic printing frame. By exposing this combination to the action of light, the stippled effect is obtained on the print; the parts that are not to have this effect are shaded. The print is then toned, fixed, and washed. When the print is mounted, the stippled parts are rubbed over with pumice powder, and the process is finished.

[*Printed, 4d. No Drawings.*]

A.D. 1873, March 31.—No. 1190.

BROWNE, ALEXANDER.—(*A communication from Pierre Auguste Despaquis.*)—(*Provisional protection only.*)—" Improvements in " photography."

This invention relates to touching up enlarged photographs.

The image produced by the enlarging apparatus is received on a very finely ground glass plate, and is touched up by brushes, or otherwise, so that, when seen through the glass, the picture stands out well. Against the ground glass is placed a sensitised glass plate or sheet of paper, and the perfected image is photographically printed thereon.

[*Printed, 4d. No Drawings.*]

A.D. 1873, April 10.—No. 1333.

STEIN, SIGISMUND THEODOR.—" Improvements in the con- " struction of cameras for photographic purposes, and in the " apparatus connected therewith, together with the process " employed therein."

The purpose of this invention is to enable photographs to be taken by the wet process, without the aid of a separate and detached dark room or chamber.

The camera is provided with a recess or space to receive a hollow case, called, by the inventor, a " heliopictor;" the heliopictor is made of glass or bitumenised wood, and is used for sensitising the plate, and (after the picture is taken) for developing and fixing the image.

The focussing glass is inserted into one side of the heliopictor; this side is hermetically closed by a yellow glass frame, when the focussing glass is not there. A yellow glass wall, on the opposite side of the apparatus, extends from the top downwards, about two-thirds of the depth of the heliopictor; the space below forms an enclosed receptacle for containing chemical liquids. The stem of a glass funnel, with a tap, is inserted in the edge of the apparatus, and is capable of being automatically closed.

The glass plate for the photograph is prepared by means

of an aqueous solution containing albumen, glacial acetic acid, and iodide of potassium; it is then substituted for the focussing glass. The funnel is inserted, nitrate of silver solution is poured into the heliopictor and is caused to run over the surface of the plate by inclining the apparatus, and is then poured away. The picture is taken by transferring the heliopictor to the camera; the apparatus is then detached from the camera and receives the developing solution, which is manipulated in a similar manner to the sensitising solution; the liquor is run out, the plate is frequently washed with water, and the image is fixed by hyposulphite of soda and well washed with water. Previous to fixing, the heliopictor is ready for employment again.

[*Printed*, 6*d*. *Drawing*.]

A.D. 1873, April 12.—No. 1341.

BILLUPS, Jonathan Edwin, and LEE, Edwin Palmer.—"Permanent coloured and other photographic pictures and "delineations."

These pictures are produced on glass by transference, or by producing a photographic transparency on the surface of the glass. This transparency is rendered permanent by placing the glass, with the photograph thereon, in a furnace or muffle, and firing the same until the photograph is permanently burnt into the substance of the glass.

Colours can be applied to the delineations and coloured photographs in enamel can be obtained, in which the colours are rendered permanent by incorporating them with the substance of the photographs previously burnt into the enamel tablet or glass, and therefore forming a part of the same, by means similar to the above. Vitrifiable colours, mixed with water, are applied by a brush; the glass enamels, so treated, are then placed in a furnace and fired until the colours are thoroughly fused into the material of the glass or tablet.

[*Printed*, 4*d*. *No Drawings*.]

A.D. 1873, April 22.—No. 1449.

BILLUPS, Jonathan Edwin, and LEE, Edwin Palmer.—"Improvements in producing permanent coloured and othe "photographic pictures, delineations or devices on porcelain "and other fictile ware."

A photographic transparency is transferred to or produced on the surface of the porcelain. The whole is placed in a furnace and fired until the photograph is permanently burnt into the substance.

Permanent coloured photographs on porcelain can be produced in a similar manner. Vitrifiable colours are applied with a brush to the porcelain with the photograph thereon. The same is placed in a furnace and fired, until the colours are thoroughly fired into the material.

[*Printed, 4d. No Drawings.*]

### A.D. 1873, May 30.—No. 1954.

WOODBURY, WALTER BENTLEY.—"Improvements in photo-"mechanical printing, and in apparatus and appliances "therefor."

1st. A method of obtaining, by photography, type-printing blocks from subjects in half tones only.—A Woodburytype sheet of bichromatised gelatine is exposed under a photographic positive with a transparency of netting interposed. The transparency may be an impression from a grained stone or from fine ruled lines. This sheet of gelatine, when washed, gives a relief having the positive photograph represented by a number of lines. An impression of this, by hydraulic pressure, in soft metal is used for printing at a type press. An electrotype from this may be employed when a number of copies are required.

Sometimes a negative of the network is copied with the negative to be reproduced yielding a lined positive, from which a relief and blocks are made as above.

2nd. Improvements in Woodburytype.—The side next the glass of a bichromated gelatine film is exposed for a few seconds to daylight before exposing it under a negative. To the edges of the relief thus produced a sheet of tin foil is attached by gum; a sheet of plate paper is laid on the back of the tin foil and the whole is passed through an ordinary rolling press, thus impressing the foil into all the details of the relief. The foil and relief are backed up, by pressure, with a composition containing shellac and asphalt, lac and sulphur, lac and Venice turpentine, lac or rosin and silica.

When the composition has hardened, the tin foil adheres to it. The gelatine relief is then removed from the mould and the mould is printed from.

3rd. A method of printing Woodburytype by machinery.—In a cylindrical hole in a solid block of iron a cylinder of soft metal fits very loosely; the cylinder has a taper hole through it lengthwise. The gelatine reliefs are inserted between the interior of the iron block and the soft metal cylinder. Into the taper hole is driven a taper spindle, which forces the soft metal against the interior of the iron cylinder, thus impressing the reliefs on the outside of the metal cylinder; the taper spindle forms a shaft for the cylinder to be used in printing. The cylinder is mounted in vertical slots in a frame which has a bed of plate glass on which the paper to be printed rests. A fixed roller may be used in place of the glass plate.

4th. A method of preparing blocks by photographic agency to be used in producing the water mark in paper.—A relief from a negative is made and a reverse is obtained from it in soft metal. From this a spongy or porous electrotype is made in the following manner:—The plate to be coated is placed face upwards in the bath in a horizontal position. At short intervals, during the whole process of depositing, fine, and afterwards, coarse granules of asphalte are sifted over the plate. On being removed from the bath, the granular material is dissolved out from the copper by a solvent, thus giving an electrotype which will drain away the water from the pulp of the paper. The electrotype may be so mounted that the air may be exhausted through it to hasten the abstraction of the water from the pulp.

[*Printed*, 8*d*. *Drawing*.]

A.D. 1873, June 5.—No. 2011.

WILLIS, William, junior.—" Improvements in photo-chemi-" cal printing."

The surface of paper or other material that is to receive the image is coated with a salt of platinum, iridium, or gold, then with a salt of another metal and with a solution containing ferric oxalate or tartrate. The surface thus obtained is exposed to light under a photographic negative and is treated with an oxalate, and washed in a dilute acid and in water.

The following are examples of the operations :—

1st method.—The paper is coated with chloro-platinite of potassium, with nitrate of lead and with ferric oxalate. After exposure, potassic oxalate and weak oxalic acid are used, and the picture is finished by immersion in hyposulphite of soda solution and washing in water.

2nd method.—Nitrate of silver is substituted for nitrate of lead, and the print is finished by immersion in a strong solution of chloride of sodium, or in a weak solution of the same, followed by weak ammonia, and washing in water.

3rd method. — Platinic bromide and ferric tartrate are used to coat the paper; after exposure, potassic oxalate is employed, and the picture is finished by immersion in weak oxalic acid and washing in water.

[*Printed, 4d. No Drawings.*]

A.D. 1873, June 12.—No. 2079.

FRÈRET, LÉON.—(*A communication from Paul Marny Godard.*) — A " method of producing and reproducing drawings, " devices, and designs on porcelain, delph-ware, and such " like materials."

A negative is produced from the design and a print is taken from the negative on to specially prepared paper. The bath for preparing the paper contains glycerine and mineral colour, gelatine, and bichromate of potash. The bath is used at a temperature of 90° to 100° Faht. One surface of a sheet of paper is wetted in this bath and then allowed to dry in a dark room. After the print is taken, by exposure to light, the paper is soaked in cold water, applied on porcelain, &c. and made to adhere thereto by pressure. The whole is placed in warm water and the paper soon detaches itself leaving the gelatine on the object. All the parts of the gelatine that have been exposed to light, and therefore form the picture, will be fixed to the porcelain. The picture is then covered with enamel or vitreous glaze and baked in an oven.

[*Printed, 4d. No Drawings.*]

A.D. 1873, June 16.—No. 2122.

HODGSON, EDWARD. — (*Provisional protection only.*) — " A " revolving camera for taking portraits and other pictures."

A hollow drum, or a skeleton frame of the shape of a polygonal prison, has top and bottom discs in which grooves are cut for the insertion therein of any convenient number of prepared glass plates on which pictures are to be photographed. The focussing ground glass plate forms one side of the frame, the other side are occupied with the prepared glass plates, each of which may be brought, in succession, by the turning of the drum, into the exact position previously occupied by the focussing glass. The drum is pivoted in a box, so that it can be freely revolved by a handle outside the box. The box is provided in front with a sliding lens, directly opposite to which is a small strip of yellow glass to focus through.

By means of a circular dial plate on the top of the box, centred on the upper pivot of the drum, and numbered at the proper distances corresponding to the number of plates or sides of the drum, the plates are brought in succession before the lens and thus receive an impression or picture.

When there are to be different views on each plate, the focussing glass is brought behind the lens after each operation, until all the views are taken.

When only one or two plates are required, the drum may be only a quadrant, and the views, as they are taken are turned into an empty space.

For a stereoscopic camera, the drum is made to revolve on a horizontal axis.

[*Printed*, 4*d*. *No Drawings*.]

### A.D. 1873, July 4.—No. 2316.

BRADLEY, Henry.—A "mode of preparing and printing " designs or objects on metal or other surfaces by the agency " of photography, such designs or objects so produced to be " made permanent by the subsequent operation of the engra- " ver, for the purpose of copper-plate or letter-press print- " ing."

This invention is a ready means for transmitting to a copper plate, the design to be engraved on the same.

The plate is electro-coated with silver, with a very thin deposit, which is afterwards oxidised by submerging the plate in an acid. On the front of this plate prepared wax is applied, then a film of collodion, which is afterwards sensitised with a solution of silver. The plate is then placed in the camera to receive the design or object and is treated with albumen and a coat of preservative, after which process it is ready for the engraver.

Designs produced by the above (or positive) process are reversed by the process of printing, to avoid which a glass reversed negative is taken and the copper plate is coated with a sensitive film, prepared with wax (as in the positive process) and collodio-chloride of silver. A positive printed from the negative, and toned, and fixed, is ready for the engraver.

By a modified process, this invention may be applied to surfaces, for engraving, of a non-metallic character.

[*Printed*, 4*d*. *No Drawings*.]

A.D. 1873, July 7.—No. 2347.

VAN TENAC, Charles Louis.—(*Provisional protection only.*) —An "optical instrument intended for the magnified pro- " jection with their natural colours of all opaque objects of " small size."

Only so far as this instrument is "a photographic instrument " to produce enlarged negative plates direct from the subject," is it entitled to be noticed here.

The light is placed at the focus of the hind reflector of the system of reflectors which forms the body of the instrument. The tube containing the optical glasses is fitted to the body of the instrument; its axis is perpendicular to the plane of the subject to be photographed, it meets the axis of the instrument at an angle of 45°.

The white board or cloth on to which the image is projected, in the case of the microscope, is presumed to be changed for the sensitive surface in the use of the instrument for taking enlarged negatives.

[*Printed*, 8*d*. *Drawing*.]

A.D. 1873, July 17.—No. 2468.

BANKS, John Henry.—"Improvements in the production " of printing surfaces, and of engraved metal surfaces appli- " cable for other purposes."

A photographic negative is obtained, in the ordinary way,

upon a glass plate, except that it is developed with bichloride of mercury, when this is washed off the image is treated with iodide of potassium, and lastly the photograph is subjected to the action of hydrosulphuret of ammonia, so as to make all the white parts in the original drawing or photograph as opaque as possible. A second glass plate is coated with a film of gelatine of some thickness, which is dried and covered with a mixture containing a bichromate and hydrate of lime. When the film is dry and hard, the first plate carrying the negative is placed in the pressure frame, and the film side of each glass plate is placed in contact. The whole is pressed tightly together and exposed to light. The white portions of the second plate are soluble and the printed portions insoluble; the soluble parts can be washed away with hot chloride of sodium solution. Acetic acid is used to harden the relief design, and a plaster cast is taken from it; a second plaster cast can be taken from the first.

Instead of using a hot solution of chloride of sodium to wash away the soluble parts of the second plate, a cold solution of iodine or of iodide of potassium may be employed, followed by a very weak ammoniacal solution.

The white portions of the plaster relief may be deepened, by rolling over the raised portions a mixture of printing ink and copal; when dry, the plaster mould is dipped into a weak bath of soap and well brushed and the plaster is hardened with a weak solution of alum and pyrogallic acid. Another way is to cut away, with a gouge, all the whites that are not sufficiently deep. For stereotyping purposes, the plaster is hardened by means of a hot bath containing size, treacle and peroxide of chromium; the plaster is then exposed to light, while wet a frame is laid upon it, and the plaster of Paris is poured in and put under pressure just before it sets. When an electrotype is required, the plaster cast from the gelatine film is soaked in cold water and its face is heated over a gas flame. A frame is put over the face and a preparation of hot wax, resin, and lard is poured on it. The wax impression may be blackleaded or coated with mercury; the latter operation is accomplished by washing the mould with spirits of wine, and, when nearly dry, allowing the fumes of mercury to settle on the face.

To produce a copper engraved plate, the inventor takes a

negative with lights and shades reversed; he adopts the same portion of the process of the first and second glasses which relates to the printed representation on the second glass, and raises the surface with acetic acid only. A plaster mould is taken; the electro-cast from this is the engraved plate required which is printed from by the copper-plate press. The mode may be modified by not taking a negative but by placing the print in contact with the second glass; it acts as the negative, and after it is reproduced on the second glass the process is as above. A plate is produced which can be printed from in the ordinary manner.

[*Printed*, 6*d*. *No Drawings*.]

A.D. 1873, July 31.—No. 2592.

BECKETT, JAMES.—(*Provisional protection only*.)—A "system "of albumenizing paper."

The object of this invention is to give greater stability to the paper used in photography, thereby rendering it less permeable to the chemicals used in washing, toning, and fixing.

Instead of coating one side of the paper only, it is immersed in the albumen solution, thereby coating the paper on both sides at once. The paper may be coated on both sides by floating, but immersion is a quicker plan.

[*Printed*, 4*d*. *No Drawings*.]

A.D. 1873, September 4.—No. 2911.

HASELTINE, GEORGE.—(*A communication from George P. Critcherson*.)—(*Provisional protection only*.)—A "machine for "burnishing photographic pictures."

The burnishing is effected by means of a heated steel burnishing bar, and a pressure roller, the photograph being placed between the burnishing bar and the roller, and moved by the latter to and fro over the bar.

A metallic plate is mounted upon a cast-iron frame. The burnishing bar is placed over the plate. The bar and the photograph being burnished are protected by the plate from the smoke or vapour arising from the lamp or heater. The pressure roller is placed above the burnishing bar and is rotated by a handle attached to its axis; its surface is roughened so that it does not slip over the back of the photograph but moves it. Set screws, springs, and adjustable bearings allow the pressure of the roller to be regulated.

[*Printed, 4d. No Drawings.*]

A.D. 1873, September 22.—No. 3106.

BARRAUD, Henry.—(*Provisional protection only.*)—"Im-"provements in the treatment of photographic surfaces and "pictures."

The photographic picture is printed from a negative, and a lined, dotted, stippled, or varied ground is applied thereto. By this means some of the effects of a line or stippled engraving may be imparted to the picture.

Upon the sensitised or other surface on which the picture is to be produced is obtained an impression on from one or more plates, stones, or other suitable printing surfaces upon which lined, dotted, stippled, or otherwise varied designs have been produced. The character of the ground and the colour of the pigment in which it is printed may be varied to suit the subject of the picture or of the different parts of the picture to be produced.

"Instead of applying the said ground to the surface before "the photograph is printed, it may be applied after the "picture is fixed, either by printing as before from a plate, "stone, or other surface or by transferring (from a sheet of "paper or other material upon which the said ground has "been printed) an impression thereof to such parts of the "picture as may require it."

[*Printed, 4d. No Drawings.*]

A.D. 1873, October 10.—No. 3288.

SELL, Eugen Ernst Johannes.—"Means of and apparatus "for taking photographs at night."

This invention consists in producing a new photographic light by burning bisulphide of carbon with peroxide of nitrogen.

An ordinary petroleum lamp, with a circular wick, contains bisulphide of carbon instead of oil. Water surrounds the chamber that contains the bisulphide, so as to keep it cool. The peroxide is led to the centre of the wick of the lamp, and is uniformly mixed with the bisulphide by impinging against a cone placed with its apex downwards within the circular wick.

The pressure of the peroxide is regulated by a stop cock and means are provided to carry away the products of combustion. A reflector is used, when necessary, to direct and condense the light.

[*Printed, 4d. No Drawings.*]

A.D. 1873, November 20.—No. 3782.

KENNETT, Richard.—A "substance to be used instead of "collodion and other emulsions for photographic purposes."

This substance is produced, for use, in a dry or solid condition and will keep good for any length of time.

The compound essentially consists of an aqueous solution of gelatine together with a bromide chloride or iodide, and nitrate of silver. The compound is cleared of certain salts, which are formed during the mixing, and then dried. These processes must be conducted in non-actinic light.

The bromides chlorides or iodides that may be used are those of potassium, cadmium or ammonium. When the compound is mixed, and thoroughly incorporated, it contains a free bromide, and nitrate of potash; these are removed or eliminated by pouring the compound into a dish, letting it cool and cutting it into small strips which are washed with many changes of water until all the free salts are dissolved out. The subsequent drying process is accomplished by heating the compound in flat dishes, until it is reduced to a thick paste. When cold, it is stripped from the dishes and placed in suitable frames in a drying closet in which a circulation of dry air is maintained.

[*Printed, 4d. No Drawings.*]

A.D. 1873, December 1.—No. 3938.

JOHNSON, John Henry. — (*A communication from Honoré Muhé.*) — (*Provisional protection only.*) — "Improvements in "polishing and glazing photographic proofs and in the ap- "paratus employed therein."

Photographs are submitted to the action of a peculiar press and burnisher; the latter may be heated by a lamp during the polishing process. The press consists of a roller, the surface of which is roughened to ensure the drawing in of the mounted photograph when presented thereto. Beneath this roller is fitted an adjustable table, mounted on hinges at one end, whilst the other end is supported on powerful helical springs mounted on adjusting screws. The table is recessed, immediately beneath the pressing roller. to receive the burnisher.

A slightly projecting collar, on each end of the cylinder, rotates in contact with the surface of the burnisher at each end and thereby prevents the photograph from being unduly pressed upon by the roller, which is rotated by a winch handle on its axis. The surface of the table at the front part, is ribbed to facilitate the removal of the photograph therefrom on its leaving the burnisher.

[*Printed*, 4*d*. *No Drawings*.]

---

## 1874.

A.D. 1874, February 27.—No. 731.

BROWN, Joseph.—(*Provisional protection only.*)—" Improve-" ments in the ornamentation of vitreous, ceramic, glazed, " and metallic surfaces."

This invention consists in reproducing photographs upon these surfaces, by the employment, in conjunction with an oleaginous or fatty substance, of fusible fluxes together with metallic oxides.

A photographic image having been obtained on a film of bichromated gelatine, and the soluble portions having been washed away, the surface is ready to be submitted to the ordinary printing process. For the production of impressions upon paper, the inventor employs a printing or transfer material or medium consisting of the above-named substances, fluxes, and materials, which, when submitted to the action of an elevated temperature, are capable of forming a glaze or fused surface, whereby the image may be burnt in and rendered permanent upon the surface which it is desired to ornament.

The paper suitable for the purposes above-mentioned is made by coating the surface of paper with a solution containing a gelatinous and an amylaceous substance.

[*Printed*, 4*d*. *No Drawings*.]

A.D. 1874, March 31.—No. 1124.

WIRTH, Frank.—(*A communication from Adolph Steinheil, and Edward Steinheil.*)—" Improvements in photographical portrait " apparatuses and other optical instruments."

These improvements relate especially to photographic lenses or combinations of lenses used for the purpose of taking portraits.

Two compound lenses form the entire combination. Each compound lens consists of two parts cemented together, and is at an adjustable distance from the other compound lens. Of the individual or component lenses, the two exterior lenses are flint and the two interior lenses of crown glass. The interior surfaces of each compound lens, namely those that face each other in the combination, are of equal curvature. The external curves of the crown glass lenses, and therefore the internal curves of the flint lenses, are unequal, the surface next the camera having the shorter radius. The external curves of the flint lenses are unequal, the surface next the camera having the longer radius.

[*Printed*, 8*d*. *Drawings*.]

A.D. 1874, April 28.—No. 1481.

LAKE, William Robert. — (*A communication from Joseph Parker Bass.*)—A " machine for burnishing photographs, cards, " and other like articles."

The general principle upon which these burnishing machines operate is described in No. 2684, A.D. 1872; the present invention relates particularly to certain modifications of these machines.

This invention consists "in giving to the feeding roll a " slight endwise vibration over the burnishing tool. The " surface of this feeding roll being slightly rough this motion " is communicated to the card or picture, giving to it a " rubbing or to-and-fro motion over the burnishing tool in "addition to its movement across it." The roller has cams or cam surfaces cut at each end, against which bear pins attached to the standards of the machine, so that, as the roller revolves, the pins force it into a vibratory motion in addition to its rotary motion. The same result may be accomplished by a grooved cam cut either in the periphery of the roll or of its arbor, a pin (attached to the stationary part of the machine) working in the groove.

[*Printed*, 6*d*. *Drawing*.]

## A.D. 1874, May 8.—No. 1634.

CLARK, ALEXANDER MELVILLE.—(*A communication from Claude Léon Lambert.*) — A "method of retouching photographic "negatives and positives."

This invention relates to a method of retouching photographic negatives, whether they are taken from nature or are enlarged reproductions, whereby all retouching of the positive proof on paper is rendered unnecessary.

The invention consists in the application of a semi-translucid sheet on each side of a negative or positive, and in a method of quickly and readily retouching by operating on these surfaces.

The large negative, when fixed and finished, is covered, on both sides, with a sheet of thin paraffin paper of the grain known as "papier pelure végétal." The retouching is effected upon these two surfaces, by applying, wherever necessary, an impalpable galvano-plastic powder or other suitable finely-divided substance; a stump is used to apply the powder. The sharpness of the lines may be ensured by means of a black lead pencil. A positive image is then obtained from the negative by means of the printing frame.

If the lines of the primary cliché be too strong, a sheet of very thin glass may be interposed between the negative and the paper during the last part of the time of printing.

The complete or partial transformation of the background, &c. may be effected by placing, on the positive, a sheet of extra thin glass at $\frac{1}{25}$th of an inch from it. Another glass is then placed on the mask of yellow paper, at about $\frac{1}{12}$th of an inch therefrom.

This process may be applied to transparent positives of all

dimensions, which will then serve for the reproduction of negatives without further retouching.

[*Printed*, 4*d*. *No Drawings*.]

A.D. 1874, May 21.—No. 1807.

PLACET, Paul Emile.—" Improvements in engraving."

The processes which are the subject of this invention give engravings with the grain which is necessary for lithographic or typographic printing.

An animal or vegetable substance is dipped into a liquid which has the property of dissolving it, then into a second liquid having a tanning or contracting property, thus covering the surface of the substance with the required grain.

The following are examples of the modes of operating :—

1st. Bichromated gelatine paper is exposed to light under a negative and dipped into a saturated solution of bichromate of potash. The excess of liquid is removed by blotting paper, and the gelatine paper is plunged into an aqueous solution containing sulphate of iron and acetic acid; instantly the design is developed and grained. The dried engraving may be pressed, so as to give, at once, a plate for impression.

2nd. After exposing the gelatine paper under a negative, it is placed on another surface, so as to be able to develop the image on the reverse side. The reversed proof is then developed in a solution of bichromate of potash, and is plunged into the sulphate of iron solution.

3rd. A carbonised gelatine proof may be developed by means of solutions of ammonia, potash and hypochlorite of lime; it s then plunged into the solutions of bichromate and of sulphate of iron, when the grain appears.

4th. The gelatine may be run over plates of glass or metal and then sensitised, or it may be dried and sensitised.

5th. The tanning material may be applied in the solid state. Solvents may also be used in the form of powders on the gelatine slightly wetted.

Other examples are given, which are modifications of the above.

To be usefully applied, the design should be reproduced by electrotypy.

[*Printed*, 4*d*. *No Drawings*.]

A.D. 1874, June 4.—No. 1935.

DILLON, Thomas Arthur.—A "method of and apparatus for "indexing, copying, multiplying and securing public and "other records and documents."

The object of this invention is to obtain an efficient record system, by means of an adaptation of scientific processes comprised in photography, type printing and stereotype, combined with a method of printing on thin metallic plates.

Only the adaptation of photographic processes and dependent processes can be noticed here.

For a type printer to read from, a photograph of a given document is taken on prepared paper by flattening it down with a clear sheet of glass. A microscopic copy may be taken for subsequent enlargement whenever a fac-simile of the original document is required. The original document in the Record Office, say, for instance, the Registry of Deeds Office, Ireland, need never be removed from the custody of the Registrar.

When a metallic fac-simile of the document is required, any photo-mechanical process may be used to impress the photographic copy into the metallic plates.

To impress the metallic plates, a stereotype is taken from the set up type, in the ordinary way. The metallic sheets, or brass plates are impressed in ink and indented by the stereotype by means of a fly press.

Copies are taken from the reversed or embossed side of the plates on white writing paper. The paper is moistened with nitrate of silver, laid on the plate and covered with blotting paper and then with a sheet of india rubber. A moderate pressure blackens the nitrate of silver whenever it comes into contact with the paper.

Type ink may be used to print from the reversed sheets; or manifold paper, or electro-printing on chemically prepared paper may be employed.

[*Printed*, 6*d*. *No Drawings*.]

A.D. 1874, July 16.—No. 2490.

DEBENHAM, William Elliott.—"Improvements in fixing "colours on photographs and drawings, and in the method

" of preparing the paper or other material to receive " colour."

To fix the colour on a coloured photograph, it is immersed in a solution of gluten in alcohol, or in an alcoholic solution of gelatine or kindred substance prepared by dissolving the gelatine in water and adding as much alcohol as the solution will bear without precipitating the gelatine. The gelatine coating may be rendered insoluble by treatment with tannin or chrome alum; the chrome alum is afterwards exposed to light. Instead of the above immersion process, the solution may be allowed to flow over the photograph.

To prepare a photograph in order that colour may adhere thereto, either of the above solutions may be applied, or a solution of glycerine or sugar, or a mixture of these. If the work can be coloured before being mounted, as in the case of a photograph to be enamelled, the preparation liquid is applied to the back of the paper. The alcohol makes it penetrate to the front, and the colour is taken in a very even and fine manner. The fixing solution may be used to attach the photograph to the collodionised glass in enamelling.

[*Printed, 4d. No Drawings.*]

A.D. 1874, July 27.—No. 2617.

LESTER, THOMAS.—" Improvements in photographic ap- " paratus."

This invention is intended to facilitate the production of photographic negatives in open daylight.

The invention consists in the application to the camera of a box containing three parallel baths, into which the collodionised plate may be successively plunged without exposure to daylight, then transferred to a plate box and finished at leisure. The first bath contains nitrate of silver, the second, acidified protosulphate of iron, and the third a mixture of honey, iodide of potassium, and water for washing the developed picture.

Screws, passing through the sliding top of the bath box and pressing on the bath covers, make the baths watertight.

The collodionised plate is plunged into the baths by means of a dipper, which holds the plate at one edge, and is passed through a slot formed in the upper part of the body of the camera. The slot is covered by a slide with an aperture for the admission of the dipper.

In the dipper, the glass plate can be held firmly by one edge, by means of a clamp screw pressing against a stay at the lower end of the dipper. A stop on the lower end of the dipper cuts off light.

" The bath box is fitted to the camera by means of a slide, " which is screwed on to its lower edges and works in a groove " in the top of the bath box."

[*Printed*, 8*d*. *Drawing*.]

A.D. 1874, July 31.—No. 2667.

BROOKES, ALFRED GEORGE.—(*A communication from Louis Alexandre Farjon.*)—" Improvements in copying presses, a " part of which improvements is also applicable to stamping " and punching machines."

Descriptions and drawings are given of stamping or embossing presses for giving a convex or concave shape to photographic cards.

A pressure bar is united to the base plate of the press by means of screw bolts or pins, which also form guides for the travelling plate. The pressure is applied to the travelling plate by means of a cam lever, which bears against the pressure box and the travelling plate, a sliding piston being placed between the lever and the travelling plate. Helical springs encircling the bolts may be used to raise the travelling plate after the pressure has been applied, or the springs may surround short tubes on the bolts.

A portion of the lower surface of the travelling plate is convex and of an oval form to fit a corresponding oval hole in the base plate. The convex portion of the plate may be rectangular or the press may be made to emboss two or more cards at once, or to emboss two or more parts of the same card at the same time.

[*Printed*, 10*d*. *Drawing*.]

A.D. 1874, October 15.—No. 3541.

CLEARY, WILLIAM.—(*Provisional protection only.*)—A " press " for glazing or polishing photographs, cards, or other similar " articles."

This invention relates to that kind of press wherein the photograph is passed between a revolving roller and a fixed burnishing bar.

The invention consists principally in giving to the burnishing bar a to-and-fro motion, in the direction of the axis of the roller, whilst the latter revolves. The roller is draw filed in the direction of its length, so that it may take a firm hold of the back of the photograph. The burnishing bar, underneath the roller, is fitted into slides. On the axis of the roller is a winch handle, and a cam or other equivalent device, to impart the to-and-fro motion to the bar.

The steps in which the axis of the roller revolves are provided with pressure screws, and the machine is kept heated, whilst in operation, by means of a lamp.

[*Printed, 4d. No Drawings.*]

A.D. 1874, October 21.—No. 3633.

CLARK, ALEXANDER MELVILLE.—(*A communication from Claude Léon Lambert.*)—"Improvements in producing carbon photo-"graphs."

Permanent carbon photographs are produced with salts of chromium, in their true aspect and gelatinised at the same operation.

A glass plate first receives a greasy coating, then a layer of normal collodion, and is immersed in cold water. The pigmented papers are then removed from their printing frames and one is applied on the coated glass plate whilst it is in the bath. Both are removed and subjected to slight pressure. The positive print is developed in warm water, and may either serve as a transparency or as a picture to be seen by reflected light; this point is an important feature in the invention.

The positive print, thus developed, is immersed in gelatine solution and the card mount is applied thereto in the bath. When the picture is dry, it is detached from the glass and is brilliant and glossy.

Portraits face to face, protected by mica, may be obtained by an analogous process to that described above.

The gradual strengthening which is applied to salts of chromium for obtaining clichés consists of an aqueous solution containing ammonia, sugar and permanganate of potash. This olution may be used to strengthen transparent positives as well as negatives.

Borders to prints in salts of chromium on albumenised paper may be obtained, in a tint different from that of the picture by employing a yellow mask and a negative with an ornamental border; the image is developed by hypo-sulphite and fixed.

At the back of the gelatine picture, tinted powder may be applied by a stump to impart opacity.

[*Printed*, 4*d*. *No Drawings*.]

### A.D. 1874, October 29.—No. 3740.

SAWYER, John Robert.—" Improvements in the production " of photographic prints."

This invention relates to the preparation of a temporary support to serve during the development of the image obtained on a sensitised pellicle of gelatine and carbon.

The invention consists in the preparation of a temporary support or surface which becomes adhesive while moist, and in that state, holds the picture. This surface becomes contractile when dry, and allows the developed print to leave it when mounted upon the final support.

When made of paper, the temporary support is coated with a solution of gelatine containing chrome alum; when dry, it is pressed powerfully by rollers or by hydraulic pressure. The sheets of gelatinised paper are next coated with a solution of lac and borax containing sub-carbonate of soda, dried, and milled by being passed through a rolling press. The surface is then made repellent to water by rubbing over with a solution of resin in turpentine together with white wax.

The employment of the temporary support is as follows:—The prepared surface of the support is brought into contact with the surface of the print. Both are lifted out together and dried; they are then immersed in warm water and the paper at the back of the picture is skinned off. The print is then developed, rinsed in cold water, placed in a solution of alum, then in cold water, and is transferred to its final support under water, the surface of the final support being prepared with an adhesive preparation. When the surfaces are dry, the temporary support may be stripped from the finished picture.

[*Printed*, 4*d*. *No Drawings*.]

A.D. 1874, November 17.—No. 3947.

DESPAQUIS, PIERRE AUGUSTE.—"Improvements in the pre-"paration of gelatine and albumen photographic plates for "printing from with lithographic inks."

The basis of this invention is:—"That a layer of bichro-"mated gelatine intended to furnish a plate for impressions "with lithographic inks however thin it is may be divided "into three parts or thicknesses."

In the Poitevin process (Albertype) for instance, the total sensitive film consists of sensitised albumen next the glass, covered with sensitised gelatine. The albuminous layer is rendered insoluble, but between it and the exterior of the gelatine layer is a soluble or hygrometric layer which sometimes loses its adherence to the layer of albumen and is dragged off by the inking roller.

According to the present invention, the intermediate and soluble layer is absent, for the two solarizations, one on the back and the other on the face of the compound film, penetrate into the parts of the film which ought to furnish the blacks and deep half tones. In the case of an insoluble support, the solarization of the blacks and deep half tones penetrates up to the support, so as to render them adherent to it.

A cylinder press for taking off proofs consists of two horizontal rollers, at a regulatable distance apart, with an endless cloth stretched so as to encircle them. The cloth carries the proofs, which are damped and inked by upper longitudinal rollers, and pressed against the descending sheet of paper by an endless cloth that passes round a third adjustable roller. In the damping, the water is distributed by a spongy roller and a hard roller. An endless sheet of zinc may be substituted for the cloth that passes round the third roller. The paper may be continuous.

[*Printed*, 8*d*. *Drawing*.]

A.D. 1875, May 4.—No. 1663.

JOHNSON, John Henry.—(*A communication from La Société Anonyme de Publications Periodiques.*)—A " process for printing " photochromatic images."

By means of this process, there are obtained polychromatic images, visible either as a reflection or as a transparency. These images are produced by the superposition of a series of photographic impressions. The number of negatives by which the total piecture is formed depends upon the number of monochromes in the resulting picture. Each negative is covered with opaque colour, so as only to give one colour in the resulting picture. The negatives are printed from separately upon a mixture of the colour required. The image is developed upon each film as in the carbon process, and the films are superposed by immersion in water, pressure and desiccation being used as each layer is applied. The whole picture is then immersed in an alum bath, and washed with water.

The ordinary mechanical means for printing in colours may be employed; either previously to the photographic impression or subsequently to the same, so as to superpose various tints upon the photographically printed image. According to another plan a photographic impression may be followed by a lithographic impression, and so on alternately until the final picture is produced.

[*Printed, 4d. No Drawings.*]

A.D. 1875, May 7.—No. 1699.

HARE, George.—A " photographic changing box and dark " slide."

The chief features of this invention are :—

1st. The inserting, by means of morticing or otherwise, in the body of the lid of the changing box, a metallic shutter with its springs and parallel guide pieces.

2nd. A flexible cover for the changing box.

3rd. The hinging of the back of the dark slide at one end, together with a method of fastening in the photographic plate by means of a spring catch lock, brought into action by pressing forward the back.

By means of the springs and parallel guides, the shutter is

self-acting, being opened only by the attaching of the camera slide to which the plate is to be transferred and closed by the act of removing it.

The flexible cover is used instead of the rigid sliding top peculiar to an ordinary transfer box. The cover is preferably made of a series of wooden laths glued to a band of leather.

In the present invention, the plates are inserted from the bottom.

The dark slide, used in connection with this box, has its end aperture closed, by means of a piece of metal projecting from the back and working on an axis at the further end. To permit of the egress of the plate, a bolt is lifted and the end thrown open.

[*Printed*, 4*d*. *No Drawings*.]

A.D. 1875, May 10.—No. 1734.

TUCKER, John Bone.—(*Provisional protection only.*)—" Im-" provements in mounting papers, either printed, ornamented, " or plain, on zinc, iron, tin and other metal plates."

Photographs, so mounted, are said to be weatherproof and dirt proof, and to bear washing without injury.

The process of mounting consists in coating both sides of the paper with size and allowing it to dry. The metallic plate is then coated with a mixture of copal varnish and lithographic varnish and the back of the coated paper is placed on the varnished face of the plate. The whole is subjected to pressure, so as to cause the plate and paper to adhere firmly together, and they are dried spontaneously or stoved. Finally the paper is coated with copal varnish and the whole is finished by again stoving.

[*Printed*, 4*d*. *No Drawings*.]

A.D. 1875, May 13.—No. 1777.

JOHNSON, Charles.—(*Provisional protection only.*)—" An " improvement in photographers' head rests, applicable also " to stands used for other purposes."

Usually the upright rod that carries the rest is fixed at the required height by means of a set screw and a socket, a separate joint being provided to adjust the inclination of the rest.

This invention consists in providing for the vertical and lateral adjustment by one and the same means. A ball-and-socket joint is employed of the following construction :—" The " ball through which the upright rod passes is made in two " halves, clamping the upright rod between them so as to grip " the rod tightly when the socket in which the ball works is " tightened up. This socket is also made in two parts hinged " together and provided with a thumbscrew by which the " socket may be tightened up or loosened as required. By " loosening this thumbscrew the upright rod may be slid " through the ball, and at the same time inclined by the ball " working in its socket, whereby the rest is adjusted both as " regards height and position at one and the same action, the " whole being firmly secured by tightening the screw."

[*Printed*, 4*d*. *No Drawings*.]

A.D. 1875, May 14.—No. 1791.

MALFROY, AUGUSTE.—(*A communication from Jean Baptiste David.*)—A "photographic apparatus with portable camera " obscura."

By means of the " portable camera obscura or laboratory " (which is a fixed box placed under the ordinary photographic apparatus that contains the lenses and dark slide), the photographic plate is submitted to the developing, fixing, and rinsing solutions necessary to complete the picture without having a dark room for the purpose.

In the fixed box is a tin box without a cover, containing three gutta percha vessels. The first vessel contains the nitrate bath, the second the sulphate of iron bath, and the third the hyposulphite bath ; the second and third vessels have a second compartment for rinsing water.

To immerse the photographic plate in succession in these baths, a rack and pinion movement brings an aperture (in the photographic apparatus beneath the plate) successively over the vessel to be used. The plate is moved up or down by means of nippers.

The gutta percha vessels are filled up by means of receptacles that communicate with them by india-rubber tubes having closing taps. By raising the receptacles by hand, the compartments are filled, the respective taps being open ; the taps are then closed. By the lowering of the receptacles (the taps being open) the vessels are emptied.

[*Printed*, 8*d*. *Drawing*.]

A.D. 1875, May 17.—No. 1825

CLARK, Alexander Melville.—(*A communication from Claude Léon Lambert.*)—"Improvements in photographic printing "frames, and in other apparatus used in connection there-"with."

1st. Printing frames.—These frames are for printing a number of positives at once with borders, and with or without devices thereon. A main frame contains a glass plate, which is obscured or masked, with the exception of any convenient number of apertures. The negatives are placed upon the apertures, and are held securely by a frame hinged to the main frame of the press. Masks, in connection with the hinged frame, have apertures rather smaller than those named above; the sensitised papers are placed in these apertures, and hinged flaps are folded down upon them and fastened by springs.

The frame for printing the border is similar to that above described. The picture is protected by a mask fixed to the glass in the middle of a larger aperture of rectangular form. To obtain a tinted border with a device thereon, a tissue of collodion and gelatine is used.

2nd. Paper cutting apparatus.—Two frames hinged together have between them the sensitised paper to be cut. One frame carries a glass plate, the other has a number of bars with transverse slits. By this means the paper is cut of the correct size for the above printing frames.

3rd. A registering photometer.—A rectangular oblong box has a number of dials provided with ratchet teeth, between which are holes that form stops for an index, and thus a definite time of exposure is given to the paper. By means of a strip of sensitised paper on a roller on the axis of the index, a record of the printing done each day is obtained.

[*Printed*, 1*s*. 4*d*. *Drawings*.]

A.D. 1875, May 21.—No. 1863.

SABUNJIE, John Louis.—(*Provisional protection only.*)—"Im-"provements in apparatus employed in the 'printing' of "photographs."

The printing frame consists of a metal frame fitted with glass and provided with metal bars or clasps, which are hinged by one end to the frame, and are formed at the other end to clip the edge of the frame tightly. The photograph to be copied is laid on the glass, and is covered with a piece of prepared paper, and then with a number of pieces of thick paper. The clasps are then shut down, and the frame is exposed to the light. This printing frame may also be employed when printing from glass negatives.

In a more simple form of the printing apparatus, metal clasps are made to clip the glass without the aid of a metal frame.

[*Printed*, 4*d*. *No Drawings.*]

A.D. 1875, July 22.—No. 2613.

DILLON, Thomas Arthur.—(*Provisional protection only.*)—A "method of embossing thin metal sheets with copies of manu-"script writings, engravings, printed matter, and pictorial "subjects by means of photography."

From the sheets or from moulds or castings taken from the embossed surface, copies may be made.

A negative on bichromated gelatine is dried and pressed in contact with a sheet of thin metal over which is laid a sheet of lead, a rolling press being used for the pressure. A copy of the gelatine surface will be found indented into the brass and lead.

A sheet of prepared paper, according to the method of the Ordnance Survey when adopting photo-zinco-graphic copying may be used. A transfer being made on to the stone or zinc, fine resin is dusted on the greasy ink; biting in with acid allows the letters or lines to stand up in relief. Thin metal covered with lead is then laid upon this relief, and is indented as described above. From the indented metal copies may be taken by the method described in No. 1935, A.D. 1874. Stereotype plates may be cast by covering them with thin paper and a thin coating of stereotype composition such as a preparation of glue paste and whiting on paper.

[*Printed*, 4*d*. *No Drawings.*]

A.D. 1875, August 31.—No. 3059.

HUGHES, Edward Thomas.—(*A communication from Adolf Hommel.*)—"Improvements in photography."

A clean glass plate is covered, on both sides, with a sensitive collodion film. When a copy of the object to be photographed is taken, in the camera, upon this plate, two images are produced, which are developed, fixed, and covered with a preserving varnish.

This double negative, produced at the same time upon both sides of a glass plate, is called a "photo-plasto-graphical" negative; copies from it (positives) are called "photo-plasto-"graphs."

By this invention, the light that has passed through the front film is utilised to produce a perfect picture. The refraction of the rays of light in traversing the glass plate tends to neutralise the sharp outlines on the front of the plate, and thus to produce very soft copies of perfect images. Photoplastographs are obtained by the action of dispersed and nearly vertical light upon two or more negatives prepared and superposed in the herein-before described manner.

[*Printed*, 4*d*. *No Drawings.*]

A.D. 1875, September 24.—No. 3352.

SLINGSBY, Robert.—"Improvements in the construction "of photographic studios."

This invention consists "in constructing the 'sky-light' as "well as the 'side light' of a photographic studio with a "double pitch or inclination inwards from the two ends to "the middle of its length, where the two inclined or sloping "surfaces meet at an angle of about 120 degrees." Thus advantage is secured at whichever end of the studio the sitter may be placed.

Instead of the sky and side lights meeting at an angle, they may be separated a few feet by an unglazed or darkened portion of straight roof and side, under which the camera may be placed in comparative obscurity.

Instead of the side light being vertical, it may be sloped outwards from top to bottom, the sky light being at one slope and the side light at another slope. Or, they may both be at the same slope and combined in one in the form of a "Man-"sard" roof, but with the double pitch or inclination inwards at the centre.

The inward pitch or inclination of the sky and side lights

may be adapted, in a single form, to studios in which the sittings are made at one end only.

[*Printed, 8d. Drawing.*]

A.D. 1875, October 5.—No. 3453.

HADDAN, Herbert John.—(*A communication from Ernest Edwards.*)—A " method of printing from plates of gelatine or " equivalent material."

This invention is the same as that set forth in the Provisional Specification, No. 1362, A.D. 1875.

[*Printed, 4d. No Drawings.*]

A.D. 1875, November 3.—No. 3815.

TAYLOR, Thomas Lyon.—(*Provisional protection only.*)—A " press for burnishing or polishing, photographs, cards, or " other similar articles."

This invention relates to that kind of press which has a revolving cylinder or roller and burnishing bar.

The bearings of the cylinder can be raised or lowered by means of set screws; a piece of solid india-rubber is inserted at the back of the bearing pieces. By this arrangement, if the card or mount should be thicker than usual, the cylinder gives way slightly and the work is not injured.

The cylinder is of such a diameter that one turn thereof will take the photograph through, and the burnishing bar is mounted, in a transverse slide, beneath the cylinder, and thus receives an endwise motion during the passage of the photograph through the machine. Lines across the photograph are thus avoided.

A cam on the axis of the cylinder gives the endwise motion to the burnishing bar. By the time that the bar has reached the highest point of the cam, the photograph will have passed completely through the machine; at this moment, when there is no pressure, the highest point of the cam passes, and the burnishing bar is forced back by a spring into its original position against the lowest part of the cam, ready to commence another operation. Thus each turn of the handle carries the photograph through and moves the burnisher once across the surface; the burnisher then returns ready for the next photograph, and so on.

[*Printed, 4d. No Drawings.*]

A.D. 1875, December 8.—No. 4247.

ARMANINO, Leopoldo.—(*Provisional protection only.*)—A " process for coloring photographs or photo-prints."

A photograph or photo-print is rendered transparent by varnish or other means, and a sheet, painted or printed in colours prepared according to the effect required to be produced on the photograph or photo-print, is affixed to the back of the print. According to another method, the print is printed directly upon coloured, painted, or printed sheets by means of registering points or marks.

[*Printed, 4d. No Drawings.*]

---

## 1876.

A.D. 1876, March 10.—No. 1050.

CLARK, Alexander Melville. — (*A communication from William Alexander Brice.*)—" Photographic apparatus."

This apparatus dispenses with the dark room and is able to be folded up when not in use so as to be portable.

" The invention consists of a central casing, with lens and " sliding focussing frame, to the larger base part of which " are hinged side casings that may be thrown down and sus- " pended in a horizontal position in suitable manner. The " central casing is recessed at the lower part, and shorter " boxes or covers hinged thereto to be swung down on the " outer side parts. Trays for the sensitizing and developing " solutions are moved by means of strings from the side cases " into the central casing to plunge the glass plate, which is " supported on holders of a frame pivoted to the sliding " focussing frame, first in the silver bath, and after exposure " to light into the developing bath. The water bath is finally " introduced through the lower hinged door of the rear " part to fix the picture, after which it is removed for " printing."

Besides those set forth above, there are the following points which constitute the invention:—The central casing has lower side openings to admit the entrance of the trays into the central compartment. Hinged covers are fitted on the opened side cases, so as to enclose the trays and produce the dark room with the central compartment. The central casing has a detachable rear shutter or wall with a hinged door at the lower part to introduce the water tray.

[*Printed*, 6*d*. *Drawing.*]

A.D. 1876, March 17.—No. 1148.

HENDERSON, ALEXANDER LAMONT.—"Improvements in the "production of ceramic photographs."

The object of this invention is to produce photographs in vitrifiable pigments, which may be burnt in upon enamel tablets, glass, porcelain, or metal.

If the surface upon which the picture is to be placed, termed the enamel, be light in colour, a transparency of the subject is treated with such iodides, bromides or fluorides as will produce a dark image. The iodides, &c., may be combined or not with chlorides and mordants. Subsequently the picture may be submitted to the action of silicates, such as those of sodium or potassium, so as to produce the deposited metal as a silicate, and then transferred to the enamel, which is then fired so as to vitrify the silicate.

The picture may be taken direct upon the enamel and treated with fluoric acid by which an intaglio is produced. This may be used as an engraved plate for printing, or it may be filled up with enamel colour and again fired.

Instead of converting the image into a silicate, it may be fluxed by means of an emulsion containing borate of lead suspended in collodion, oil of spike, fat oil, or similar media.

If the enamel be dark in colour, a negative is employed instead of a transparency, the iodides or bromides of the metals used being such as will produce a light image.

[*Printed*, 4*d*. *No Drawings.*]

A.D. 1876, April 25.—No. 1747.

VANDER WEYDE, HENRY.—"An improvement in photo-"graphy and apparatus used therein."

A large lens is provided in the photographer's studio and is placed between the source of light and the object to be photographed. An annular plano-convex Fresnel lens, about five feet in diameter, is preferred, but a lens of another kind may be used. The source of light may be that from a window, or reflected from a mirror or other reflecting surface illuminated by daylight or artificial light. In one instance, a reflector placed outside of the studio reflects the vertical rays from the sky, through a window, on to the lens. In another instance, the electric or magnesium light is placed in the focus of a paraboloidal reflector, so as to reflect the rays on to the lens in a parallel direction. A sheet of ground glass may be introduced between the source of light and the lens.

The effect of the converging rays from the lens is an increase in the actinic power of the light, and greater softness in the effects produced.

The lens may be made "by casting the glass in moulds and " afterwards submitting the angular surface to the direct ac- " tion of a furnace until the surface runs and becomes smooth " and glossy." Each annulus of which the lens is composed may be cast in segments and joined together.

[*Printed*, 6*d*. *Drawing*.]

A.D. 1876, May 11.—No. 1991.

SARONY, Oliver.—"Improvements in the production of " borders, titles, and other devices or ornamentation on " photographic pictures, and in means or apparatus employed " therefor."

This invention is chiefly designed for use in carbon printing. The border or ornamentation is obtained, by two successive exposures through two different openings on the same negative. By this method a print like a chromotype can be produced, in the ordinary printing frame, at one printing. The two openings may be contained in a sliding frame fitted to the plate holder, or in separate frames pivoted within the camera.

When the openings are used for a single portrait, one opening, in form and size, corresponds exactly to the style and dimensions of the picture, the other is of a size to include both the picture and the border and in it is fixed a plate of colourless glass having a mask corresponding in size, shape, and position to the first opening. The clear portion of the plate bears upon it the device to be produced beneath or around the picture.

The sensitised plate is exposed, to take the portrait, through the first-mentionea opening. Then the lens being capped, and the dark slide closed, the second opening takes the place of the first, and the border is exposed and printed. The second exposure is for a few seconds with the aid of a cap that is covered with tissue paper, ground glass being within the tube.

Modifications of the above described arrangement enable more than one portrait to be taken on one plate.

[*Printed*, 4*d*. *No Drawings*.]

A.D. 1876, May 30.—No. 2274.

BURROWS, Frederic Thomas, and COLTON, Joseph Dufais.—"Improvements in desks for use in retouching photographic "negatives."

1st. A series of screens, of varying degrees of transparency, so as to show the negative more or less in detail.—The screens are glass plates fitted into a disc that is able to rotate on a pivot at the back of the carrier on which the negative is mounted; any one of the screens can thus be brought quickly to coincide with the opening in the carrier over which the negative is fixed.

2nd. The negative is secured by clamps in a circular carrier fitted to rotate in the desk, so that the operator can turn the negative round, to suit his convenience, as the work proceeds, without removing the negative.

3rd. A concave reflector is mounted upon adjustable supports behind the desk. For night work, a lamp is combined with the reflector, and a plate of violet glass is fitted in the revolving carrier.

4th. The magnifier is so mounted upon a fixed standard at the left hand that it can be adjusted without interfering with the right arm of the operator.

5th. A pencil sharpener is fitted to slide within the desk. This instrument consists of a block on which a sheet of emery paper is clamped by a metal frame; the block draws out from the case, at one side, and is retained by a spring.

[*Printed*, 6*d*. *Drawings*.]

A.D. 1876, June 6.—No. 2367.

BROWN, Richard, and BARNES, Robert William.—" Im-" provements in and applicable to photography."

The object of this invention is to produce a mezzotint effect in photographic vignette portraits.

The stippled granular effect is produced direct, in the process of printing, from a negative from a groundwork of morocco leather or other reticulated surface. The print is treated by the artist, with crayon or pastil, so as to produce an effect according to taste; from the prepared print a second negative is taken which bears the reticulated surface and also the artistic manipulation.

A mask cut from the vignette is placed on the reticulated negative, or on the second negative. Only the lighter and inner outlines of the photograph are fastened down, the outer vignetted portion of the figure being curled or turned over in a graduated manner. This prepared negative and mask is then placed on the vignetted photograph as ordinarily produced, and the whole combination is placed on a revolving table and submitted to the action of daylight until the granular effect, with or without the background, is printed on the vignetted picture. The fixing and completing of the photograph is as usual.

[*Printed*, 4*d*. *No Drawings*.]

A.D. 1876, July 21.—No. 2963.

BROOKES, Alfred George.—(*A communication from Charles Eugène Farjon.*)—" Improvements in copying presses, which " improvements are also applicable to other presses."

The application of this invention to embossing photographs entitles it to be placed here.

The travelling plate of these presses is adjustable to articles of various thickness by means of screws, sliding tubes and springs. The pressure is exerted by means of levers in connection with screws or cams.

For embossing photographs an oval opening is made in the travelling plate to allow of the bulging of the cardboard, which is placed upon an india-rubber bed on the base plate when the pressure is applied.

[*Printed*, 6*d*. *Drawing*.]

A.D. 1876, July 22.—No. 2973.

MORGAN-BROWN, William.—(*A communication from Louis Ducos Duhauron.*)—" Improvements in photography in colours, " and in the apparatus for that purpose."

Three negatives of the same subject are obtained; one by green light, the second by orange light, the third by violet light. Three positives or monochromes are respectively printed from these negatives; the positives are on gelatinised semitransparent paper prepared with the complementary colours. The three semitransparent monochromes, being superposed, represent, by the blending of the tints, the total colour of the object taken.

The camera used to obtain the negatives may be triple; it contains three lenses. Images are simultaneously directed to the three lenses by reflection and refraction, unsilvered reflecting glasses being used. A correcting lens (concave in form) is used in this camera and is interposed between the object lens and the sensitive surface, very near to the latter.

The coloured glasses to qualify the light to each negative are placed either close to the sensitive plate or immediately outside the object lens. Coloured varnishes are used to tint the glasses.

Aurine or eosine is used in sensitising the negative from green light, and chlorophyll in that from orange light; these adjuncts are only necessary when wet processes are employed; no accelerating substance is used with dry processes.

[*Printed*, 8*d*. *No Drawings.*]

A.D. 1876, August 22.—No. 3298.

JOHNSON, John Henry.— (*A communication from Marcel Charbonnier and Robert Charbonnier.*)—A " process and apparatus for producing photographic images upon or ornamenting glass, or glazed, or enamelled articles."

A polished copper plate is uniformly coated with an aqueous mixture containing bichromate of potash and honey; the coating is then dried and exposed under a positive photograph or an engraving. The copper plate is then taken away from the positive, slightly heated, and brushed over with vitrifiable metallic powder. The picture develops during the brushing of the powder, and is treated with collodion. Upon being washed in water until the coloured bichromate is removed, the picture is formed by the powder which has been deposited.

A small frame or holder is provided. This instrument is made of a flexible material such as cardboard, caoutchouc, &c. The flexible frame is laid upon the washed copper plate, takes up the collodion film carrying the picture, and deposits it on the glass or glazed article which it is required to ornament, care being taken that the metallic powder is not in contact with the glaze of the article. The collodion is then cut out round the frame and the frame removed. The picture, thus fixed, is dried, covered with oil of petroleum and baked.

These pictures may be transferred to ceramic articles in the biscuit state, the article being maintained in the wet condition during the transfer."

[*Printed*, 4*d*. *No Drawings*.]

A.D. 1876, August 28.—No. 3374.

McLIESH, William. — (*Provisional protection only.*) —" Im-" provements in photographic apparatus."

In this invention, by the arrangement of photographic lenses, " pictures in stereoscopic or natural relief are ob-" tained."

"This relief is obtained by employing lenses of larger " diameter with a short focus, the margin of the lens forming " a parallax corresponding to the distance between the human " eyes, the diaphragm or stop being so placed as not to cut " off the marginal rays. The diameter of the lenses employed " must not be more than four inches or less than two inches, " and the back focus (not the equivalent focus) must not be " less than one or more than three diameters of the front " lense. Any form of lense or combination of lenses may be " employed for the purposes of this invention as above de-" scribed, except that known as the ordinary portrait lense, " which from its limited field and want of penetration is quite " unsuitable."

[*Printed*, 2*d*. *No Drawings*.]

A.D. 1876, August 30.—No. 3417.

BRICE, William Alexander.—A " photographic test-plate or " actinometer."

The object of this invention is to furnish a testing plate that, by being applied to the camera, determines the quality of the chemicals employed, and the quick or slow working of the lens; the testing plate also affords information as to the cause of fog in the picture.

The invention consists of "a frame with a sliding glass " plate to which are applied fixed pieces of transparent " material superposed in layers of one, two, three and more " in regular succession to produce a greater or less obstacle " to the passage of the light." The centre part of the glass plate is left free; the adjoining parts are covered with one layer, the next adjoining parts with two layers, and so on to the extremity of the plate.

The testing plate being set up between the lens and the sensitised plate, the picture is taken as usual. When developed, the whole picture is visible, but is divided into sections of unequal intensity. The absence of fog where the light has passed through several layers, indicates that the chemicals are pure, that there is no diffused light and that the nitrate bath is of the proper acidity. Where the light has passed through a less number of layers, evidence is given respecting the proper time of exposure. The middle or uncovered section indicates, by the evident over exposure, that the lens is good and rapid, that the chemicals are in good condition, and that the actinic power can produce pictures with rapid exposure.

[*Printed*, 4*d*. *Drawing*.]

A.D. 1876, Sertember 18.—Ne. 3644.

NESBITT, George.—(*Provisional protection only*.)—A "photo- " graph tinting apparatus."

The apparatus is made of a frame, with a back of the exact size of the photograph to be tinted; the back is hinged, so as to open and shut to receive the picture, and, when the picture is in, to shut and press down tightly on the tinting plate. It also has a rabbet to receive the tinting plate, or name plate, and a groove in front of the rabbet for the introduction of a glass with any size or shape of mask to cover the parts of the portrait to be tinted. Thus instead of a separate frame for each size or shape of picture it is only necessary to have a separate glass with a different size or shape of mask.

[*Printed*, 2*d*. *No Drawings*.]

A.D. 1876, October 20.—No. 4060.

GEDGE, WILLIAM EDWARD.—(*A communication from Jules Alphonse Terpereau.*)—" Improvements in portable photo-" graphic apparatus."

This invention relates to a portable camera for tourists and photographers.

This invention consists in the arrangement of the box, which is small, but opens to a size sufficient for forming a dark chamber to take any kind of photograph on plates up to about $10\frac{1}{2}$ inches.

Upon the sides of the box is nailed a curtain, composed of three pieces so as to form a chamber. A window, of glazed linen is in one of the panels or pieces. The curtain is cut of a shape to permit of the immediate closing of the camera and to prevent interference by the wind.

The second part of the invention consists in the curtains being held by an arrangement of rods in two parts.

Instead of a metal bottle to contain water, a pocket, or india-rubber bag is provided, at the top of which is a funnel through which the water is introduced. There is an india-rubber pipe for letting out the water; the end of the pipe has a stopper which is hermetically self-closing. " When operating " this little receiver is hooked on to one of the uprights in " such a manner that the water may come directly into the " basin placed in the middle of the camera."

The apparatus can be enclosed in a very small space, because the setting up rods separate and the sides fall into the box, which is portable. It rests upon a tripod.

[*Printed*, 6*d*. *Drawings*.]

A.D. 1876, October 24.—No. 4117.

NEWTON, HENRY EDWARD.—(*A communication from Cheri Rousseau.*)—A " process for obtaining pictures in colors."

A photographic carbon print of the object to be reproduced is mounted on a temporary supporting paper in a reversed position and is tinted with oil colours. When the colours are dry, the whole surface of the picture is coated with an adhesive substance and transferred to the canvas or other permanent backing, being caused to adhere thereto by gentle pressure. The supporting paper which still covers the face of the carbon print is now removed, presenting the coloured picture to view with the colours blended with the tints and shadows of the photographic print.

The coloured pictures may be retouched on the face by a skilled artist.

[*Printed*, 2*d*. *No Drawings.*]

A.D. 1876, November 9.—No. 4344.

DONISTHORPE, WORDSWORTH.—(*Provisional protection only.*) —"Improvements in apparatus for taking a succession of "photographic pictures and for exhibiting such pictures."

The pictures are taken at equal intervals of time, to record the changes taking place in or the movements of the object being photographed. By a succession of pictures, so taken, of any moving object, the eye may be impressed with a representation of the object in continuous movement as it appeared when being photographed.

The camera is constructed to receive a number of prepared plates, one behind the other. By mechanism, the series of plates moves forward progressively, so that as soon as the foremost plate has been exposed for the required length of time, the whole series moves forward a certain distance; the foremost plate is then dropped into a receiver. The plate which is then foremost is exposed and dropped, and so on in succession, until the requisite number of plates have been exposed.

The admission of light to the lens is controlled by a movable screen, which is only brought in front of the lens during the shifting of the plates. The mechanism for moving the plates and the screen is driven at a uniform rate. After exposure, the plates are developed; prints from these may be obtained in the ordinary manner.

The series of pictures may be used in the phenakistoscope.

[*Printed*, 2*d*. *No Drawings.*]

A.D. 1876, November 11.—No. 4385.

SLINGSBY, ROBERT.—"An improvement in the construction "of photographic studios."

This invention relates to an improvement upon the photographic studios described in No. 3352, A.D. 1875.

This improvement consists in the addition to the studio constructed as in the former invention "of a vertical light

" having likewise a double inward pitch or inclination at the " centre corresponding to that of the sky and side lights, this " vertical light meeting the highest part of the skylight and " rising therefrom to the ridge of the roof of the studio."

The results of this arrangement are said to be "a still " shorter time of exposure, a softening of the shadows, and " greater delicacy of modelling."

[*Printed*, 4*d*. *Drawing*.]

A.D. 1876, December 30.—No. 5065.

WILKINSON, James, HARDMAN, James, and BIRTWISTLE, Richard. — (*Provisional protection only*.) — " Improvements in the mode or method of burnishing photo- " graphs, and in the machinery or apparatus used therefor."

This invention consists in passing the photograph, when mounted on paper or cardboard, through or between a pair of metal rollers heated by gas or steam.

The rollers are mounted in adjustable bearings, a handle, at one side, gives them motion. At the opposite side to the handle, a gas pipe perforated with a number of holes is passed into each roller. The gas being lighted and the photograph placed between the rollers, the bottom roller is rotated by the handle and gives motion to the upper roller by wheel gearing. Both sides of the photograph come out with a highly glazed surface.

One heated roller suffices when the back part of the card is not required to be glazed.

[*Printed*, 2*d*. *No Drawings*.]

# INDEX OF SUBJECT MATTER.

[The numbers refer to the pages on which the Abridgments commence. The names printed in *Italic* are those of the persons by whom the inventions have been communicated to the Applicants for Letters Patent.]

# THE SOURCES OF MODERN PHOTOGRAPHY

*An Arno Press Collection*

## THE PREHISTORY OF PHOTOGRAPHY

Sobieszek, Robert, editor. **The Prehistory of Photography.** 1979

## THE DAGUERREOTYPE IN EUROPE

Sobieszek, Robert, editor. **The Daguerreotype in Germany.** 1979

Chevalier, Charles. **Nouvelles Instructions sur l'Usage du Daguerréotype** *and* **Mélanges Photographiques.** 1841/1844

Mentienne, [Adrien]. **La Découverte de la Photographie en 1839.** 1892

Thierry, J. **Daguerréotype.** 1847

## HISTORIES OF PHOTOGRAPHY

Bunnell, Peter C., editor. **The Universal Exposition of 1900.** 1979

Blanquart-Évrard, [Louis Désiré]. **La Photographie.** 1870

Colson, R[ené], editor. **Mémoires Originaux des Créateurs de la Photographie.** 1898

Eder, Josef Maria. **Geschichte der Photographie.** 1932

Eder, Josef Maria, editor. **Quellenschriften Zu Den Frühesten Anfängen Der Photographie Bis Zum XVIII. Jahrhundert.** 1913

Lécuyer, Raymond. **Histoire de la Photographie.** 1945

Stenger, Erich. **Die beginnende Photographie.** 1943

Stenger, Erich. **The History of Photography.** Translated by Edward Epstean. 1939

## MANUALS AND DICTIONARIES AND CATALOGUES

Davanne, A[lphonse]. **La Photographie.** Two vols. in one. 1886/1888

Dillaye, Frédéric. **La Théorie, la Pratique et l'Art en Photographie.** 1891

Eder, J[osef]-M[aria]. **La Photographie Instantanée.** Translated from the German by O. Campo. 1888

Guerronnan, Anthonny. **Dictionnaire Synonymique.** 1895

Kodak, Ltd. **Price List of General Photographic Apparatus and Materials.** 1904

Martin, A[nton Georg C.]. **Handbuch der gesammten Photographie.** 1854

Saint-Victor, [Claude Marie François] Niépce de. **Recherches Photographiques.** 1855

Snelling, H[enry] H[unt]/Anthony, E. **A Dictionary of the Photographic Art/ A Comprehensive and Systematic Catalogue.** Two vols. in one. 1854 van Monckhoven, D[ésiré]. **Photographic Optics.** 1867

Vogel, H[ermann] W[ilhelm]. **Die Photographie farbiger Gegenstände.** 1885

von Rohr, Moritz. **Theorie und Geschichte des Photographischen Objektivs.** 1899

Woodbury, Walter E. **The Encyclopaedic Dictionary of Photography.** 1898

## PHOTOGRAPHIC PATENTS AND INDUSTRY

Great Britain, Patent Office. **Patents for Inventions.** Two vols. 1861-1903

Engrand, Bernard. **L'Industrie Photographique en France.** 1934

Kühn, Willy. **Die Photographische Industrie Deutschlands.** 1929

## MONOGRAPHIC STUDIES

Lo Duca, [Joseph-Marie]. **Bayard.** 1943

Nadar [Gaspard Félix Tournachon]. **Quand J'Étais Photographe.** [1900]

Bunnell, Peter C., editor. **The Photography of O.G. Rejlander.** 1979

## FRENCH PRIMITIVE PHOTOGRAPHY

Figuier, Louis. **La Photographie au Salon de 1859** *and* **"La Photographie" & "Le Stéréoscope".** Two vols. in one. 1860/[c. 1868]

Ken, Alexandre. **Dissertations Historiques, Artistiques et Scientifiques sur la Photographie.** 1864

[Kravets, Torichan Pavlovich, editor. **Documents on the History of the Invention of Photography].** 1949

Lacan, Ernest. **Esquisses Photographiques.** 1856

Mayer & [Louis] Pierson. **La Photographie.** 1862

## EARLY COLOR PHOTOGRAPHY

Sobieszek, Robert, editor. **Two Pioneers of Color Photography: Cros & Du Hauron.** 1979

Sobieszek, Robert, editor. **Early Experiments With Direct Color Photography.** 1979

## AESTHETICISM AND FIN DE SIÈCLE PHOTOGRAPHY

Adams, W[ashington] I[rving] Lincoln. **Sunlight and Shadow.** 1897

Bunnell, Peter C., editor. **The Aesthetics of French Photography.** 1979

Demachy, R[obert] and C. Puyo. **Les Procédés d'Art en Photographie.** 1906

Holme, Charles, editor. **Art in Photography.** 1905

Matthies-Masuren, Fr[itz]. **Künstlerische Photographie.** 1907

Poore, H[enry] R[ankin]. **Pictorial Composition.** 1903

## THE MODERN PERIOD

Genthe, Arnold. **As I Remember.** 1936

Gernsheim, Helmut, editor. **The Man Behind the Camera.** 1948

Gräff, Werner. **Es Kommt Der Neue Fotograf!** 1929

Stenger, Erich. **Die Geschichte der Kleinbildkamera bis zur Leica.** 1949

Stotz, Gustaf, et al. **Film und Foto.** 1929

Whiting, John R. **Photography is a Language.** 1946

Bunnell, Peter C., editor. **Willi Warstat on the Aesthetics of Art Photography.** 1979